Data Communications
Gigabit Ethernet
Handbook

Data Communications Gigabit Ethernet Handbook

Stephen Saunders

McGraw-Hill

New York   San Francisco   Washington, D.C.   Auckland   Bogotá
Caracas   Lisbon   London   Madrid   Mexico City   Milan
Montreal   New Delhi   San Juan   Singapore
Sydney   Tokyo   Toronto

Library of Congress Cataloging-in-Publication Data

Saunders, Stephen.
    Data communications gigabit Ethernet handbook / Stephen Saunders.
      p.   cm. — (McGraw-Hill series on computer communications)
    Includes index.
    ISBN 0-07-057971-7
    1. Ethernet (Local area network system) 2. Gigabit
communications. I. Title. II. Series.
TK5105.8.E83S28 1998
004.6'8—dc21   98-14848
                CIP

McGraw-Hill

*A Division of The **McGraw·Hill** Companies*

1 2 3 4 5 6 7 8 9 0 DOC/DOC 9 0 3 2 1 0 9 8

ISBN 0-07-057971-7

The sponsoring editor for this book was Steve Elliot, and the production supervisor was Pamela A. Pelton. It was set in Century Schoolbook by North Market Street Graphics.

Printed and bound by R. R. Donnelley & Sons Company.

This book is printed on recycled, acid-free paper containing a minimum of 50% recycled, de-inked fiber.

Dedication

For my grandmother, Sylvia.
And for Dawn.

Contents

Part 6   Gigabit Ethernet Design Tutorials

Chapter 20. How to Build a Gigabit Ethernet Network

Chapter 21. Choosing Server Components for a Gigabit Ethernet Network

Part 7   Available Gigabit Ethernet Hardware   313

Acknowledgments

Having an idea for a book is one thing. Getting it into print is quite another. Quite simply, the book you are now holding could never have been published were it not for the enthusiasm and expertise of the authors of the chapters that follow. These are people who are not familiar with the concept of free time, but they contributed to this project anyway. They all have my deepest gratitude.

This is also a good opportunity for me to thank my colleagues at *Data Communications* magazine. I'm grateful to my boss, Lee Keough, for letting me write yet *another* book—and for her encouragement. Also a big thank you to David Newman, *Data Comm's* senior technology editor, for his advice on content and technical accuracy. Special thanks go to Ken Surabian, *Data Comm's* design director, for his stunning cover design. Thanks too to senior editor Erica Roberts for contributing (unwittingly) insightful analysis on the first generation of gigabit Ethernet products. And as always, a big thank you to executive editor Aaron "The Beard" Fischer for just being himself.

I would also like to thank Steve Elliot, my book editor at McGraw-Hill, for first giving me the idea to write this book, and then riding me mercilessly until I delivered it (six months late). Steve, you are unique.

Thanks also to Frank Kotowski, senior editing supervisor at McGraw-Hill Professional Publishing division, and to Stephanie Landis and the other staff at North Market Street Graphics, for making the process of turning raw copy into a finished product so easy.

Finally, I would like to express my deep gratitude to the following PR folk, marketing gurus, and freelance writers. They are all consummate professionals. Without their anonymous (but much appreciated!) hard work this project would never have gotten off the ground:

Robert Barlow, Barbara Bates, Marilyn Callaghan, David Callisch, Erin Curtis, Peter Davé, Ana Fiejeski, Kurt Foeller, Kevin Gallagher,

Alice Homolka, Barbara Hurst, Amanda Jaramillo, Diane Kennedy, Yechiel Kurtz, Mark Levenson, Maureen Liberty, Susan Lider, Lori Lux, Molly Miller, Nancy Prater, Linda Pugliano, Dawn Slusher, Jeanne Talbot, and Donna Woznicki.

Stephen Saunders

Editor's Introduction

Back at the beginning of 1997, when I first had the idea for a book about gigabit Ethernet, I was pretty sure it would be timely. But in the time between starting this project and getting the book into print, the excitement about this new technology has exceeded even my expectations. Gigabit Ethernet has made the transition from its previous status as a promising new networking option, to its current position as the main choice of network managers looking to build high-speed backbone networks, at lightening speed. Indeed, this change has been almost as speedy as the technology itself.

Why the change? One reason for gigabit Ethernet's unexpected popularity is the renaissance in packet-based networking. Of course, packet-based LAN technologies never went out of style. But with the arrival of ATM in the early 1990s, there was a widely held belief that there were limits to what could be accomplished in a network without the use of ATM's fixed-length cells or its highly evolved quality-of-service mechanisms.

Not any more. Beyond adding raw performance, the networking community has taken a two-pronged approach to retrofitting Ethernet with the facilities it needs to handle today's applications—namely, prioritization and QOS. On one side, bodies like the IEEE and the IETF have been hard at work on prioritization standards; examples include the resource reservation protocol (RSVP) as well as emerging protocols such as 802.1Q and 802.1p and integrated services over specific link layers (ISSLL). Vendors also have been working hard, building QOS into their hardware at a fundamental level using advanced queuing and buffering techniques.

Both methods (standards-based and vendor-devised) are relatively new, and problems need to be ironed out. But network managers now know that ATM is not the only technology that can prioritize traffic or ensure bandwidth availability for time-sensitive applications.

ATM's faltering fortunes are not the only factor behind gigabit

Ethernet's popularity. Another more pressing reason is the bandwidth shortage on corporate networks. While almost everyone knows that the gigabit Ethernet revolution is being driven by a capacity shortfall, most people are only now grasping that the crisis has just begun.

Consider a few of the facts. As of this writing, 10/100-Mbit/s Ethernet adapters are only just starting to outsell 10-Mbit/s cards for the first time. But only about 20 percent of those 10/100 cards are currently being used at 100-Mbit/s speeds.

When net managers get around to upgrading their workgroup hubs and switches from 10-Mbit/s to 100-Mbit/s hardware, these dual-speed adapters will automatically switch over to transmitting at the higher rate. And that's when the bandwidth fun will really start, as the 10-fold increase in speed creates a glut of workgroup traffic that will push its way out onto corporate backbones, creating a whole new level of capacity crisis. (How long will it take for the change from 10 to 100 to come about? Not long at all if the all-out price-cutting war between leading switch and hub manufacturers continues.)

Clearly, gigabit Ethernet will be network managers' strongest ally in the fight against bandwidth shortages. The gigabit Ethernet market is burgeoning, with scores of vendors rolling out affordable products. And beyond boosting bandwidth, a lot of these players are building multi-layer routing capabilities into their switches—making the switches smart enough to prevent workgroup traffic from getting onto the backbone in the first place. Gigabit Ethernet will also play a key roll elsewhere on the network, for instance by providing high-speed connections into servers as well as delivering fat-pipe connections to power workers.

Sounds great. But as with any new technology, deploying gigabit Ethernet will not be all smooth sailing. Corporations that deploy gigabit Ethernet will face several problems. As noted, gigabit Ethernet QOS and prioritization features are still in their infancy. Further, there is still no consensus on how best to deploy routing in a gigabit Ethernet network. Congestion control also is an area where progress needs to be made. And net managers will have to master an entirely new set of topology and design rules. Finally, getting gigabit Ethernet to run safely and reliably over the installed base of Category 5 UTP copper cabling will be a real challenge. Beyond these technological challenges, network managers also will need to carve away the hype and marketing-speak in order to find the best products for their networks.

That's where this book comes in. Its rationale is to provide network managers, consultants, and designers with the knowledge they need to migrate to gigabit Ethernet without getting burned.

To that end, this book is organized into seven parts. Each part deals with a different aspect of the migration, and the parts are placed in the order in which net managers themselves will have to deal with these issues—from learning the fundamentals of the technology to designing and rolling out a network.

Thus Part 1 is designed to provide the reader with a comprehensive understanding of the gigabit Ethernet standard—an essential first step. Part 2 goes on to explain how the gigabit Ethernet standard works best when used in conjunction with several other technologies such as 802.1p, 802.1Q, and 802.1x. Part 3 deals with the critical issue of how routing intelligence will be deployed in a gigabit Ethernet network. Part 4 compares gigabit Ethernet to the other available high-speed LAN technologies, such as ATM and Fibre Channel. Part 5 provides tutorials on how best to undertake the migration from today's installed LAN technologies to gigabit Ethernet with the least disruption possible. Part 6 provides six design lessons detailing how to get the best possible performance out of the gigabit Ethernet network and the equipment attached to it. Finally, Part 7 offers a detailed account of what to look for when buying gigabit Ethernet switches, hubs, and adapters. It also contains evaluations of the merits and downsides of the first gigabit Ethernet products.

To make sure this handbook contains the best, the most prescient, the most accurate, and the most timely information possible, I chose not to write it myself (readers will please forgive this jest in such a serious book. I can assure them this will be the last attempt at levity in this volume). Rather, I went back to the people who are closest to the technology and therefore ideally placed to provide the right answers: namely, the technology mavens at the leading gigabit Ethernet equipment manufacturers. Their response to the idea of a book on gigabit Ethernet was overwhelmingly positive. Never before have so many acknowledged networking experts—and, it should be noted, competitors—come together and collaborated on such a work. The result is a collection of tutorials written by the very brightest minds in networking about one of networking's brightest technologies.

Finally, one caveat. Every effort has been made both by myself and by the contributors to make sure that the information given in this book is accurate. However, gigabit Ethernet is a fast-moving technology in all senses of the word. At press time the standard defining gigabit Ethernet was stable, but readers should note that it is conceivable that some minor changes may be made subsequent to publication.

Stephen Saunders
Executive Editor, Data Communications *magazine*
saunders@data.com

Understanding Gigabit Ethernet

Purpose

To allow readers to obtain a detailed understanding of gigabit Ethernet technology as defined in the IEEE 802.3z standard.

What Is Covered

Part 1 covers every aspect of the gigabit Ethernet spec via five technical tutorials. These tutorials go into varying degrees of detail. Chapter 1 provides an overview of the gigabit Ethernet standard, highlighting its advantages and disadvantages. Chapter 2 goes into more detail on the different elements that make up the overall 802.3z specification.

The next three chapters each focus exclusively on one significant aspect of the standard: Chapter 3 deals with packet bursting, the modified version of the CSMA/CD algorithm that forms the basis of shared-media gigabit Ethernet networks. Chapter 4 examines how elements of the gigabit Ethernet standard can be used to build full-duplex repeaters, which deliver better performance than a hub at lower cost than a switch. Finally, Chapter 5 deals with the crucial issue of cabling, including the distance goals for the various transceivers and cabling types covered in the spec.

Contributors

SHL Systemhouse Corp.; Packet Engines Inc.; Nbase Communications Inc.; 3Com Corp.

Chapter

1

Gigabit Ethernet: A Cautious Overview

Al Lounsbury*
MCI Systemhouse

1.1 Introduction

Gigabit Ethernet? Heed the hype, and the high-speed technology sounds like the answer to most net managers' prayers. Simple and familiar, and a pipe fat enough to handle any app: switched workgroup, corporate intranet, real-time multimedia, whatever.

Filter out the industry buzz, however, and giga*bit* starts sounding a little like giga*but*. True, the technology looks a lot like the Ethernet that fast Ethernet corporate networkers know and love. And at 1 billion bits per second, it has the potential to blow through virtually any bandwidth bottleneck.

But (that word again) gigabit Ethernet isn't simply straight Ethernet running at a supersonic clock speed. In fact, the ways in which it differs from its predecessors may be more important than its similarities, at least when it comes to deployment and network design.

For one thing, gigabit Ethernet relies on a modified media access control (MAC) layer, which affects network size and utilization. For another, its cabling requirements are very different. For the most part,

---

* As director of technology strategies for MCI Systemhouse's sister company, SHL Systemhouse (Canada), Lounsbury is responsible for monitoring and correlating the relationship between various information technologies, their corresponding market trends, and their technical viability and suitability to business process. Lounsbury is an electrical engineer and a member of the IEEE and the Professional Engineers of Ontario. His in-depth engineering understanding and extensive experience (15 years) in designing, implementing, and integrating various computer and networking technologies allow him to evaluate information technologies effectively. Lounsbury's unique combination of business management and technical design skills coupled with his hands-on experience form the basis for his expertise as a senior technology strategist.

think fiber—at least in the short term. At press time, the specification for Category 5 was still in the works, and a commercially available UTP product will probably not be available until early 1999.

Further, many installed servers can't churn through data at gigabit speeds, and some gigabit adapters are similarly performance-bound. And treating gigabit Ethernet like a one-size-fits-all scheme could actually contribute to network slowdowns. Finally, gigabit Ethernet (like its forerunners) doesn't now deliver the quality-of-service (QOS) guarantees needed by multimedia applications.

So much for the reality check. The trick with any technology is understanding what it can and can't do and designing networks accordingly. The best place for gigabit Ethernet is on interswitch links between campuses and wiring closets.

Corporate networkers who want to get the goods on gigabit should start digging into the details now. With the vendor marketing machines going into overdrive, the answer is to fight back with the facts. Table 1.1 will point you in the right direction.

1.2   Sizing Up Gigabit

One of the key issues to understand about gigabit Ethernet is the maximum size of the network. The best way to understand the limits—and the reasons for them—is to take a quick look at some Ethernet history. Back in 1980, the original IEEE 802.3 Ethernet spec

---

TABLE 1.1   Gigabit Q&A

Getting the goods on gigabit Ethernet means sorting a few facts out of an enormous amount of fiction. Here's a quick review of some of the more common questions—and the correct answers.

Is gigabit Ethernet an ATM killer?
No. Gigabit Ethernet will probably slow early adoption of ATM, since it will be cheaper per network adapter and per switch port. But the technologies are intended to solve different problems. ATM was designed to deliver various classes of service for data and real-time traffic. Ethernet—straight, fast, and gigabit—is a data-oriented transport. Further, the next-generation Internet infrastructure (Internet 2) is being designed around ATM, and the forthcoming Network Device Interface Specification version 5 (NDIS5) furnishes critical application program interfaces (APIs) for negotiating QOS over native ATM.

Will gigabit Ethernet ever support QOS?
It depends on how QOS is defined and what parameters are specified. Resource reservation protocol (RSVP) provides a very basic QOS for local networks; however, across the switched public network, it acts only on a best-effort basis. Actually, two QOS camps are emerging. ATM delivers true QOS for real-time services. But companies have thus far shown little interest in real-time services, which has slowed or stalled ATM rollouts. Ethernet vendors, meanwhile, think they can take some ATM QOS features and append them to protocols like TCP/IP. This is not an approach that yields optimal solutions. The real question then becomes: are the features being developed close enough to true QOS for the market to adopt them? The answer remains to be seen.

---

defined a mechanism called carrier-sense multiple access with collision detection (CSMA/CD). This scheme ensures that all stations are granted access on a first-come, first-served basis. Since Ethernet was intended only to carry data, no provisions were made for quality of service or prioritization. CSMA/CD simply ensures that the same access rules apply equally to all network nodes.

The designers of Ethernet worked it out so that stations up to 2 km apart could sense a collision. This distance limitation results from the relationship between the time required to transmit a minimum-sized Ethernet frame (64 bytes) and the ability to detect a collision (a limit known as the round-trip propagation delay). When a collision occurs, the MAC layer detects it and sends a jam signal, telling the transmitting stations to back off and then retry.

That's fine at 10 Mbit/s. When the IEEE defined 802.3u (100BaseT) in 1994, it maintained the Ethernet framing format and raised the speed limit to 100 Mbit/s. But increasing the clock rate 10-fold means that the time needed to transmit a frame is reduced by a factor of 10. That, in turn, directly affects network diameter, shrinking it from 2 km for 10BaseT to 200 m for 100BaseT.

Since gigabit Ethernet represents another tenfold increase in clock speed, it should require another 10-fold reduction in network diameter. But a 20-m network is clearly impractical, so the 802.3z working committee has come up with a mechanism that preserves the 200-m collision domain of 100BaseT. In essence, the committee has redefined the MAC layer for gigabit Ethernet. This is necessary because at gigabit speeds, two stations 200 m apart will not detect a collision when both simultaneously transmit 64-byte frames. And the inability to detect collisions leads to network instability.

The mechanism that makes a 200-m network diameter possible is known as *carrier extension*. Here's how it works. Whenever a gigabit network adapter transmits a frame less than 512 bytes long, the gigabit MAC sends out a special signal (while continuing to monitor for collisions).

All told, the frame and carrier extension will last for a minimum of 512 bytes, which is equivalent in time to transmitting 64 bytes at 100 Mbit/s. If the gigabit MAC detects a collision during this period, it reacts just like its conventional counterparts, sending a jam signal and telling the offending stations to back off and try again.

Sharp-eyed readers will notice that the math doesn't work here: gigabit Ethernet should be able to transmit 640 bytes, not 512, in the time it takes a fast Ethernet interface to transmit 64 bytes. The IEEE 802.3z working committee decided that a 640-byte extension was too inefficient and shaved the extension down to 512 bytes. First, the group cut the number of repeater hops permitted in gigabit Ethernet to one, down from two in 100BaseT. This provides some timing relief in detecting collisions. Second, earlier Ethernet implementations have a safety margin built into all engineering specs. This margin was basically eliminated in gigabit Ethernet, putting the onus on manufacturers to adhere strictly to the final specification when it is ratified. Cheating on gigabit Ethernet timing, even a little, will cause network instability.

Carrier extension lets gigabit Ethernet scale to a usable size, but it has a downside. No user data is loaded into the carrier extension portion of any frame shorter than 512 bytes. This doesn't exactly make for efficient use of the bandwidth.

How bad can it get? In worst-case calculations, with traffic consisting completely of 64-byte frames, gigabit Ethernet's effective throughput would drop to 120 Mbit/s. That's only 12 percent of total capacity, just slightly better than 100BaseT.

Of course, no network carries small frames exclusively. Frame size distribution varies, but the average on most Ethernets is somewhere in the range of 200 to 500 bytes. Under those conditions, gigabit Ethernet would deliver anywhere from 390 to 977 Mbit/s, enough of a bandwidth boost to keep things clipping along on most corporate nets.

What's the bottom line for net managers? Analyze, analyze, analyze: take a close look at what size frames are being carried on the network now, and then assess how much of a capacity increase gigabit Ethernet is likely to offer.

1.3 Bursting the Bubble

To improve gigabit efficiency, the IEEE also has defined an advanced version of carrier extension called *packet bursting*. With this technique, the gigabit network adapter transmits the first frame by the existing gigabit rules. That is, it includes carrier extension (padding) for packets of less than 512 bytes. At the same time, it starts a packet burst timer set for a transmit duration of 64 Kbits. After the first packet has been transmitted, if there is another packet waiting to be transmitted, and if the packet burst timer has not expired, it transmits 96 bits of carrier extension (padding), followed by the next packet within the same frame. This process is repeated until there are no more packets waiting to be transmitted or the packet burst timer has expired, whichever occurs first.

At first glance, the scheme looks a lot more effective than simply padding short frames. But for packet bursting to work, various applications must stream their short frames for consolidation at the adapter. Trouble is, most apps aren't aware of packet bursting and would have to be rewritten to take advantage of the technique—and that's going to involve a huge redevelopment effort.

Consider, for example, what happens when the client portion of a client-server application commits to an online transaction. The client sends a small commit frame and waits for an acknowledgment before continuing. Thus, there are no other frames from the client to consolidate for packet bursting. Similarly, network file open and close requests and subsequent acknowledgments tend to use small frames. This is a significant end-node concern only since network servers could readily consolidate gigabit Ethernet frames with other user traffic in order to benefit from packet bursting.

In addition, many software vendors argue that developers don't want to be burdened by these sorts of low-level issues. They have a point: developers historically ignore what goes on in the protocol stack, a willful ignorance that has sparked off countless finger-pointing sessions. Trying to convince the software wizards to rewrite their applications to exploit packet bursting—or even explaining why their tried-and-true apps need to be recoded—is likely to be an exercise in futility.

Of course, fixes like carrier extension and packet bursting are only needed when gigabit runs in half-duplex mode. Going full-duplex eliminates the need for CSMA/CD. Stations transmit and relieve data on different wire pairs, so there are never any collisions and thus no need to wait before transmitting. But full-duplex mode generally works only in point-to-point configurations, and implementations will need to comply with the recently ratified full-duplex spec from the IEEE 802.3x working group.

1.4 Buffer or Suffer?

There's another way around carrier extension and its inherent limitations: a new class of gigabit device called a buffered distributor. These boxes meld features found on repeaters and switches and are priced somewhere between the two. They're generally being marketed as low-cost alternatives to full-blown switches that give net managers a way to experiment with gigabit Ethernet without taking out a second mortgage on their glass houses.

A gigabit buffered distributor uses full-duplex links and the flow control mechanisms defined by IEEE 802.3x. Like an Ethernet repeater, it transmits each packet received to all other connected nodes, furnishing shared bandwidth for workgroup devices. Like a switch, it can simultaneously receive on multiple ports, storing the frames in local memory. When memory begins to fill, the buffered distributor invokes 802.3x flow control to inform the transmitting node to stop sending while it empties its buffers. With this approach, it's possible to achieve nearly 100 percent throughput in a shared gigabit domain, even with 1- to 2-km gigabit fiber links. But it's important to understand that this scheme assumes all transmitting nodes comply with 802.3x as well as 802.3z.

1.5   High in Fiber

Cabling is another key way that gigabit Ethernet differs from its slower forerunners. First and foremost, the technology will be teamed with fiber. Multimode (Lx) will be able to transmit at gigabit rates to at least 550 m, with single-mode runs extending to 5 km.

But fiber is typically more expensive than copper, especially in terms of termination and installation. In order to reduce associated cost in the wiring closet, the 802.3z working group proposes using twinax or short-haul copper for distances to 25 m.

What about plain copper? The simple answer is that they're still working on it in a new IEEE working group called 802.3ab. In late 1997, after reviewing a handful of proposals, the IEEE selected five-level pulse amplitude modulation (PAM-5) as the line code transmission scheme to deliver gigabit Ethernet over the required 100-m distance of Category 5 (unshielded twisted-pair) copper cabling. The scheme will require four pairs of copper cabling and will operate in both half- and full-duplex environments. The group hopes to get a standard ratified by late 1998 and to have commercial products available in 1999.

Why is sending gigabit data over copper such a challenge? The real issue will be reflections and corresponding echo cancellation. Gigabit Ethernet can theoretically travel over Cat 5 copper—as long as there are no objects in the actual signal path. Once any object is inserted into the path—such as an RJ-45 jack or a punch-down block—signal reflections start to occur. These reflections can easily destabilize the entire network and are almost impossible to pinpoint. Electrical reflection and the crosstalk interference are not restricted to gigabit transmissions. At 10 Mbit/s over copper, the problems are irrelevant; at 100 Mbit/s, they become more serious, which is why some 100BaseTX implementations fail.

Although PAM-5 will certainly help, sending gigabit Ethernet over UTP will still require the use of high-end digital signal processors to cancel reflections. These processors are expensive in comparison to standard 100BaseTX Ethernet cards. If this theoretical design approach works as anticipated, the DSP algorithms will actually provide a more reliable connection (with a lower error rate) than the existing 100BaseTX installations. However, in order to reduce the cost of these UTP gigabit adapters, mass volumes will be required. Whether UTP gigabit to the desktop has the potential to generate those volumes is in doubt.

1.6   Server Slowdowns

It's not just the wire itself that can put a damper on gigabit speeds. The stations on the wire can also slow things down seriously—particularly in the case of UNIX and Windows NT servers from Intel Corp. (Santa

Clara, California). These operating systems run in protected mode, and raw throughput at the server typically tops out between 100 and 200 Mbit/s. Tying these servers into a gigabit connection is a recipe for disaster. The server and its local gigabit network adapters would be overwhelmed, and the resultant dumped data and retransmissions would actually slow network performance.

The small (less than 1-Mbyte) memory caches and slow main memories on most Intel-based servers create another limiting factor. A 60-ns DRAM may sound like it really sizzles, but the theoretical maximum rate at which a system can read and write data to a network adapter, after local cache burnthrough or CPU starvation, is approximately 128 Mbit/s, assuming a 32-bit data structure. (Cache burnthrough indicates that the CPU has used up all the data waiting for it. CPU starvation means there is no data waiting to be processed.) Here again, slotting a gigabit Ethernet network adapter into one of these servers would degrade performance rather than pump it up.

The outlook is a lot brighter for higher-end UNIX servers from Hewlett-Packard Co. (Palo Alto, California), IBM, and Sun Microsystems Inc. (Mountain View, California). Enterprise servers typically process data from the network at anywhere from 200 to 400 Mbit/s. If they're being fed by fast Ethernet connections, enterprise servers may actually be underutilized. Deploying gigabit Ethernet adapters in this scenario can significantly improve performance.

1.7 Going Gigabit

Where gigabit Ethernet really comes into its own, however, is on fiber internetworking switch links (ISLs). Simply stated, an ISL is a data link between local switches and hubs that consolidates multiple 10- and 100-Mbit/s traffic streams between wiring closets or campuses (see Fig. 1.1).

Let's take a closer look. For the majority of desktop applications, a switched 10-Mbit/s connection supplies more than enough bandwidth. The 10-Mbit/s switch that feeds the local desktops would be connected to a gigabit Ethernet switch.

For workstations that execute data-intensive applications, a switched 100-Mbit/s connection is more appropriate. The 100-Mbit/s switch would in turn be connected to a gigabit switch. And the gigabit switches are tied together by an ISL.

When deploying gigabit buffered distributors or switches, it's critical to offload traffic from the ISL in increments small enough for downstream devices to handle. Consider a server equipped with a single 100-Mbit/s network adapter. Since most downstream devices don't support 802.3x flow control, it's all too easy for a server's buffers to overflow. When this happens, overall network performance will slow down,

Gigabit Ethernet comes into its own as an internetworking switch link (ISL) that aggregates 10- and 100-Mbit/s feeds from desktops and servers.

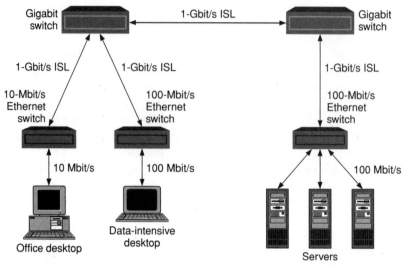

Figure 1.1   Going gigabit.

since dropped data must be recovered using conventional application time-out/resend logic.

Routers also could slow gigabit Ethernet. Conventional routers are already at or near their performance limits, which is one reason router makers are scrambling to speed up their boxes. A new crop of routers and hybrid switches/routers is starting to appear, and some of these handle gigabit speeds.

So much for what gigabit Ethernet does and does well. Having a clear sense of what the technology *can't* accomplish is equally important: real-time services like multimedia are simply not in the picture (or the frame) for now.

Unfortunately, this seems be one of the biggest misconceptions about gigabit Ethernet. Many people in the industry argue that support for real-time services is a protocol issue rather than a physical-layer problem. That's not completely correct. The physical layer of a network that furnishes quality-of-service guarantees should offer connection admission control (CAC) and predictable packet arrival. ATM (asynchronous transfer mode) does both: It is a connection-oriented scheme that uses uniform-length cells. But gigabit Ethernet is a connectionless technology that transmits variable-length frames. As such, it simply can't guarantee that real-time packets get the preferential treatment they require. Gigabit Ethernet may be a very fat pipe, but it's not a magic bullet.

2

The Gigabit Ethernet Standard

Bernard Daines*
President, CEO, and founder, Packet Engines Inc.
With additional work by Bryan Osborn and Matt Johnson[†]

2.1 Introduction: Why a Standard?

In the world of networking, there is one primary objective: to ensure that information is delivered to its destination. In the world of Ethernet, this objective is achieved by way of the data frame, or packet, which carries data in a prearranged format. This format allows the transfer of data between software applications on network-connected devices (see Fig. 2.1).

Suppose one network-connected device is sending data packets to another network-connected device, except that the devices were made by different vendors (see Fig. 2.2). Since each vendor has its own methods of implementation, these devices may not be completely compatible, making it impossible for them to understand and communicate with each other. So how would these two devices ever be able to work together seamlessly, as if they were indeed created by the same vendor? The answer lies in standards.

A standard is a set of guidelines that describes how a product must function in order to achieve its objective (in this case, interconnectivity).

---

* Bernard Daines is president, CEO, and founder of Packet Engines Inc., a privately held company headquartered in Spokane, Washington. Daines is a graduate of Brigham Young University and received the first degree in computer science issued by that institution. Following a move to Northern California, Daines worked for IBM and Hewlett-Packard prior to founding his first company, Tidewater Associates. His clients there included Northern Telecom, 3Com, Texas Instruments, Pacific Bell, and Cisco. In 1992, Daines cofounded Grand Junction Networks, which later was acquired by Cisco Systems. In 1994, Daines established Packet Engines to develop leading-edge gigabit Ethernet products based on that technology. During his career, he has created 38 successful ASIC designs as well as numerous circuit boards, systems, and test equipment products.

† The following staff at Packet Engines also contributed material to this chapter: P. J. Singh, Steve Ramberg, Kevin Daines, Howard Johnson, and Brian MacLeod.

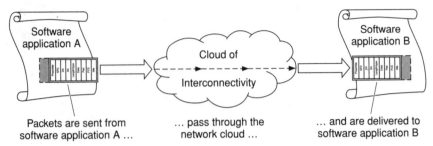

Figure 2.1   Communication between software applications is achieved through the use of a standardized data packet structure.

These guidelines are determined by standards-making bodies made up of industry leaders. A standard does not state how the internal details of a product must be constructed but merely serves as a set of rules or an outline that must be followed to achieve an overall end result. This brings into play the concept of interoperability, which is needed when many different types of network-connected devices are expected to work together seamlessly, as if they were produced by the same vendor. A good standard allows each vendor to accomplish the same task in different ways while providing harmonious interoperation.

All adaptations of a standard must be able to connect and operate alongside each other, because many different product variations may be used in a network. In Fig. 2.3, each stack of OSI Model layers (the OSI Model is described in more detail later in this chapter) represents one network interface, while the cloud of interconnectivity represents the connections between the interfaces. The arrows represent the medium, or wiring, used to connect each interface to the network. Each link in this diagram has been designed by a different vendor, yet a standard allows all the links to operate with one another.

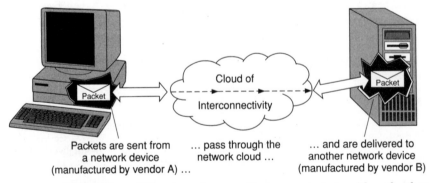

Figure 2.2   Products from different vendors are able to communicate with each other through the use of standardized data packets.

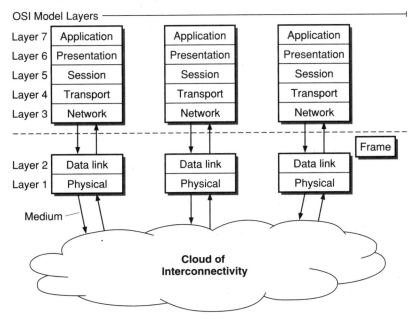

Figure 2.3 The standardized frame is the vehicle of communication between network components that mesh in the cloud of interconnectivity.

Gigabit Ethernet is an example of a good network standard. The gigabit Ethernet standard is concerned only with what occurs below the dashed line in Fig. 2.3; that is, with standardized frame size and content, as well as the data link layer, the physical layer, and medium functions. When all of these components operate according to standard specifications, they can then mesh in the cloud of interconnectivity.

The overall Ethernet standard, as determined by the Institute of Electrical and Electronic Engineers (IEEE), is recognized as IEEE Standard 802.3. Gigabit Ethernet, one member of the large family of Ethernet standards, is known as IEEE Standard 802.3z.

2.2 Data Frames

At the heart of the Ethernet standard is the data frame, a universal convention for the exchange of data. The frame format is a constant around which all versions of Ethernet revolve. The frame, of course, contains data, but it also includes essential information needed to ensure that the data frame stays as pristine as possible and ends up where it is supposed to. Integral to this format are limits on the minimum and maximum frame length, which assist in preserving network efficiency. The 802.3 standard frame definition is an essential component of interoperability. Frames will be explained in more detail later in this chapter.

2.3   What Is Gigabit Ethernet?

Gigabit Ethernet is simply an extension of earlier Ethernet technologies, only gigabit is much faster. Instead of traveling at speeds of 10 or 100 million bits per second (Mbits/s), gigabit Ethernet sends data through a network at the rate of 1 gigabit per second (1000 Mbits/s). Gigabit Ethernet uses the same frame format as all other existing, ubiquitous Ethernet technologies. By utilizing the expanded bandwidth of gigabit Ethernet technology, LANs can better accommodate multimedia and other heavy traffic.

The cloud of interconnectivity for gigabit Ethernet is built from two types of devices: repeaters and switches. Repeaters are single-stream devices (meaning communication takes place along one path) that cost less than switches because they simply regenerate signals and repeat *all* data transmissions to *all* connected ports. Switches are multiple-stream devices (meaning that many ports can receive and transmit simultaneously) that not only rejuvenate signals but also use address filtering to send data packets to *only* the needed destinations.

IEEE Standard 802.3z also addresses backward compatibility with existing Ethernet technologies. Most importantly, it uses the same 802.3 Ethernet frame format. The best way to understand how gigabit Ethernet works is by first examining the OSI Model.

2.4   The OSI Model

The majority of action specific to gigabit Ethernet occurs in the lower two layers of the Open Systems Interconnection (OSI) Model, a seven-layer representation of network systems. Standardized in 1984, the OSI Model specifies the kinds of functions performed in each layer, but not the details of how each layer works (see Fig. 2.4).

An operating system such as Windows NT communicates only with the top layer of the OSI stack. The application, presentation, session, transport, and network layers prepare the data for communication through a physical medium. The data link layer controls access to the physical medium, while the physical (PHY) layer controls actual transmission through it. The bottom two layers of the OSI Model—the data link and physical layers—are the focus of this chapter.

2.4.1   The data link layer

The data link layer controls access to the physical medium. It schedules the transmitting and receiving of data. The main component of this layer is the media access control (MAC), which transforms data

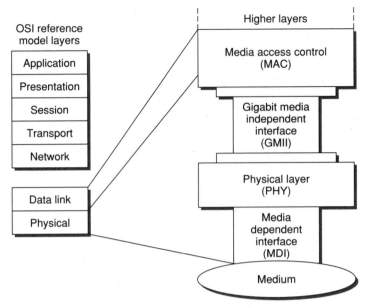

Figure 2.4 Close-up of the data link and physical layers.

units into frames and determines how a network device schedules, transmits, and receives data. The MAC is also responsible for ensuring reliable data transmission across the medium. Sublayers within the MAC include the MAC client, the MAC control, and the central MAC unit. These sublayers are explained in more detail later in this chapter.

2.4.2 The physical layer

The physical layer (PHY) encodes frames with delimiters needed for transmission and then sends them along the appropriate medium. More specifically, the PHY, sometimes known as a transceiver, defines the electrical or optical signaling, line states, clocking guidelines, data encoding, and circuitry needed for data transmission and reception. Contained within the PHY are several sublayers that perform these tasks, including the physical coding sublayer (PCS), the serializer/ deserializer (or physical media attachment sublayer), the optical transceiver (also known as a physical media dependent sublayer) for fiber mediums, and the media dependent interface (MDI). These sublayers are explained in more detail later.

2.5 The Ethernet Frame

Obviously, a computer does not speak English, nor can it understand the groupings of information found on a spreadsheet or any other application. At the most basic level, the computer's world is digital; it can only understand 1s and 0s. By combining infinitely complex patterns of these 1s and 0s, called *bits,* the computer can create representations of millions of numbers, any number of words, and many thousands of colors and shapes. These bits are then organized by the user's choice of protocols, until a code is generated that the computer can understand (more often than not, the user is not aware that these protocols even exist).

Even though the information is encoded, the computer can't take an entire page from a document on the desktop and cram it all down the wire at once. In Ethernet, there is a special process called packetization that must occur first. Packetization refers to the process of breaking data into smaller chunks in preparation for transmission across the network.* As the data stream flows down the OSI stack from the application layer to the network layer, it is broken down by degrees into successively refined groupings. Eventually, groupings of bytes reach the data link layer, where they are packaged into specific groupings called frames.

The core of the Ethernet system is the Ethernet frame, as defined by IEEE Standard 802.3. A frame is the unit used to deliver information from one device to another and is one of the concepts that helps define the essence of Ethernet. The Ethernet frame is composed of a bundle of bytes organized into very specific fields or sections (see Fig. 2.5). These

---

* Unfortunately, the term *packet* is rather vague in its usage. In general, a packet refers to any small group of data that is part of a larger division of data. In Ethernet, a packet is a frame with transmission delimiters.

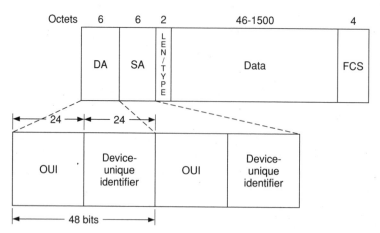

Figure 2.5 The 802.3 Ethernet frame structure with the makeup of the DA and SA.

sections include one field each for the destination address (DA) and source address (SA), a small length/type field, a variable-length data field, and the frame check sequence (FCS), which is tacked onto the end of the frame. At the receiving node, all the packetization steps are reversed. Therefore, a large unit of data, broken down by one OSI stack for transmission in bits across the network, is reconstituted by the receiving OSI stack into the same recognizable unit of data.

2.5.1 Destination address

The first field encountered in the frame is known as the destination address (DA). This field is present to help each device determine where the frame needs to go. The DA is further broken down into two parts that make it unique to each device. These two parts include the organizationally unique identifier (OUI) and the device-unique ID. The IEEE assigns a specific OUI to each vendor that builds Ethernet devices; IBM, Cisco, and 3Com, for example, each have multiple different OUIs. In turn, vendors typically complete the DA on manufacture by combining a device-unique identifier with the company OUI. This two-step process ensures that no two Ethernet devices in the world could ever accidentally share the same destination address. Now that the DA is complete, it is known as a physical address or MAC address.

2.5.2 Source address

Following the DA field is the source address (SA) field. The SA is identical in format to the DA except for the position in which it appears in the frame. It is present to identify where the frame came from in case frame errors occur or the source device needs to be contacted for some reason.

2.5.3 Length/type

Sandwiched between the data field and the SA is a tiny field called the length/type field. Originally, this field contained information that identified the type of protocol used in the frame to the hardware it encountered. Later, the IEEE 802.3 committee changed this field to indicate the actual length of the frame or packet. The committee decided that the redundancy of length indication would be used as an additional means to ensure frame integrity. In gigabit Ethernet, this field is a length field if the value is less than 1500 or a type field if the value is 1536 or larger.

2.5.4 Data

The largest and obviously most useful field of the frame is the data field. User data and other internal information is packaged into this portion. In other words, this is where the meat of the information is contained, while the other sections are merely headers and footers to the encoded information. This field is limited to 46 bytes minimum and 1500 bytes maximum. All user data must be proportioned into bundles that fit within the data field of the Ethernet frame.

Breaking user data up into smaller portions makes the information more manageable. Since the system can expect the traffic to be of predictable length, the latency (or delay) of the frames remains fairly constant and smooths out transmission. Frames are delivered at a relatively constant, even rate. In addition, since frame size is known, the size and number of buffers in a given device can be reduced to be proportionate to the amount of traffic expected, which helps control the cost of hardware.

There are also other advantages to limiting the data field length. Since Ethernet creates a network by linking devices together across a shared medium (or system element), when the data is broken up into short bundles, each device will have only a short waiting period before being able to transmit. This method is very efficient; the first few frames of a data block can be received by the end device before the last packet has even been transmitted by the source device.

The fixed maximum frame size is also useful when it comes to error control. In the rare event that a packet is dropped or is found to contain a coding error, the amount of data that needs to be retransmitted remains fairly small. The fixed minimum packet size is present for historical reasons associated with the timing characteristics of CSMA/CD. Packets smaller than the minimum size are padded until they reach minimum size status.

2.5.5 Frame check sequence

Perhaps one of the most important properties of the frame is its ability to indicate when errors have occurred in the coding sequence. In Ethernet, the frame check sequence or FCS provides the error indication. The FCS, also referred to as a cyclic redundancy check (CRC), appears at the end of the frame. On receiving a new frame, the receiving device performs a sequence of mathematical equations on the incoming frame, compares its result to the residual value (CRC polynomial), and determines whether any errors have occurred.

2.6 Ready for Transmission

As described earlier, the data link and physical layers are the channels though which data frames pass to be prepared and sent across the transmission medium to their destination. Having already dealt with the OSI layers and frames, the next step in understanding the gigabit Ethernet standard is to look at how the data link and physical entities prepare data to be transmitted across the network.

Timing is an important attribute that makes gigabit Ethernet work by ensuring the reliable delivery of frames to their destination. The data link layer is mainly concerned with scheduling frames for transmission, ensuring that they travel across the medium in pristine condition. As frames are lined up to be sent on their way, the data link layer uses the MAC protocol to determine who can access the broadcast medium and when. The MAC is made up of two sublayers that perform these scheduling duties (see Fig. 2.6).

The MAC receives frames from the MAC client (formerly known as the logical link layer), which contains formatted data from the software application. As described earlier, the MAC formats all data into frames that contain a destination address, source address, length/type field, data field, and frame check sequence.

The gigabit Ethernet MAC has two operating modes: half-duplex (CSMA/CD) and full-duplex. In half-duplex mode, gigabit Ethernet uses the traditional Ethernet carrier sense multiple access with collision detection (CSMA/CD) feature. With CSMA/CD, the same channel can only transmit or receive at one time. A traditional repeater combines nodes together into a collision domain, where only one device may successfully transmit at any given instance.

CSMA/CD is much like placing your ear to a train track to determine whether a train is coming. (NOTE: *Do not* try this at home!) The MAC uses this protocol to determine whether a packet is coming toward the node before it transmits. If the medium appears to be clear, the MAC client will hand the PHY a packet; the PHY then transmits the packet.

Frames often proceed across the medium without encountering a collision. If so, the MAC client will prepare the next frame for transmission. Before a device can transmit, however, it must listen to the wire again. This behavior allows the network to detect and recover from collisions.

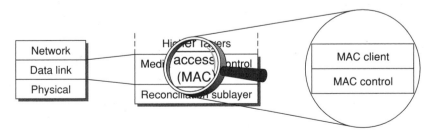

Figure 2.6 Sublayers in the MAC include the MAC client and MAC control.

A collision results when a frame sent from one end of the network collides into another frame sent from the opposite end of the network (see Fig. 2.7). Timing becomes especially crucial if and when a collision occurs. If a collision occurs *during* the transmission of a packet, the MAC will stop transmitting and resend the packet when the transmission medium is clear. If the collision occurs *after* a packet has been sent, then the packet is lost, because the MAC has already discarded the packet and started to prepare the next packet for transmission. In all cases, the rest of the network must wait for the collision to dissipate before any other devices can transmit.

In a full-duplex network, collisions do not occur. Full-duplex means that there are two channels for each connection: one for transmitting and one for receiving. This setup eliminates most of the need for CSMA/CD, because there is no need to determine whether the connection is already being used. Full-duplex networks are very efficient, because data can be sent or received at any given time.

Gigabit Ethernet NICs use buffers when operating in full-duplex mode. The buffers store incoming and outgoing data frames until the MAC has time to push them higher up the OSI protocol stack. During heavy traffic transmissions, the buffers may fill up with data faster than the MAC can process them. When this occurs, the MAC control sublayer holds off the upper layers from sending until the buffer has room to store more frames; otherwise, frames would be lost, since there would be no place to store them.

In the event that the receive buffers approach their maximum capacity, a high-water mark alarms the MAC control and sends a signal to the source of data transmission to halt sending frames until the buffer can catch up. The high-water mark is set to guarantee that enough buffer capacity remains to give the MAC time to tell the other device to shut down the flow of data before the buffer capacity overflows. Likewise, there is a low-water mark that will notify the MAC control when there is enough open capacity in the buffer to restart the flow of incom-

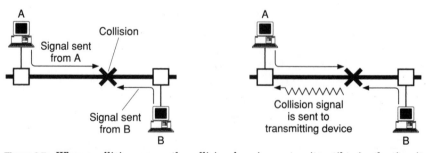

Figure 2.7   When a collision occurs, the collision domain must wait until twice the time it takes for a frame to reach its destination before any other devices can transmit.

ing data. The MAC control restarts the data by sending a resume signal to the transmission source.

When the MAC sends a frame on its way to the physical coding sublayer of the PHY, it first travels through the reconciliation sublayer, also called the gigabit medium independent interface (GMII). The GMII is designed to enable the Ethernet MAC chip to hook up to any one of a number of different PHY chips in a standard way.

2.7 The Encapsulation Process

As mentioned earlier, the physical layer is that portion of the OSI Reference Model that provides an interface between an Ethernet device and its medium. But it is much more than just an interface; it also contributes the mechanical, electrical, functional, and logical means to establish, maintain, and disengage physical connections over the given transmission medium. In the PHY (generically referred to as a transceiver) and its associated sublayers, raw data packets are prepared to be transmitted physically on some form of medium such as fiberoptic cable. The packets are encoded, translated into electrical impulses, converted to photons (in the case of fiber), and then physically sent down the wire. This process of preparation begins in the physical coding sublayer. The PHY is made up of three sublayers that perform these encoding functions (see Fig. 2.8).

2.8 Packets

During the process of encoding, the frames have additional information added to them for transmission across the network. The basic data frame (discussed in Sec. 2.5) is coupled with other control functions, including a preamble and a start delimiter. Once these addenda have been made, the bundle is then known as a *packet*. The packet

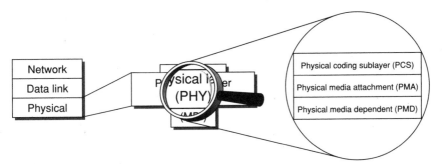

Figure 2.8 Sublayers in the PHY include the physical coding sublayer, physical media attachment, and the physical media dependent.

length in Ethernet protocol varies, but it is limited to a minimum of 72 bytes and a maximum of 1526 bytes. With these final additions, the packet nears completion.

The portion of the frame that acts as an engine, pulling the rest of the packet along, is known as the *preamble*. As the name suggests, this portion is an introduction to the data contained within the frame. It is sent to allow the receiver at the destination device and all receivers in between to synchronize their clocks or other circuitry with the incoming transmission. The last octet in the preamble is called the start frame delimiter (SFD). The SFD octet is used to indicate that the data field of the MAC frame is about to begin.

The 1000BaseX frame has two other elements that encapsulate the packet. The end packet delimiter (EPD), like its counterpart, the start packet delimiter (SPD), is another addition made to the frame by the transmitter. The EPD is appended after the frame check sequence and is stripped off by the receiver. The first octet in the eight-octet preamble is replaced with a section known as the SPD pattern. This pattern is substituted by the transmitter and acts as a kind of flag to all receiving devices to help find the beginning of the preamble. Later, the SPD pattern is also stripped by the receiver, which returns a normal preamble to the MAC.

In addition, there is one other component associated with the packet: the interpacket gap (IPG), or idle interval. The Ethernet transmitter is always on. When packets are not being sent, the transmitter sends a pattern known as *idle*. Idle is sent mainly to aid synchronization at the receiver. The Ethernet IPG rule provides that the minimum span of idle between the end of the FCS on one packet and the beginning of the preamble on the next is to be 12 octets.

Now that the packet and its associated components have been discussed, Fig. 2.9 gives a visual representation of an 802.3 packet, including the number of octets in each component.

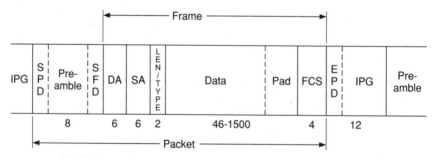

Figure 2.9 The Ethernet packet and frame structures.

When packets have been formatted by the MAC to include all essential components, they can then be prepared for network transmission. This is accomplished in the physical layer (PHY), where the packets undergo a series of transformations that allow them to be sent over the transmission medium. Figure 2.10 is a magnification of the PHY pipe that gives an overview of the internal functions of the PHY.

2.9 Physical Coding Sublayer: 8B/10B Coding

The physical coding sublayer is the first division of the PHY and is responsible for the actual encoding of transmitted data to a form suitable for the physical medium. In gigabit Ethernet, the coding process used is called 8B/10B (8-bit/10-bit) coding. 8B/10B, which arose from Fiber Channel and other fiber protocols, is a good coding scheme for transmission on optical fiber, and, since gigabit Ethernet initially functions over fiber, it is a natural choice.

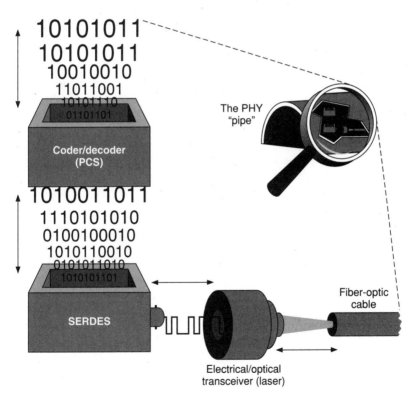

Figure 2.10 The PHY processes incoming and outgoing data packets.

When the PHY, or, more specifically, the PCS (also known as the coder/decoder), is handed an 8-bit group of data from the MAC, it takes the data and reencodes it into a larger group of 10 signal bits (see Fig. 2.11). That is, the entering group of 8 bits is taken together and translated into a different, larger bit pattern before it is even transmitted on the wire. This aids in the process of interpreting frames.

Generally, the relationship of an 8-bit data code is one to one with octets, meaning that every combination of bits is representative of a particular octet. When 8 bits are translated into 10, additional unique combinations are made available for purposes other than denoting data octets. These liberated code combinations are unique in that they do not have a corresponding data octet value and can thus be used to indicate the end and beginning of packets and frames and signal control instructions between devices or provide other functions.

In addition to providing auxiliary code combinations, 8B/10B coding also ensures accuracy of transmission. Only bit combinations that include several alternations between 1s and 0s are allowed. The resulting code therefore contains an abundance of transitions: at least two per group. This assists the decoding process by aiding in the synchronization of the timing clock. A transition-rich bit stream does not allow the receiver's clock to wander. Instead, the receiver is required to align its internal clock constantly with the incoming bit stream.

Coding also accomplishes another feature called DC balancing. Since data is represented by swings in voltage, it is possible that the system

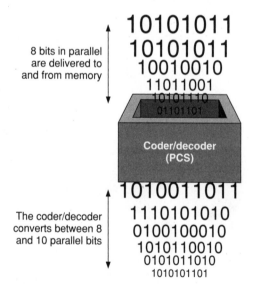

8 bits in parallel are delivered to and from memory

10101011
10101011
10010010
11011001
10101110
01101101

Coder/decoder (PCS)

The coder/decoder converts between 8 and 10 parallel bits

1010011011
1110101010
0100100010
1010110010
0101011010
1010101101

Figure 2.11 The PCS encodes 8 outgoing parallel bits into 10-bit groupings and decodes incoming 10-bit groupings back into 8 parallel bits.

could spend more time, on average, in either a positive or negative state. If this were the case, AC-coupled components in the chain of signaling could eventually develop a DC charge, which would tend to distort the transmitted signal. This interference is referred to as *baseline wander.* 8B/10B coding precludes any form of baseline wander by inherently balancing the number of 1s and 0s within each transmitted data code group so that no DC charge can accumulate. Perhaps the single largest advantage to using the 8B/10B encoding scheme is its simplicity. 8B/10B coding is widely used and easy to design, and carries with it many benefits.

2.10   Serializer/Deserializer

In digital communication, some juxtaposition needs to take place in order to transmit encoded information across a wire. The objective, of course, is to transfer encoded packets from one device to another. At the output of the PCS sublayer, there is now a packet encoded with 10 bits per octet. It contains a DA, an SA, and various other pieces of information to help the system determine where to send the signal, but the problem lies in the fact that the packet needs to be delivered across a single pair of wires or a fiber-optic cable. So how can a grouping of 10 bits (nothing more than a bundle of 1s and 0s) be sent down the medium? In order for this to occur, the group of outgoing bits must be broken down into even smaller bit groupings, perhaps only 1 or 2 bits at a time, and chained together. The process of splitting parallel bits up and sending them one right after the other is called *serialization.*

In Ethernet, this function takes place in a device known as the *serializer/deserializer* (SERDES) (classified as the PHY sublayer physical media attachment or PMA). As the name indicates, the SERDES has the function of serializing the packets that are handed to it for transmission and then deserializing the incoming information (see Fig. 2.12). Deserializing is merely the reverse process; the bit stream enters, 1 bit at a time, and the SERDES assembles the bits back into 10-bit groupings. Special 8B/10B codes allow the deserializer to align to received data.

2.11   Optical Transceiver

While gigabit Ethernet utilizes the same wiring methods as other Ethernet standards, IEEE Standard 802.3z's initial emphasis is on fiber-optic cables, which offer the highest performance of any frame-transferring medium. Fiber uses a single thread of glass to carry light signals. A primary reason for fiber's popularity is that outside electri-

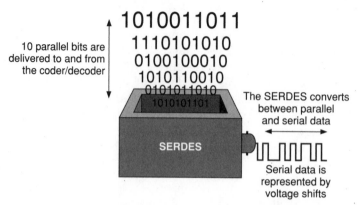

10 parallel bits are delivered to and from the coder/decoder

1010011011
1110101010
0100100010
1010110010
0101011010
1010101101

SERDES

The SERDES converts between parallel and serial data

Serial data is represented by voltage shifts

Figure 2.12   The SERDES transforms outgoing parallel bits into serial data and then converts incoming serial data back into parallel bits.

cal currents cannot interfere with light signals carried over glass. Another reason fiber is so popular is that it is the fastest, most reliable data transmission medium available. Because light signals can travel for extremely long distances without losing much of their strength, and because there is no interference, fiber offers the most data throughput of any medium. Although fiber is the most expensive medium, it is clearly the most reliable.

Fiber-optic technology is analogous to light communication between ships. A communicating sailor opens and closes metal slats covering a strong light in a series of coded patterns that represent the information in a message.

In fiber-optic data transmission, the sailor and the light are replaced by a laser and glass fiber. Once the serial bit stream has been converted into electrical form, it is passed on to a converter. The converter is able to read the electrical signals and drives a laser that focuses light into the glass fibers. Much like the sailor's message, light is pulsed in patterns that represent in optical form what was translated into electrical form. When the light reaches the receiving device, the pulses are examined for state changes and timing and are then changed back into electrical impulses.

In gigabit Ethernet, it is the electrical/optical transceiver (classified as the sublayer called physical media dependent or PMD) that transforms the incoming pattern of changing voltages into light waves or pulses so that it can be sent across the fiber cabling (see Fig. 2.13).

Another sublayer for the PHY is the autonegotiation function, which allows pairs of devices to communicate with each other regarding the configuration of each connection. Both connected devices can then determine the most efficient operating mode to use when exchanging data.

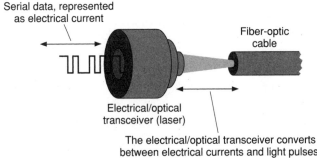

Figure 2.13 Outgoing electrical currents are converted into light pulses by the electrical/optical transceiver, which also converts incoming light pulses back into electrical currents.

2.12   Media

Like earlier versions of Ethernet, gigabit Ethernet will operate over a variety of physical media. As stated earlier, gigabit Ethernet will initially support optical fiber cabling with a maximum distance of up to 550 m on multimode fiber and 5 km on single-mode fiber.

Gigabit Ethernet will eventually be able to operate over other mediums that use copper as the wiring conductor. These include unshielded twisted-pair (UTP) wire and twinaxial wire such as coaxial cable. Unshielded twisted-pair wire is the least expensive and slowest medium and consists of four twisted wire pairs. Electrical noise from nearby pairs is canceled out because each wire pair is braided. Shielded twisted-pair, which can also be used, is a faster and more expensive medium that operates the same way as UTP but provides a higher level of protection from electrical current interference. Along with the twisting of the wire pairs, it also utilizes braided copper shielding and foil shielding around the wire pairs. The IEEE has identified the initial maximum copper link distance for twinaxial wire as 25 m but is currently investigating other methods to allow a copper link distance of up to 100 m on UTP.

Gigabit Ethernet sends frames of data across a medium at 1000 Mbits/s. The PHY conveys these frames to the cabling medium by way of a media dependent interface (MDI) connector, which transfers the data to the medium, and packets are then on their way to their destination.

2.13   The Reverse Process

When packets reach their end destination, all the processes in the OSI layer stack are reversed. Light pulses are transferred back into electrical impulses by the optical transceiver. The SERDES receives the electrical impulses serially and then assembles, or deserializes, them back into 10-bit packets. The packets then are sent into the physical coding sublayer, which transforms the packets into 8-bit groupings (the opposite of the 8B/10B encoding process). These groupings are given to the MAC, which deposits them in the appropriate memory buffers for later processing.

3

Enhanced CSMA/CD: Packet Bursting

Moti Weizman*

Director of hardware engineering, Nbase Communications Inc.

3.1 Introduction

Scaling Ethernet to run at 1 Gbit/s in a shared-media environment is not a trivial undertaking. In theory, the fastest and safest way to define a shared-media gigabit Ethernet standard would be simply to take the well-known CSMA/CD algorithm used in 100BaseT and run it 10 times faster, in just the same way as fast Ethernet relies on a version of CSMA/CD that runs 10 times faster than that used in 10-Mbit/s Ethernets. However, attempting to do this would result in a network with a maximum diameter of a few meters, which is undoubtedly too small for most network configurations.

In 1996 NBase Communications (Chatsworth, California) proposed a solution called *packet bursting*. The technology, which is essentially a slightly modified version of CSMA/CD, was adopted as part of the gigabit Ethernet standard in November 1996. Packet bursting is itself based on another solution to the diameter dilemma called *carrier extension*, developed by Sun Microsystems Inc. (Mountain View, California). Crucially, packet bursting provides better network utilization compared to pure carrier extension, especially with smaller frames.

Packet bursting is only designed for use in shared-media (or half-duplex) networks, which use repeaters to share a gigabit of capacity among multiple stations. It is not part of the full-duplex specification

---

* Moti Weizman is director of hardware engineering at NBase Communications Inc., Chatsworth, California. He has a BSEE from the Technical Institute in Israel. Moti has 10 years experience in the development of Ethernet products and ASICs, and he is an active participant in the IEEE 802.3z group. He can be reached at motiw@nbase.com.

used by some gigabit Ethernet switches and adapters. This is because, in a full-duplex environment, each attached station has access to a dedicated (as opposed to shared) 1-Gbit/s connection, and the CSMA/CD algorithm is disabled altogether, thus allowing nodes to send as well as receive data simultaneously.

In order to ensure interoperability, the IEEE gigabit Ethernet standard mandates that any device designed for operation in a shared-media (half-duplex) network must implement packet bursting in the receiver component of the hardware. Implementing it on the transmitter, however, is optional. This means that any shared media device will be able to receive packet bursting traffic from any other device, regardless of its manufacturer. However, only those devices that support the spec on their transmitters can take advantage of the performance enhancement of packet bursting when sending data.

This chapter discusses the difficulties involved in scaling CSMA/CD, describes packet bursting in detail, and presents the benefits of packet bursting.

3.2   Background

So why would a shared-media gigabit Ethernet LAN that used the original Ethernet algorithm have such a small diameter? The reason is the CSMA/CD mechanism, which requires that the worst-case round-trip delay on the network be less than or equal to the transmission time of the shortest legal frame. (See Fig. 3.1.) This minimum frame size is 64 bytes for existing 10 and 100 Ethernet. Since the time it takes to transmit a 64-byte frame at gigabit speed is one-tenth the time it takes at 100 Mbit/s, the maximum network diameter shrinks to 20 m. (Note that the maximum diameter is 2 km for 10BaseT, and it shrinks to 200 m for 100BaseT.)

This number is even less if delays in active components such as repeaters are considered. These delays cannot be scaled down to one-tenth of the delays in a 100-Mbit/s repeater with today's silicon technology. In other words, it's not yet feasible for manufacturers of repeater chips operating with a 25-MHz clock to scale them to operate with a 250-MHz clock.

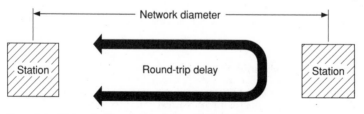

Figure 3.1   Network diameter and round-trip delay.

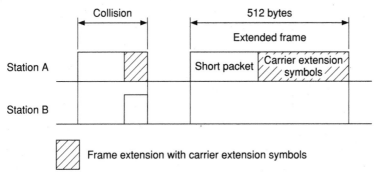

Figure 3.2   Carrier extension.

3.3   Carrier Extension

The carrier extension scheme developed by Sun Microsystems provides a simple way to extend the network diameter to 200 m while leaving the current CSMA/CD algorithm unchanged. With carrier extension, the minimum slot time is increased to 512 bytes as opposed to 64 bytes. Frames that are shorter than 512 bytes are transmitted in a 512-byte window. A carrier extension symbol is used to mark frames shorter than 512 bytes and fill the remaining time in the 512-byte window. (See Fig. 3.2.)

Carrier extension is a simple antidote to the diameter problem. In fact, it is too simple. A quick analysis reveals two concerns: a very low network utilization for frames shorter than 512 bytes and a higher collision probability that may increase the number of lost frames and amplify some other known CSMA/CD deficiencies.

3.4   Low Utilization for Short Frames

Obviously, transmitting carrier extension signals instead of real data will result in a low utilization for short frames. For example, a 64-byte frame will have 448 bytes of wasted (nondata) carrier extension symbols attached to it.

Network utilization (U) is described in Eq. (3.1)—network utilization, carrier extension, no collision.

$$U = \frac{FL - ND}{FL + PRM + IFG + CE} \times 100 = \frac{FL - 18}{FL + 20 + CE} \times 100 \quad (3.1)$$

where FL = frame length
  ND = nondata information that has to be transmitted per frame; composed of 18 bytes of MAC addresses, type fields, and CRC fields
  PRM = preamble, transmitted at the beginning of each frame—8 bytes

IFG = interframe gap—12 bytes
CE = carrier extension size if (FL < 512) then (512-FL) else 0

It's clear from Fig. 3.3 that, with 64-byte frames, the utilization of a gigabit network is less than 10 percent, or 100 Mbit/s. That's only double that of a 100-Mbit/s Ethernet, which typically see 50 percent utilization. However, gigabit Ethernet closes the gap as the frame becomes longer, and above 512 bytes it approaches ten times the performance of fast Ethernet.

Should net managers really care about short frames? Yes. Consider control frames. Most are short. Typically, they are transmitted every few data frames; in extreme cases, they are transmitted on every data frame. Even though these frames may not be the majority of the traffic in a typical network, they will have a noticeable effect on the network utilization.

3.5   Frame Loss

Carrier extension increases the collision window from 64 bytes to 512 bytes. Therefore, it also increases the probability of a collision. This effect will amplify some of the deficiencies already present in CSMA/CD, notably the capture effect, frame loss due to 16 consecutive collisions, and variable delay.

A somewhat extreme example will illustrate the higher probability of a collision. Consider a case where station A continually transmits a

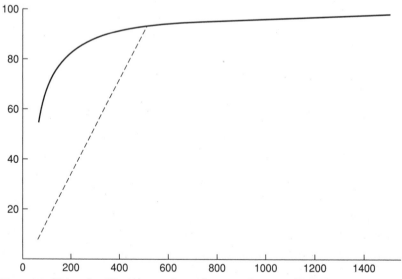

Figure 3.3   Network utilization graph, carrier extension, no collisions.

64-byte frame every slot time, and station B has one 64-byte frame to transmit at a random time. At 100 Mbit/s, network station A will capture the channel for only 64/512 or 12.5 percent of the time. Thus, the probability that the single frame from B will collide with the stream from A can be written as the collision window per frame window. This value is 64/512 = 0.125, or 12.5 percent.

In a gigabit network with carrier extension, however, station A will capture the channel 100 percent of the time, since every 64-byte frame will be extended to capture a 512-byte window. Thus, the probability that the single frame from B will collide with that stream is 1 (i.e., it will always collide on the first attempt). The binary exponential back-off algorithm will, of course, eventually resolve the contention. The question to ask is how this affects the behavior of the network. It really depends on the traffic patterns; however, the main result will be that frame loss due to 16 consecutive collisions will start at a lower network utilization compared to fast Ethernet. The gigabit network, then, will not give the expected 10-fold increase in performance, since it will be unable to operate at the same percentage levels of traffic as fast Ethernet.

3.6 Packet Bursting

Packet bursting is an addition to carrier extension—an extension to the extension, if you will.

In a heavily loaded gigabit Ethernet network where carrier extension is in use, packet bursting will improve bandwidth utilization for short frames and decrease the probability of collisions.

Essentially, packet bursting allows users to gain back some of the performance lost due to carrier extension itself, which is essential, as noted earlier. Packet bursting accomplishes this in a simple way that minimizes the impact to the existing standard. The concept is to transmit a burst of frames every time the first frame has successfully passed the collision window of 512 bytes, applying carrier extension to the first frame in the burst only. This will effectively average the wasted time (in carrier extension symbols) over the few frames that are transmitted. (See Fig. 3.4.)

The size of the burst varies depending on the number and size of the frames being sent by the station. Frames are added to the burst during transmission in real time. To decide how many frames may be added to the burst, the number of bytes already transmitted is totaled after each frame sent. If less than 1500 bytes have already been transmitted, another frame can be appended to the same burst. Conversely, if 1500 bytes or more have been sent, the burst is terminated.

If the initial frame has a size greater than 1500 bytes, it will end up being the only frame in the burst, since the frame size already exceeds

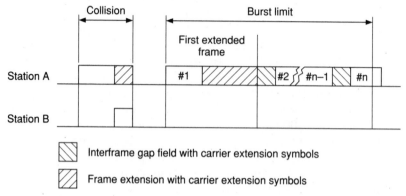

Figure 3.4   Packet bursting.

the 1500-byte burst limit. The maximum number of frames a burst can contain is 13 (if all are 64 bytes in length). The maximum length of a burst will occur when the first frame length is 1499 bytes (which does not exceed the burst limit) and the next frame has the maximum legal size of 1518 bytes. Thus, the maximum burst length is 1518 + 1499, or 3017 bytes.

3.7   Higher Utilization for Short Frames

Packet bursting averages the carrier extension symbols over several frames as opposed to just one frame. Thus, it improves the utilization when short frames are transmitted. To figure out the extent of the improvement, we need to calculate the network utilization with packet bursting (U) described in Eq. (3.2)—network utilization, packet bursting, carrier extension, no collision.

NF is the number of frames in a burst calculated as

$$NF = \text{Ceiling}\left(\frac{BL - CE}{FL + PRM + IFG}\right)$$

and

$$U = \frac{(FL - ND) \times NF}{(FL + PRM + IFG) \times NF + CE} \times 100$$

$$= \frac{(FL - 18) \times NF}{(FL + 20) \times NF + CE} \times 100 \tag{3.2}$$

where NF = number of frames in a burst
BL = burst limit—1500 bytes
CE = carrier extension—if (FL < 512) then (512-FL) else 0

FL = frame length

PRM = preamble, transmitted at the beginning of each frame—
8 bytes

IFG = interframe gap—12 bytes

ND = Nondata information that has to be transmitted per frame,
composed of MAC addresses, type fields, and CRC fields—
18 bytes

Examining the graph in Fig. 3.5, we can see that the network utilization for pure 64-byte frames has improved from 10 percent before packet bursting to almost 40 percent of 1 Gbit/s after. Compare that to 50 percent utilization out of 100 Mbit/s (~50 Mbits) in fast Ethernet, and the result is approximately eight times the bandwidth typically available on fast Ethernet, a considerable performance improvement. (See Table 3.1.)

3.8 Less Frame Loss

Packet bursting also alleviates problems with collisions, since the burst of frames may collide only during the first frame, as opposed to conditions with carrier extension, where each one of the frames may suffer a collision. If the probability of one frame colliding is the same as before packet bursting but fewer frames are subject to collision, the total collision probability is less. Trying to evaluate the actual improvement is difficult and depends on the traffic patterns. Packet bursting will be

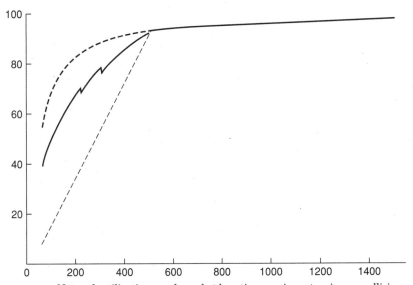

Figure 3.5 Network utilization graph, packet bursting, carrier extension, no collisions.

TABLE 3.1   Maximum Network Utilization Comparison for a 64-Byte Frame Stream
with No Collisions

	Percent utilization	Actual bandwidth
Fast Ethernet	50	50 Mbit/s
Giga + carrier extension, no packet bursting	10	100 Mbit/s
Giga + carrier extension + packet bursting	40	400 Mbit/s

much more efficient for a traffic mix of short frames, and the improvement will be less significant as the traffic mix includes longer frames. However, since short packets are part of every network, it's logical to expect frame loss to start at a higher network load under the packet bursting scheme, creating an overall higher throughput on the network.

3.9   Late Collisions

Packet bursting does affect the behavior of nonlegal networks where the round-trip delay is longer than the allowed 512 byte-times. In such a network, late collisions may occur after the first frame. However, in this situation, management information should include information about late collisions, alerting the network manager that something is wrong with the network.

3.10   Conclusion

Packet bursting addresses all the concerns about short frames in gigabit Ethernet with one simple solution. It retrieves a substantial portion of the performance loss caused by the need to add the carrier extension. It does not change the dynamics of the CSMA/CD algorithm at all. It just makes the whole system more efficient. Finally, packet bursting requires very small changes to the existing MAC definition and no changes whatsoever to the repeater concept.

4

The Full-Duplex Repeater

Bernard Daines*

President, CEO, and founder, Packet Engines Inc.

With additional work by Bryan Osborn and Matt Johnson[†]

4.1 Introduction

With the advent of gigabit Ethernet and even faster network speeds on the horizon, network devices must be able to support the increased bandwidth and speed at which data is transmitted. While networks achieve faster and faster speeds, traditional repeaters may bog the network down rather than relieve it, and, consequently, traditional repeaters will be moved further and further down the network hierarchy. In fact, given the difficulty of scaling CSMA/CD to gigabit speeds and the complexity of schemes such as packet bursting that have been devised to compensate for low performance, half-duplex repeaters may not even make it to store shelves. Full-duplex repeaters, however, provide more intelligence than a traditional repeater at a lower cost than a switch, bridging the gap between switches and CSMA/CD repeaters (see Fig. 4.1). So, how do they compare to traditional repeaters and switches?

---

* Bernard Daines is president, CEO, and founder of Packet Engines Inc., a privately held company headquartered in Spokane, Washington. Daines is a graduate of Brigham Young University and received the first degree in computer science issued by that institution. Following a move to Northern California, Daines worked for IBM and Hewlett-Packard prior to founding his first company, Tidewater Associates. His clients there included Northern Telecom, 3Com, Texas Instruments, Pacific Bell, and Cisco. In 1992, Daines cofounded Grand Junction Networks, which later was acquired by Cisco Systems. In 1994, Daines established Packet Engines to develop leading-edge gigabit Ethernet products based on that technology. During his career, he has created 38 successful ASIC designs as well as numerous circuit boards, systems, and test equipment products.

† The following staff at Packet Engines also contributed material to this chapter: P. J. Singh, Steve Ramberg, Kevin Daines, Howard Johnson, and Brian MacLeod.

Figure 4.1 Full-duplex repeaters fill the gap in cost and performance between CSMA/CD repeaters and switches.

Traditional repeaters have historically been essential components in the Ethernet network topology. Compared to other connectivity devices, they are relatively cost efficient (a switch can cost as much as four to five times more than a repeater with the same number of ports) and provide for easy integration of smaller network segments. But they require the transmission of useless padding, which means constrained performance at different packet sizes and the wasting of precious bandwidth. To combat this problem, elaborate schemes like packet bursting, which essentially chains smaller packets back to back, have been proposed.

Switches, in contrast to repeaters, are much more than simple signal boosters or packet duplicators. They are among the most intelligent and powerful network devices available. Switches limit traffic on segments by forwarding frames to only those ports that need them. They provide sophisticated, full-duplex performance while allowing wide versatility in building a network, especially at the backbone* level. However, switches do have one major drawback: They are expensive.

It's clear that there's a need for some in-between device that can provide cost-efficient, full-duplex performance in a network hierarchy, and this is where full-duplex repeaters enter the picture—especially at gigabit rates. As Fig. 4.2 shows, a full-duplex repeater offers a much higher performance level than a CSMA/CD repeater at a much lower cost than a switch. In fact, early estimates indicate that a gigabit Ethernet full-duplex repeater can cost three to four times less than a gigabit switch.

However, the overall performance of repeaters is severely weakened in gigabit implementations. Concerns include limited throughput under heavy traffic, constrained performance with various packet sizes, and carrier extension—which wastes throughput. Another concern is lack of

---

* A backbone is the highest level of a LAN hierarchy, which acts as the primary path for LAN traffic.

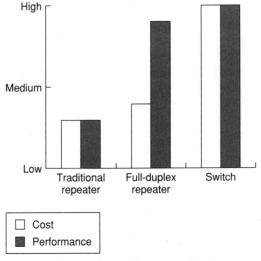

Figure 4.2 Comparison of the cost and performance of network devices.

versatility, since repeaters must be placed within collision domains. A collision domain combines Ethernet devices in one area where only one device may transmit at any given instance. Each collision domain has a timing constraint, which essentially limits the round-trip time (the time it takes for signals to reach their destination and come back again). This constraint limits the network diameter (the total length of cabling that can be run between devices).

Traditional repeaters may work fine for the lower levels of a network hierarchy, but it has become increasingly clear that they will not provide the performance needed as network speeds increase. In order to maintain a practical diameter of 200 m, the minimum effective packet size would have to be increased from 64 to 512 bytes. Packets smaller than 512 bytes must subsequently be padded (or carrier-extended) to the 512-byte length to meet this requirement.

4.2 From Repeaters to Switches

A traditional, CSMA/CD repeater is a very simple device. It contains no logic but is merely present to forward incoming signals. When the bit stream enters the port of a repeater, the bits are quickly replicated. Then, like a speaker who drones on and on for hours, the repeater spurts the stream out to everyone connected to it except the port of ingress. There is no delay to the packet, because there is no need for the repeater to decode or read the incoming packet. The repeater blindly duplicates *all* signals, be they packets, jam signals, or even line noise.

The attributes of the traditional repeater may not sound very attractive, but repeaters do have their advantages. Attenuation,* a problem that has always plagued transmitted signals, is combated in part by the repeater. The signal that reaches the repeater is usually mutilated and weak from its trip across the wire. By duplicating the incoming packets, the repeater restores the signals back to their original shape and strength. This rejuvenation is beneficial, primarily because it extends the practical length over which a signal can be sent and still recovered reliably.

The CSMA/CD repeater is also very fast because it does not have to worry about routing problems or decoding. Many manufacturers specify that their repeaters have a port-to-port latency (delay) of only 20 bit-times. When a packet arrives, it is immediately copied and sent to all outbound ports, regardless of its final destination. There are no long delays or unnecessary operations to perform.

Since a traditional repeater does not have any complicated processing devices or logic, it consequently does not have a very high price tag. For example, a fast Ethernet switch in 1996 had an average per-port price of $785, whereas the average price of a shared hub was $137 per port. This is perhaps the repeater's greatest advantage over other devices, but you get what you pay for. A typical CSMA/CD repeater does not contain much memory space, usually along the lines of a few bits at most. Since it cannot store packets, its potential for causing bottlenecks in the network is increased.

Another drawback of a CSMA/CD repeater is that it operates under a shared environment where only one client can talk at any given moment. Therefore, an inherent risk of collisions exists, since all connected clients are competing for the network's attention. Packets involved in a collision cancel each other out, and a signal is sent back to the transmission sources, alerting them that there has been a collision. This slows down the network and wastes bandwidth. It is truly a shared environment; everyone must listen while one person talks. This means that, realistically, everyone serviced by a repeater collectively shares the bandwidth from instant to instant. This shared bandwidth dilemma, however, is not experienced by a switch.

As described earlier, switches are the most high-profile of all network devices. Unlike traditional repeaters, they are intelligent, full-duplex units that have the ability to send a packet only to the

---

* Attenuation is the decrease in signal strength of an electrical or optical transmission as it travels between points.

appropriate port instead of wasting bandwidth by replicating the packet to all ports. The difference is the processing speed. Since the switch takes the time to read packets, it is limited not only by line speed but also by the time taken to process the packet.

Switches can also receive multiple packets simultaneously, and each port can transmit and receive at the same time. This means that, with full-duplex operation, the effective bandwidth of each port is equal to twice the line speed (i.e., for a 100-Mbit/s full-duplex port, the throughput is 200 Mbit/s).

Switches have more logic than a repeater; they have the intelligence to determine exactly where a packet is going and then send the packet only to the necessary port or ports. Therefore, a switch ensures network efficiency and preserves bandwidth, since it can filter through many more packets than a repeater can. In addition, a switch, like a repeater, rejuvenates signals.

Another important advantage of a switch is its ability to extend the range of a network by separating collision domains. As explained earlier, traditional repeaters, which have collision domains in order to limit the stress of a network, can only process one packet at a time. Switches, however, do not have this problem and can process many packets at any given moment, thus reducing the risk of some bottlenecks and also extending network range.

4.3   Enter the Full-Duplex Repeater

Now that both extremes of the spectrum have been examined, there is a network device that can bridge the gap between unintelligent repeaters and expensive switches: the full-duplex repeater. As its name suggests, it is a repeater that operates in full-duplex mode, offering higher performance than a repeater at a lower cost than a switch. Offering more versatility than a traditional repeater because of their full-duplex operation, full-duplex repeaters have no link length restrictions, since they have no need to be placed in collision domains. Full-duplex repeaters provide network managers with another essential option in the progression of scalability to gigabit Ethernet.

Since full-duplex media, PHYs, and MACs already existed, combining these entities with a repeater was relatively simple. In addition, IEEE Standard 802.3x, which defines how Ethernet is to operate in full-duplex mode (including flow control), was ratified in early 1997. These facts made developing the full-duplex repeater simpler, since virtually no new technology needed to be invented.

As for switches before them, a description of full-duplex repeaters is not included in the gigabit Ethernet standard. This is because the full-duplex repeater is not a new standard but merely an extension of existing standards. The extensions to the 802.3 standard are not concerned with the internal workings of such devices as switches, bridges, and repeaters. The IEEE standards relating to gigabit Ethernet, such as 802.3x, define only the behavior of the link, the PHY, and the MAC. The full-duplex repeater needs only to adhere to these related standards and their defined characteristics. Other specific internal functional details can be modified to fit the needs of each vendor and its customers.

The full-duplex repeater does, however, bring three new characteristics to repeaters: a MAC for each port, first in-first out (FIFO) memory buffers, and a completely self-contained arbitration mechanism. The MAC on each port allows for the management of the FIFO buffers. As packets are received, they are temporarily stored in buffers until the arbitration mechanism can transmit them to the other ports. Arbitration is performed to see which packet will be transmitted first. Because this arbitration takes place entirely inside the repeater box itself and not on the link, link lengths are limited only by the restrictions of the medium itself. Arbitration now occurs between the buffers on each port instead of between end nodes in a collision domain. This process will be described in more detail later in this chapter.

4.4 Full-Duplex Repeater Applications

So why should a full-duplex repeater be used, and where can it be used? Full-duplex repeaters provide many options to network managers who are looking for a faster response time than can be achieved by a 100-Mbit/s fast Ethernet network. In addition, while many are concerned about buying prestandard network devices that don't have universally compatible Layer 3 capabilities (e.g., VLAN, QOS) full-duplex repeaters provide a gigabit Ethernet solution today that is transparent to all pre- and poststandard schemes.

Full-duplex repeaters can be used in a myriad of applications. One such application is in power workgroups, where a group of workstations connect to a server, legacy LAN, or other network segment via a full-duplex repeater as shown in Fig. 4.3.

Another potent full-duplex repeater application is the server farm. A major source of bottlenecks in fast Ethernet networks can be found in servers. Using a full-duplex repeater as a gigabit uplink from fast Ethernet switches to a server farm (Fig. 4.4) can relieve much of the congestion found in fast Ethernet networks. The full-duplex repeater is able to handle more traffic to and from the server, while the fast Ethernet

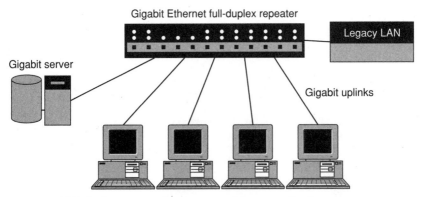

Figure 4.3 Full-duplex repeater in a power workgroup application.

switches protect the desktop machines from being flooded by gigabit traffic.

Full-duplex repeaters can be used to aggregate small-area backbone traffic. In smaller networks, most of the traffic stays within workgroups and rarely floods onto the backbone, making full-duplex repeaters an ideal device for this sort of application (Fig. 4.5).

Even when gigabit Ethernet switches are introduced into the market in force, full-duplex repeaters can continue to be used in various applications. One prime example is use of a full-duplex repeater as a port expander in front of a gigabit switch port to connect many gigabit Ethernet ports together, as shown in Fig. 4.6.

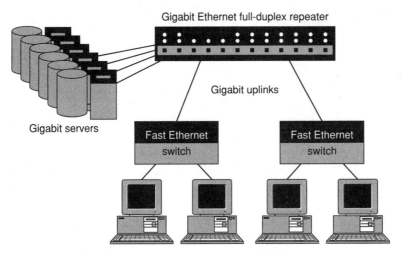

Figure 4.4 Full-duplex repeater in a server farm application.

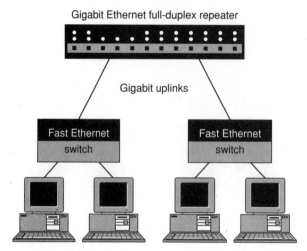

Figure 4.5 Full-duplex repeater in a small backbone application.

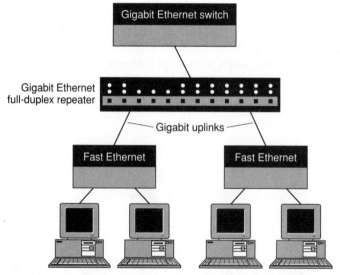

Figure 4.6 Full-duplex repeater in a future application with gigabit Ethernet switches.

4.5   A Look Inside

The major differences that separate a full-duplex repeater from a traditional repeater (besides full-duplex performance) are FIFO buffers, flow control, and the movement of the arbitration mechanism from the link to the *inside* of the repeater (see Fig. 4.7). These changes ensure

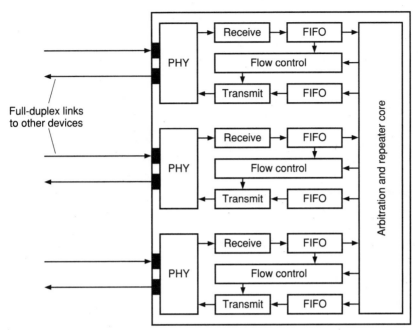

Figure 4.7 The full-duplex repeater uses FIFO buffers for data storage and an arbitration core placed inside the device.

that a repeater can work at gigabit speeds. To understand fully how the inside of the full-duplex repeater works, the best place to start is by taking a closer look at the FIFO buffers.

4.6   FIFO Buffers

A full-duplex repeater is tied to the CSMA/CD repeater in that all users connected to each are tied together in a shared-bandwidth arrangement but not a shared-media architecture. That is, packets are still rebroadcast to all other ports: but that is where the similarities end. Full-duplex repeaters, in contrast to CSMA/CD repeaters, receive and transmit packets simultaneously, using the help of small FIFO buffers on each port.

When a port receives a packet, it is stored in the port's FIFO. In networks that are relatively free of congestion, the packet is immediately forwarded to the repeater's internal bus, where it is then repeated to all other ports. When packets arrive at several ports while the repeater is busy, the FIFOs are responsible for remembering the order in which packets arrived. Packets can then be transmitted in the same order they were received on a first in-first out (FIFO) basis.

In some instances, several packets may arrive at virtually the same time. When this occurs, it is all but impossible to determine transmit order. This situation would be analogous to an Ethernet collision. However, the full-duplex repeater has an edge over other devices in these situations. When collisions occur, arbitration is performed to determine fair access to the internal bus. Since the FIFO buffers are able to store incoming packets, the domain where arbitration is performed is reduced from a point-to-point situation to a port-to-port situation. This eliminates collisions, since the full-duplex repeater does not perform arbitration on the link but instead confines the arbitration logic to the internal hardware. Arbitration is discussed in more detail later in this chapter.

In networks where traffic is extremely heavy, there may come a point where the FIFO is so saturated that it cannot store any more packets. Full-duplex repeaters have limited buffer space in order to make them more cost effective. Due to this limited buffer space, when there is no longer any room for storage, incoming packets will be spilled onto the floor, so to speak, and be lost. To prevent such situations, the full-duplex repeater uses flow-control frames as defined by IEEE Standard 802.3x.

4.7   The Leaky-Bucket Model

In order to explain how the FIFO buffers work, it's helpful to use the analogy of a leaky bucket. Picture a small bucket being filled by a long length of garden hose. The hose is connected to a water tap that is turned all the way on. The bucket has a small hole in the bottom, which is allowing water to seep out slowly. If the tap were turned off, the bucket would soon be drained of its contents. If water were flowing into the bucket at a low enough rate, the influx and efflux would be about the same. However, in this example, a tap running at full capacity is used.

Since the flow of water into the bucket is greater than the amount of water flowing out, the bucket will soon be filled to capacity and begin to overflow, an undesirable situation. In order to prevent this, the tap must obviously be shut off. The only problem is that the tap is a good distance from the bucket. Waiting until the bucket is full and then walking over to the spigot to turn off the water would result in a spillover before the source could be reached. Conversely, leaving the tap turned off for too long would cause the bucket to run dry.

In order to keep the seepage of water consistent, high- and low-water marks must be employed, as seen, both figuratively and literally, in Figs. 4.8 and 4.9. The high-water mark is set at a point where the bucket can hold only the amount of water that would continue to flow during the process of shutting the tap off. Similarly, when the water has drained down to the low-water mark, the water can be allowed to flow again.

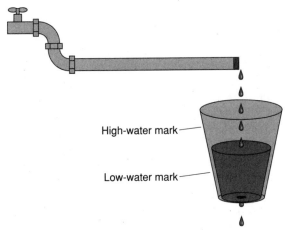

Figure 4.8 High- and low-water marks are utilized on FIFO buffers (bucket) to determine whether they are becoming full or empty.

Having set up the model, it's now possible to explain what the items represent. The bucket in the example is a FIFO buffer with an influx and outflow of data (water). The leak is a connection to the internal bus, whereas the tap and hose represent a link to a source end node, or network device. Since it takes some time to notify the sending device that it must halt transmission of packets until later, the MAC controlling the FIFO waits only until the high-water mark is reached; then it pauses that particular link partner. The full-duplex repeater, in the meantime, can continue broadcasting packets and begin emptying the burdened FIFO. The data in the buffer will eventually recede until it reaches the low-water mark. At this point, the MAC will indicate that its buffer is ready for more data to be sent, and transmissions will resume. Now that the issue of how data is regulated in the individual FIFOs has been examined, it's time to look at how fair access to the medium is maintained through internal arbitration.

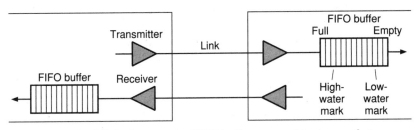

Figure 4.9   In the full-duplex repeater, FIFO buffers are used to store packets.

4.8   Arbitration

Arbitration is used by network devices to determine which packets are sent first. Traditional repeaters use arbitration between computers at the end of links to decide which DTE can transmit packets at any given time. But a full-duplex repeater moves the arbitration mechanism inside, arbitrating between buffers *within* the repeater. The FIFO buffers assist the arbitration logic in ensuring that the incoming packets are transmitted in the order that they were received.

Removing the arbitration mechanism from the link eliminates the time delay associated with transmissions that is typically found in traditional repeaters. This time delay occurs as a result of the repeater accepting and sending only one packet at a time. Previously, if the network was busy, sending stations would wait for a random time before sending their packets again. Full-duplex repeaters, on the other hand, can store packets from several ports in FIFO buffers until the arbitration mechanism selects a packet for transmission. Because of these buffers, multiple packets can be accepted almost simultaneously and stored in buffers, allowing for much greater packet throughput than a traditional repeater.

Another method is priority-based arbitration, which sends out packets that are distinguished as most important. This method is crucial to video-based applications, which require low latency. *Latency,* also known as jerkiness, is the amount of delay before the arrival of packets, and can be primarily noticed during video transmission. This is also tied into the concept of quality of service (QOS), a term that describes the arrival of data when and where the recipient needs it. QOS, largely determined by the amount of latency present, is especially important for multimedia applications. Therefore, priority-based arbitration is concerned with transmitting those packets that are most important and helping to prevent an increase in latency.

4.9   Asymmetric Flow Control

Flow control is the method of suspending packet transmission by sending PAUSE frames from one device to another to stop the transmission of data when a receiving device's buffers are full.

If everything is functioning normally, and traffic in the network is medial, packet flow is consistent. Figure 4.10 shows an example of requested data being transmitted by a server, duplicated by a full-duplex repeater, and processed by a 100-Mbit/s switch. Once the switch has determined how to route the packet, it quickly speeds the packet along.

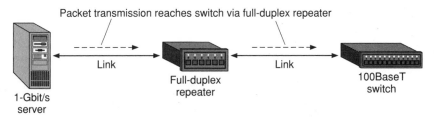

Figure 4.10   Usually, packet transmission will be sent through the network without delay.

In a congested network, however, the outlook could be much more bleak. Suppose a switch is being overburdened by traffic from another network device such as a server. The switch waits only until the buffer's high-water mark for that particular port is reached to temporarily halt transmission from the server. Likewise, if the server cannot handle traffic from the switch, then it is allowed to pause that switch. This arrangement is called *symmetric flow control*.

The instrument for suspending transmission is called a PAUSE frame. The receiving device can send a PAUSE frame, if necessary, which contains a PAUSE quanta, or 512 bit-times, that tells the transmitting device to suspend packet broadcast for a certain time. This time can be extended or canceled with further PAUSE frames. However, if no more PAUSE frames are sent by the time the most recent one expires, the transmitting device will assume the receiving device is ready and will resume sending packets.

Imagine flow control as network devices having a conversation with each other. The network devices use these conversations to tell each other the status of their buffers: for example, "I can receive packets," or "Stop sending packets, I'm full." This helps to avoid packet loss in the event the network becomes congested.

Full-duplex repeaters, because they have limited buffer space, use a form of flow control known as *asymmetric flow control* (defined in IEEE Standard 802.3z). This method of regulating packet destination sends flow-control messages in only one direction: to the end device, but not vice versa (see Fig. 4.11). Since the full-duplex repeater "repeats" only one packet to all connections at any given time, the entire communication process would therefore be held up if one end device were to send a PAUSE frame to the full-duplex repeater. This means that an end device cannot be allowed to send PAUSE frames to a full-duplex repeater. Because of this, the end device is responsible for the incoming packets, and it may be forced to lose some packets if it cannot keep up with the network traffic.

While a full-duplex repeater will usually not receive flow control instructions from end devices, there are special conditions under which

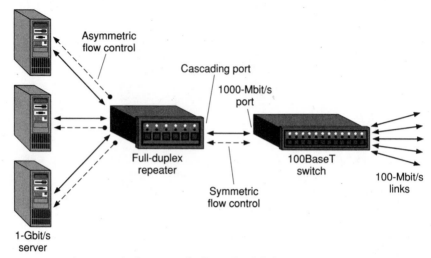

Figure 4.11 Asymmetric flow control allows the full-duplex repeater to send PAUSE frames to end devices, but not vice versa.

it may receive and act on PAUSE frames from other devices. As stated earlier, most ports on the full-duplex repeater provide a minimal amount of buffer space. There is, however, one exception: a cascading port. This port can be seen as a type of expansion port that allows the full-duplex repeater to operate with more versatility, given its increased buffer capacity (see Fig. 4.11). In other words, the term *cascading port* refers to a switched port in a full-duplex repeater that provides more buffer capacity in order to act on flow-control frames.

4.10 Link Partners

Now that asymmetric flow control has been discussed, two scenarios can be used to show how the network would benefit from its use and then what could happen if it were not implemented. The following examples assume links to an end device (a server in this example) and not other routing devices that could be connected to a cascading port.

In Ethernet communication, traffic is very bursty; a network may be flooded with frames at one moment and be idle the next. Consider a server broadcasting frames to the network via a full-duplex repeater and full-duplex links (see Fig. 4.12). In transmission and reception of packets, there are only two possible outcomes: packets will be hurried along with little or no delay, or, in high-congestion situations, PAUSE frames may be sent to suspend transmission while packets are stored in buffers.

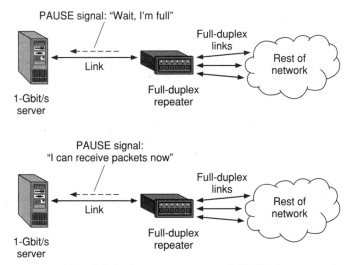

Figure 4.12   The full-duplex repeater can send PAUSE frames to other devices as part of asymmetric flow control, but not vice versa.

In the same manner as other network devices, full-duplex repeaters send PAUSE frames to suspend transmission from their link partners. As discussed earlier, the FIFO buffers on noncascading ports are limited in order to reduce the purchase price of the full-duplex repeater. Typically, end devices such as servers or computers have much more memory storage capacity than do repeaters and even switches. It is therefore very undesirable that an end device be able to flow-control the full-duplex repeater. Figures 4.11 and 4.12 are examples of situations where asymmetric flow control can be utilized: the full-duplex repeater can flow-control the servers, but the servers cannot hold off packets from the full-duplex repeater.

4.11   Conclusion

Enabled with asymmetric flow control, the full-duplex repeater becomes an essential networking tool. With competitive prices and a performance that falls between those of the CSMA/CD repeater and the switch, the full-duplex repeater is ready to take its place on the front lines of networking. Net managers can expect to see the newest and brightest member of the Ethernet family stake its claim now and continue to deliver as speeds increase in the future.

Gigabit Ethernet: Physical Layer and Cabling Options

Bruce Tolley*

Business development manager, 3Com Corporation

5.1 Introduction

The biggest advantage to gigabit Ethernet is that it is Ethernet, only faster. Thus, gigabit Ethernet is an evolution of fast Ethernet, just as fast Ethernet itself was as evolutionary step for Ethernet (see Fig. 5.1). When network managers migrate parts of their networks from 10-Mbit/s Ethernet to 100-Mbit/s Ethernet and subsequently to gigabit Ethernet, there is no change in the applications that are run on the networks, very little change in the management tools used, and no change in the CSMA/CD MAC at all. What *does* change, however, is the speed of the network (which increases each time by an order of magnitude) and the cabling requirements. Network managers who want to deploy gigabit Ethernet successfully need to be familiar with the cabling options open to them. This chapter examines the cabling issues addressed by the gigabit Ethernet task force in its work on the 802.3z gigabit Ethernet specifications. Specifically, the physical layer (PHY), the optical transceivers or physical-layer devices, the copper physical-layer devices, cabling and media options, and the

* Bruce Tolley is a business development manager at 3Com with responsibility for high-speed networking and other emerging markets. He also represents 3Com as the vice chair, events, within the Gigabit Ethernet Alliance, of which 3Com is a founding member. Prior to his current position, Mr. Tolley held various marketing management positions at 3Com. Before joining 3Com in 1991, Mr. Tolley was a senior industry analyst and research director at Frost & Sullivan in Mountain View. He has also held teaching positions at Stanford University and San Francisco State University. A graduate of the University of California at Santa Cruz, Mr. Tolley holds postgraduate degrees from Stanford University in Palo Alto, California, and an M.B.A. from the Haas School of Business, University of California at Berkeley.

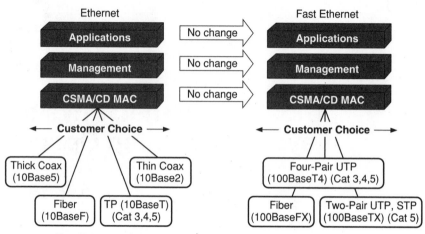

Figure 5.1   Fast Ethernet technology—meeting the current need.

distance goals for the various transceivers and cabling types are discussed.

5.2   Cabling Considerations

Network managers face different options at the transceiver and media levels depending on which speed Ethernet they are deploying. For 10-Mbit/s Ethernet, the choices are 10Base5 (thin coax), 10BaseF (multimode fiber), 10BaseT (unshielded twisted pair or UTP), and 10Base2 (thin coax). For fast Ethernet, network managers currently have three choices: 100BaseFX (multimode fiber), 100BaseT4 (four-pair UTP), and 100BaseTX (two-pair UTP).

5.3   Gigabit Ethernet Physical Layer

When it came to developing the media options and physical layer technology for gigabit Ethernet, the 802.3z task force faced some serious challenges. Category 5 UTP copper cable supports data transmissions only up to 100 MHz—well below what is required to carry gigabit Ethernet traffic. More importantly, there was no existing copper physical-layer technology that ran for distances of 100 m over Category 5 UTP cable at gigabit-per-second rates. Early in the standards process, the 802.3z gigabit Ethernet task force acknowledged that complex analog/digital circuit technology (about 250 gates) would have to be developed to allow Ethernet packets to run over unshielded twisted-pair (UTP) copper at gigabit-per-second speeds. Equally challenging, the task force acknowledged that a new physical-layer technology could take many months to develop.

This could have presented a big problem. The physical layer is a crucial part of the gigabit Ethernet specification because it provides the interface between the MAC and the transceivers in gigabit Ethernet hardware (see Fig. 5.2). The physical layer performs encoding, decoding, carrier sense, and link monitor functions. So a pragmatic decision was made, one that would prevent the unshielded twisted-pair copper issue from holding up either development of a standard or deployment of gigabit Ethernet in corporate networks. Rather than include the specification for a transceiver or PHY device for 1-Gbit/s Ethernet over unshielded twisted pair (UTP) as part of the 802.3z standards effort, the task force decided instead to focus its initial attention on developing solutions for fiber-optic cable—both multimode (MM) and single-mode (SM) fiber—and short-haul shielded copper, while at the same time concentrating the UTP efforts in a separate, dedicated task force. This was a logical choice, given that fiber is the preferred choice for most backbone connections.

To allow the standard to be developed as quickly as possible, the task force decided to borrow and modify the physical-layer signaling protocol already used by the ANSI Fibre Channel standard. This signaling protocol offers proven physical-layer technology that operates at gigabit-per-second rates. By borrowing this specification, the gigabit Ethernet 802.3z task force was able to agree quickly on the fundamental features of the PHY and reduce the time to market for gigabit Ethernet technology as a whole. The decision also enabled the networking vendors who are implementing gigabit Ethernet technology to have access to readily available, cost-effective fiber-optic components that are sold by many different suppliers.

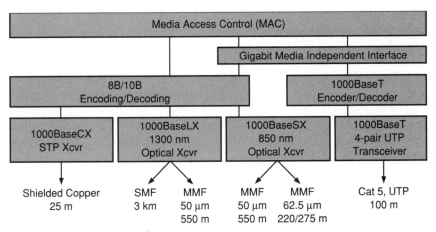

Figure 5.2 Gigabit Ethernet layers.

The Fibre Channel (FC-1) PHY signaling technology also has the advantage of supporting short (25-m) connections over shielded copper cable. These connections are for use mainly in wiring closets and between locally attached devices.

Instead of the fast Ethernet 4B/5B signaling scheme, gigabit Ethernet uses the Fibre Channel 8B/10B signaling. The gigabit Ethernet physical layer is similar to the Fibre Channel 10-bit interface, but not identical. The gigabit Ethernet serializer/deserializer (SERDES) chip is functionally equivalent to its ANSI Fibre Channel counterpart, the 10-bit SERDES chip, except for its frequency of operation. The Fibre Channel phase-locked loop (PLL) within the SERDES works at 1.0623 GHz. The gigabit Ethernet PLL has been scaled to 1.25 GHz in order to achieve true 1-Gbit/s wire speed. The SERDES is necessary because data in the media access control (MAC) layer is processed in bytes or words or double words. Data on the wire can only be sent or received 1 bit at a time. The SERDES chip converts parallel data to serial data when transmitting, and serial data to parallel data when receiving.

5.4 Optical PHYs

By relying in part on technology that had already been developed for use in another standard, the 802.3z task force has been able to speedily define two optical transceivers or physical layer devices for use with fiber cabling: short-wavelength laser, or 1000BaseSX, and long-wavelength laser, or 1000BaseLX. 1000BaseSX is Fibre Channel based, using the 8B/10B physical coding sublayer (PCS). The optics are specified at 770–860 nm, with a spectral width of 0.85 nm. They are usually referred to simply as 850-nm, or short-wavelength. The two most important facts to remember about 1000BaseSX are, first, that it is targeted at multimode fiber and, second, that the short-wavelength transceivers it uses are cheaper than those found in products implementing the long-wavelength specification.

The long-wavelength fiber laser PHY is called 1000BaseLX (L stands for long wavelength). This PHY also uses the Fibre Channel FC-8B/10B physical coding sublayer (PCS). It is specified for wavelengths of 1270–1355 nm at a spectral width of 4 nm, and is commonly referred to as 1350-nm or long-wavelength. Unlike 1000BaseSX, 1000BaseLX is specified for use on either multimode or single-mode fiber.

To review the difference between multimode and single-mode fiber, it is necessary to recall the construction and light-transmission properties of fiber-optic cable. Single-mode (SM) fiber cable reduces the radius of the core to between 5 and 10 μm, which is very close to the order of a wavelength, so that one single angle or mode of light energy passes along the fiber. Because there is a single transmission path with single-mode fiber, the light can travel long distances with little loss of signal. In contrast, multimode (MM) fiber has wider cores. Light is reflected along the core at multiple angles. The light is propagated along multiple paths; each path has a different length and hence each takes a different amount of time to traverse the fiber. These multiple modes cause the signal elements to spread out in time. Consequently, distortions occur that limit the distance over which the integrity of the light signal can be maintained.

5.5 Copper PHYs

While the 802.3z task force focused most of its efforts on fiber, it has not neglected copper cabling. In fact, two copper PHYs have been proposed. The first is a short-haul copper jumper PHY called 1000BaseCX (x for copper). It is already a part of the 802.3z specifications and will be ratified as part of the gigabit Ethernet standard by the end of June 1998. 1000BaseCX is intended for short copper connections within wiring closets. This PHY uses the Fibre Channel 8B/10B physical coding sublayer (PCS) and a different driver on shielded copper 150-Ω balanced cable, and supports wiring distances up to 25 m.

Work on the second copper PHY, which addresses support for Category 5 UTP cable, has been put into a dedicated standards effort under the 802.3ab task force. Called 1000BaseT, it is expected to be formally ratified as a standard by March CY99. The IEEE 802.3ab task force's goal is to achieve cabling runs of 100 m from the wiring closet to the network node on four-pair Category 5 cable, enabling network managers to build networks with diameters of 200 m. However, the physical coding sublayer (PCS) protocol is yet to be selected. And extensive signal processing will be required on the 1000BaseTX transceiver itself. In order to transfer information at 1 Gbit/s, serious encoding is necessary, because Category 5 UTP copper has its parameters defined only up to 100 MHz. Performance on Category 5 UTP degrades rapidly beyond that frequency. Since baseband signaling is impossible, it is much more complex to design digital signal processing for copper at gigabit-per-second data rates. Encoding helps to lower the bandwidth of the signal. However, such encoding is accompanied by the need to increase the signal power. In addition, to recover an encoded signal, equalizers need to be more sophisticated than those required to recover a baseband signal. There are also other ancillary details such as the need for echo cancellers and near-end crosstalk (NEXT) cancelers. In short, three components increase the complexity of UTP signaling logic: the more complex hardware required to encode/decode, the increase in signal power, and the need for more sophisticated receivers.

5.6   Distance Goals

Figure 5.3 shows the distance goals of the various gigabit Ethernet PHYs in comparison to Ethernet and fast Ethernet. Gigabit Ethernet simply moved the decimal point to increase speed by an order of magnitude. But increasing the data rate from 100 to 1000 Mbit/s entails some serious network distance limitations. For Category 5 UTP, the goal is to obtain horizontal runs of 100 m from the wiring closet to the end station, for a maximum network diameter of 200 m. For shielded copper, the goal is to reach distances of 25 m; thus, this cable is primarily intended for implementations within the wiring closet for short copper jumper connections for switches and servers. For multimode fiber, the goal is to reach distances of 275 m on 62.5-μm fiber and 550 m on 50-μm-diameter multimode fiber. For single-mode fiber, the goal is to obtain distances of at least 5 km.

Common sense as well as the customer research conducted by the Gigabit Ethernet Alliance and other groups indicates that the initial implementations of gigabit Ethernet will be switch-to-server, switch-to-switch, and in the backbone. Figure 5.4 shows that the initial three PHYs from 802.3z cover all the expected initial applications: campus backbones, building backbones, switch-to-switch connections, and switch-to-server connections.

Support for Category 5 UTP will only be an issue when network managers plan to take gigabit Ethernet to the desktop, which will not happen any time in the near future. 1000BaseSX can cover wiring closet connections and applications in the building backbone up to 275 m (62.5-μm diameter) or 550 m (50-μm diameter). Long-wavelength transceivers

	Ethernet	Fast Ethernet	Gigabit Ethernet Design Goals
Data Rate	10 Mbit/s	100 Mbit/s	1000 Mbit/s
Cat 5 UTP	100 m (min)	100 m	100 m
Shielded Copper	500 m	100 m	25 m
Multimode Fiber	2 km	412 km (hd) 2 km (fd)	220–550 m
Single-Mode Fiber	25 km	20 km	5 km

Source: 3Com Corp.

Figure 5.3   Gigabit Ethernet summary chart.

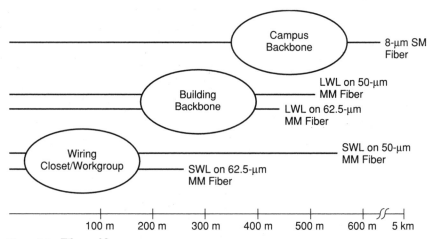

Figure 5.4   Fiber cable coverage.

(1000BaseLX) can cover building backbones and small campus back-bones with a radius of 550 m or more on multimode fiber. On single-mode fiber with a core diameter of 8 μm, 1000BaseLX transceivers can inter-connect campus backbones to support networks of up to 5 km in radius.

5.7 Conclusion

In conclusion, by borrowing a proven physical-layer signaling tech-nology from the ANSI Fiber Channel standard, the 802.3z task force has achieved rapid agreement on the specifications for the two op-tical PHYs and the short-haul copper PHY and has been able to keep to its aggressive schedule for completion of the gigabit Ethernet standard.

Consequently, gigabit Ethernet offers network managers four new choices in transceivers: 1000BaseCX for short-haul copper connections up to 25 m, 1000BaseSX for multimode fiber, 1000BaseLX for both multimode and single-mode fiber, and—although a bit further out—1000BaseTX for Category 5 UTP wire. Network managers will have the means to implement gigabit Ethernet in the campus and building back-bone with single-mode and multimode fiber media and within wiring closets with short copper jumper connections with 1000BaseCX.

Acknowledgments

My thanks to Wen-Tsung Tang of 3Com's Technology Development Center for reviewing a draft of this chapter and to Ruchi Wadhawan of

3Com's Advanced Products Group for clarifying some 1000BaseT points. Portions of this chapter were presented during a panel presentation in April 1997 at GigNet, sponsored by TTI, at the Washington Hilton in Washington, D.C. I would like to acknowledge Bob Grow, vice president of industry relations, XLNT Corp. and organizer of the panel, for his review of that presentation.

2

Gigabit Ethernet and Related Specifications

Purpose

To explain and analyze the techniques that can be deployed in gigabit Ethernet networks to control congestion problems, optimize the network for time-sensitive applications, and enable virtual LANs (VLANs).

What Is Covered

The gigabit Ethernet spec provides a high-speed transport that operates over copper and fiber cable. But that's about it. In order to make the best use of the technology, net managers must deploy it using products that also support a variety of other technologies—both standardized and proprietary.

Adding gigabit speeds to a LAN obviously creates a high risk of congestion and packet loss. Chapter 6 provides an overview of three IEEE standards that can help reduce these problems: 802.1p (multicast pruning), 802.1Q (virtual local area networks, or VLANs), and 802.3x (flow control).

Chapter 7 focuses on a proprietary specification developed by NBase Communications. Called Virtual Collision, it is an enhancement of the Ethernet CSMA/CD algorithm that addresses the needs of time-sensitive applications on shared-media gigabit Ethernet networks.

Chapter 8 deals with the knotty issue of how to enable quality of service on gigabit Ethernet. It provides an in-depth description of four technologies: Resource Reservation Protocol

(RSVP), emerging protocols such as 802.1Q and 802.1p, and Integrated Services over Specific Link Layers (ISSLL).

Finally, Chapter 9 deals with the 802.1Q draft standards so far defined for the purpose of implementing VLANs (or logical broadcast domains).

Contributors

Packet Engines Inc.; Nbase Communications Inc.; Cisco Systems Inc.; Foundry Networks Inc.

6

Associated IEEE Standards 802.1p, 802.1Q, and 802.3x

Bernard Daines*
President, CEO, and founder, Packet Engines Inc.
With additional work by Bryan Osborn and Matt Johnson†

6.1 Introduction

The Institute of Electrical and Electronics Engineers (IEEE) is an international organization that represents a broad spectrum of technical disciplines. There are many subgroups within the IEEE, including the LAN-MAN Standards Committee (LMSC). The LMSC, founded in the early 1980s by the IEEE as Project 802, is responsible for the development of LAN standards, which includes Ethernet technology (see Fig. 6.1). As described earlier in this book, a standard is a set of guidelines that describes a device and its function. A standard does not state how the product's function is to be implemented by vendors but serves instead as a set of rules or an outline that must be followed.

IEEE standards meetings are open to all participants, which ensures that everyone can have input and that all points of view are considered when standards are being set. As each IEEE standard is

---

* Bernard Daines is president, CEO, and founder of Packet Engines Inc., a privately held company headquartered in Spokane, Washington. Daines is a graduate of Brigham Young University and received the first degree in computer science issued by that institution. Following a move to Northern California, Daines worked for IBM and Hewlett-Packard prior to founding his first company, Tidewater Associates. His clients there included Northern Telecom, 3Com, Texas Instruments, Pacific Bell, and Cisco. In 1992, Daines cofounded Grand Junction Networks, which later was acquired by Cisco Systems. In 1994, Daines established Packet Engines to develop leading-edge gigabit Ethernet products based on that technology. During his career, he has created 38 successful ASIC designs as well as numerous circuit boards, systems, and test equipment products.

† The following staff at Packet Engines also contributed material to this chapter: P. J. Singh, Steve Ramberg, Kevin Daines, Howard Johnson, and Brian MacLeod.

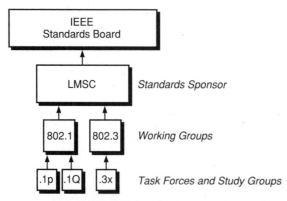

Figure 6.1 The LAN-MAN Standards Committee (LMSC),
a branch of the IEEE, manages networking standards
such as 802.1p, 802.1Q, and 802.3x.

developed, it is subjected to a repeated balloting process. During the
ballot process, individual members or observers can suggest modifications or improvements to the standard. Each and every suggestion
must be acknowledged and voted on; this may require months of effort involving literally hundreds of people. Imagine putting three
engineers in a room and getting them to agree on something. Now
imagine 300. That's the nature of the standards process. Eventually,
if the standard survives balloting, it will by definition represent a
broad technical consensus in the industry.

After a standard has been accepted by the IEEE 802 LMSC committee, it is then forwarded on to the American National Standards Institute (ANSI), an organization that accepts standards from various
bodies for ratification. ANSI then forwards the ratified standard to the
International Organization for Standardization (ISO) as a proposal for
the basis of an international standard (see Fig. 6.2). The ISO often
modifies portions of a standard to generate international support. ISO
recognition can be particularly beneficial in countries where ISO standards are given more weight than IEEE standards.

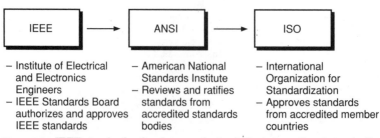

Figure 6.2 IEEE standards often become the basis for international standards.

6.2 Congestion Problems: Standard Solutions

There are basically two major criteria in evaluating the performance of a network: overall speed of the network and latency statistics, or quality of service (QOS). Using these benchmarks, it is possible to judge whether the network is supplying adequate bandwidth to its clients. More often than not, network congestion is the culprit that causes a performance decrease in both criteria and frantic frustration for a network manager.

This congestion often manifests itself in what is known as a *broadcast storm*. Broadcast storms are created by the propagation of broadcast packets in the network. All data packets in an Ethernet network contain one of three different types of MAC destination addresses: unicast, multicast, or broadcast. These addresses notify network devices of the packets' destinations and are used to help locate devices on the network. As more and more users (and therefore more and more destinations) are added to a network, the new station generates more and more broadcast packets.

To see how a broadcast storm gets started, imagine an end user sending a series of packets. Those packets are received and replicated by a repeater, which then imposes the copies of those packets on the rest of the network. As those replicated packets are being sent every which way across the network, other network devices receive them and ask the rest of the network, "Device A, are you out there?" or "Where is device B?" to locate the packets' destinations. Since packets simultaneously compete for the attention of every device in a network, and since all network devices are talking, it is easy to understand how a network can become congested. While this continues, the stress on the network builds and builds as competing packets essentially block each other's way so that nothing can get through the network. Eventually, the broadcast storm causes the network to *time out*, or collapse. Repeaters (since they repeat packets to all ports) and bridges (since they link together different LAN segments) are guilty of most of this traffic propagation.

To alleviate these congestion concerns, the IEEE has developed three standards that can aid in the operation of gigabit Ethernet: 802.1p (multicast pruning), 802.1Q (virtual local area networks, or VLANs), and 802.3x (flow control). These standards help to make networks more efficient as speeds increase, since this acceleration multiplies network congestion.

6.3 IEEE Standard 802.1p: Multicast Pruning

When someone puts an address on an envelope, they are notifying the postal service exactly where they wish that letter to be sent. On the way to its destination, the letter is routed through various arteries in the postal network, gradually bringing it closer to its destination. Eventually, the letter arrives at the listed address (see Fig. 6.3). Likewise, as noted earlier, there are three different types of MAC destination addresses that can be placed on a data packet. The first, a *unicast address,* lists only one network device as a destination. The second type, a *broadcast address,* lists all network devices as the destination and, as a result, is the main culprit in broadcast storms. The last destination address type, a *multicast address,* is used to control floods on a network by sending data only to specific end clients.

So how can a message be sent to multiple destinations without imposing unnecessary stress on a network? Unicast addresses are useless in this scenario since their data streams are being sent individually to each end user. Broadcast addresses can use up precious bandwidth, since every end user receives the data stream, whether they want it or not. Therefore, multicast addresses are used to limit the distribution of packets being sent to multiple destinations. This significantly reduces network traffic by discarding network branches not listed in a multicast group (a list of multicast addresses that all solicit the same data streams). To understand fully how multicast pruning works, it's necessary to first come to grips with the concepts behind the spanning tree configuration.

6.3.1   The network according to GARP

Multicast addresses allow a router to transmit data streams to only those areas of a network that request them. This saves bandwidth, since users who have no need for particular data streams will not

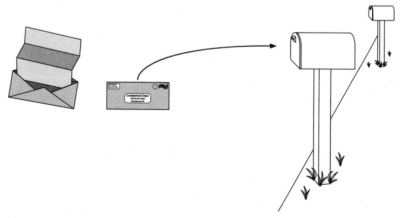

Figure 6.3 The destination addresses on packets allow them to be routed through the network similarly to the way a letter is sent through the postal system.

receive them. Furthermore, multicast addresses can be combined together into a multicast group as a collection of end users that all request the same data streams.

So how are end users assembled together into a multicast group? This is accomplished when end users who want the same data stream register with the devices that connect them to the spanning tree. The end users are in effect telling the router, "I am looking for this particular data stream, and when you come across it, send it my way."

This registration process is performed by a link-layer protocol known as GARP, or Group Address Resolution Protocol. GARP is what allows end users to register for certain data streams themselves. The router can then make a list of all the end users who have requested particular data streams and the pathway to each requestor. These connecting devices can then notify superior nodes in the hierarchy to watch for particular multicast frames.

Multicast routing, or the distribution of multicast data streams, is accomplished best when a network is configured in a "spanning tree" layout. A spanning tree is an efficient network setup constructed by routers that connect every member of a multicast group (see Fig. 6.4). In this setup, there is only one possible path between any pair of network devices, and there are no loops that might otherwise keep data streams circulating through the network indefinitely. More than one physical connection may exist between devices, but each router knows which lines are participating in the spanning tree. This allows traffic to be directed only to the necessary segments, generating minimum packet traffic.

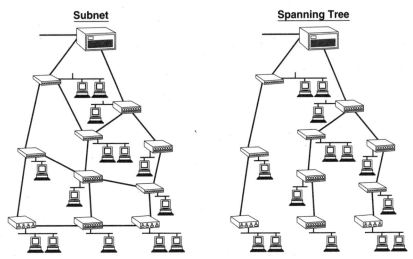

Figure 6.4 A spanning tree configuration, as opposed to a subnet setup, eliminates looping by allowing one possible connection between any two given network devices.

When the spanning tree protocol was first introduced as the 802.1d specification, packets could potentially travel in infinite loops inside the network or be sent to other connected networks. With 802.1d technology, routers were the only devices that could selectively limit the pathways traveled by multicast frames; only major branches could be excluded from receiving multicast frames. All other devices (switches and bridges included) treated multicast frames exactly like broadcast frames, sending them to all connected ports. 802.1p, on the other hand, essentially allows all network devices, except repeaters, to participate in the pruning process.

When the data stream reaches the router, it is copied and then forwarded to the appropriate network channels that lead to the end users on the list. To preserve bandwidth, the data stream is not actually replicated until the spanning tree branches—that is, until the data stream must break off in two directions (see Fig. 6.5). In this way, only the minimum number of copies is generated, since the data stream takes the shortest path toward its destination and then is replicated among the end users who have registered for it.

As an example, imagine an end user in a network requesting a particular data stream—say, a CNN feed from the Internet. The end user, using GARP, sends a request for the feed to the closest router governing that network segment. The router, in turn, seeks out the feed and sends it down the necessary network arteries of a spanning tree alignment until it reaches the registered end user. In this scenario, unnecessary network stress is avoided since only those end users who want a particular data stream are requesting and receiving the stream.

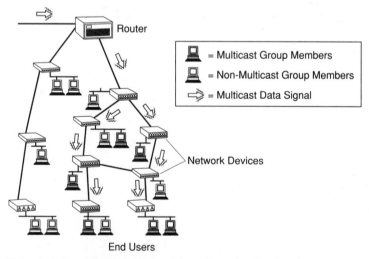

Figure 6.5 A multicast data signal is replicated only when it comes to a branch in the spanning tree.

6.3.2 Pruning

As noted previously, through the use of GARP, each device is able to communicate its desire to participate in a particular multicast transmission, informing higher-level link partners that they must keep watch for certain frames and direct them to all interested clients. The entire concept of multicast frame transmission is dedicated to controlling the flooding of frames onto the network. By using the spanning tree, the network is able to cut back even further the branches to which frames are sent.

Since users are able to change their membership in various multicast sessions at will, the spanning tree will never be stagnant. Keeping the system apprised involves a dynamic change of the specified pathways or pruning of the spanning tree to update the links that are added and rescinded. This chapter has already described what happens when a user decides to join a new group, but what happens when a user or a group decides to cancel their membership is slightly different.

In order to keep the spanning tree as current as possible while still avoiding excessive propagation of traffic, each physical network selects a governing device (for example, a router) that is responsible for the management of its network or subnetwork. The pruning process is dynamic (changing) and can be initiated in two ways.

Periodically, the governing device will make general queries of the network, calling for reports of membership enrollment. Multicast-capable devices then respond with their membership requests. The governing router listens for a specified time interval for responses to its query in order to determine what changes, if any, need to be made to the spanning tree. For the most part these queries are collected infrequently so as to keep the resultant traffic low, but often enough to keep up with changes in a large network.

Once the governing device has listened for the allocated interval, it will act on the information received. If no end users respond to the query within the time allotted, the router will assume that no one is interested in a particular multicast address. It then trims, or *prunes,* the branch from the spanning tree to increase network performance. If, however, it detects responses, it will leave the spanning tree untouched or augment it if needed.

Another manner in which pruning occurs is through *remote prompting.* An end device can send a *leave message* to request that a particular multicast session be suppressed or filtered out. However, the routing device must be careful that, when pruning this patron from the tree, it does not prune other clients who wish to continue receiving a particular service. To avoid this situation, when the designated device receives the leave message, it will make a specific query to determine whether all users have left the branch. The branch can then be pruned (see Fig. 6.6), unless clients wishing to participate are still listening.

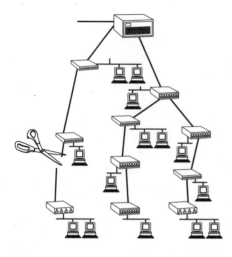

Figure 6.6 Although links are not physically disconnected from the network, multicast pruning in essence cuts away segments not participating in the multicast group.

Spanning tree pruning is much like a gardener who selectively removes branches from a tree. The liberated tree can concentrate its forces on other areas because it is no longer burdened with nurturing the removed branches. When multicast pruning is used to restrict the proliferation of multicast frames, the entire network benefits. Instead of wasting bandwidth by multiplying frames to areas that don't desire them, pruning lets the system operate with less congestion.

6.4   IEEE Standard 802.1Q: Virtual LANs and Priority-Based Arbitration

Virtual LANs, or VLANs, are smaller networks within a LAN that allow a group of end users to be tied together independently of where they are physically located (see Fig. 6.7). Previously, when an end user needed to be associated with a different workgroup, the user's network connections would have to be physically relocated and completely reconfigured. But in a VLAN, modifications (often referred to as *adds, moves,* and *changes*) can be easily executed simply by changing the end user's VLAN membership. Therefore, end users who receive the same data can be logically grouped together by giving them their own group-specific, distinct VLAN identity. Data intended for a specific VLAN will be sent *only* to network devices associated with that particular address.

6.4.1   Increased bandwidth

As explained earlier, multicast addresses are used to send data only to specific subnets, or VLANs, which in turn helps increase available bandwidth on a network. In a traditional LAN arrangement, a data

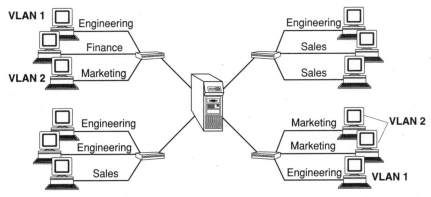

Figure 6.7   End users do not have to be in close physical proximity to participate in the same VLAN.

stream will often reach parts of a network where it is not needed and, in the process, use up valuable bandwidth. But when a LAN configuration is divided into several VLANs, data is sent specifically to the designated VLAN, and from there data can be repeated to end users (if repeaters are used at that level). This helps to relieve stress and increase bandwidth on the network, since data is being sent only to the address specifically designated on each packet. This, in turn, frees up the network for users who are accessing applications that require large amounts of bandwidth.

VLANs also assist in network management. Through the use of VLANs, a network administrator is given versatility in configuring a network. The administrator is able to group users together according to their specific bandwidth needs, relieving encumbered segments. This allows data traffic floods to be controlled so that overall bandwidth on the network is reduced and the speed with which users can receive data is increased.

As an analogy, imagine separating the course of an overflowing river into several rivers, with one river much larger than the others. The larger river represents a VLAN whose end users utilize applications with large bandwidth requirements. In contrast, the smaller rivers are VLANs whose end users don't need as much bandwidth. The backbone, and therefore the overall network, is managed much more efficiently since traffic is not going where it is not needed. Bandwidth can then, in effect, be allocated by group according to how much each VLAN needs.

6.4.2 A moving experience

Adds, moves, and changes, which involve the moving and configuring of end users to a new physical location on a network, used to be particularly exasperating for network managers. After computers have been physically moved and reconnected to the network, the network manager has to reconfigure them with new addresses, pathway information, and other user- and group-specific information. In addition, sometimes a wiring specialist has to be called in to service the connections.

When a computer is moved to another location in the network, it means that everything that pertained to the user at the old location has to be reconfigured when the computer is patched back into the network. Services connected to the user at the old location may not even be available at the new location until the network administrator loads specific applications used by the client and programs a new IP address, pathways to other devices, permission to access sensitive environments, and other concerns. The process represents an extended cost for the time of the wiring technician and network manager, and the downtime of each of the affected computers.

A study by Strategic Networks Consulting of Rockland, Massachusetts, has found that the cost of moving an employee from one location to another can range anywhere from $300 to over $1000. This figure does not include the time spent retooling the network to operate at optimum performance considering the new shift in bandwidth users. Even small companies often shift the locations of their employees in order to group departments together, expand the company, or optimize the use of bandwidth.

Although the challenge that adds, moves, and changes present has existed as long as networks have, this challenge can be easily resolved with the introduction of VLANs. The process of assigning or reassigning VLAN membership could be reduced to the act of dragging and dropping an icon on an administrator's PC, even when the end user has been physically relocated. In this situation, the administrator's application manages all configurations for each user; configurations could be contained in higher layers of intelligence and thus moved with each user's icon. As shown in Fig. 6.8, the individual connections to specific VLANs are retained.

The engineer who was moved in Fig. 6.8 is simply reconnected physically and then dragged and dropped in the administrator's program. The higher layers simply reconfigure the system in a type of autopilot manner. The engineer still has access to the systems and files used before and receives transmissions as if still at the old location, even though the engineer is in a completely different physical location. The settings are independent of the engineer's location because he or she is still a member of VLAN 1, along with the rest of engineering. This not only saves on time spent physically visiting the desk of each affected employee but also simplifies the entire process.

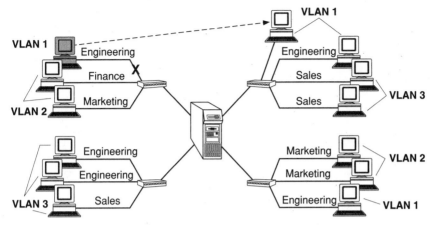

Figure 6.8 Changing the physical location of an end user does not affect membership in a VLAN.

6.4.3 Modified Ethernet frames

In order for packets to reach their destination in a VLAN, they must be encoded with pertinent information. Since VLAN technology is relatively new, modifications were made to change the original Ethernet packet so as to allow an additional 4-byte field that would enable network devices to recognize a packet as a VLAN packet. This field contains information necessary to ensure that the packet can reach its destination. The field includes information regarding the priority of each packet as well as where specifically in a VLAN the packet is

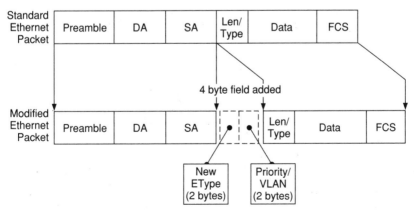

Figure 6.9 In VLANs, standard Ethernet packets are modified to include a VLAN/ priority field and a type field.

destined. Figure 6.9 compares a standard Ethernet frame with a modified Ethernet frame, including the additional 4-byte field.

6.5   Priority

In addition to indicating membership of a particular VLAN, this additional field is also used to denote the urgency of a packet. The priority field is used to address the issue of quality of service (QOS). Originally, the priority issue was managed by the .1p subcommittee but it was later handed off to the .1Q group, which then worked out the details.

Frequently, there are packets that require special attention within the network. These packets need to be forwarded without delay because they are time sensitive. This includes multimedia traffic such as real-time video, voice transmissions, or other applications that have very low latency tolerance. These types of applications not only require that data be delivered in a timely fashion but also that it arrive in the correct order. This is where QOS issues come into play.

QOS refers to the reservation of bandwidth within a network. In systems like telephone networks, this requires that a line of communication be established and remain uninterrupted until terminated. When no signals are being transmitted, the line cannot be used to carry other signals; it must remain idle (free of transmission). The signal is therefore guaranteed full utilization of the reserved bandwidth for the duration of the transfer of data.

Ethernet, which operates in a shared-medium environment, cannot set aside bandwidth for only one transmission at a time. Instead, QOS characteristics are provided through the use of priority fields. Rather than opening a connection to a particular device and tying up all resources until the connection is released, more bandwidth is effectively made available to high-priority clients.

To free up bandwidth, a packet deemed to be of high importance is tagged with a priority/VLAN field. This is almost like marking RUSH! or URGENT! on the front of an envelope. When a VLAN-capable device encounters a packet marked as having a high priority, it will suspend lower-priority transmissions and move the more critical packet to the front of the transmission line. In all, the IEEE 802.3Q standard supports eight levels of priority.

By using this system of priority tagging, Ethernet is able to deliver a measure of QOS in a shared-media environment. Just because a packet is tagged as high priority does not guarantee the exclusive use of system bandwidth. The case may arise where an even higher-priority packet could come along and usurp the right of way of another tagged packet. Despite this fact, in 90 percent of all situations the priority tag is sufficient to ensure that packets arrive in a timely fashion.

A switch, router, repeater, or other such device of any speed that is VLAN enabled can handle priority fields. VLAN and priority tagging is not a technology that is specific to any speed of Ethernet (10 Mbit/s, 100 Mbit/s, or 1 Gbit/s). However, provided that a packet does not exceed the maximum of 1518 bytes, most older hubs can handle VLAN and priority-tagged packets.

This fact is critical to the operation of VLANs and priority-based flow control. The system administrator must be keenly aware of which network clients utilize these services, since older devices may not be able to forward these types of frames. The pathway to any end device requiring VLAN or prioritization service must be composed entirely of devices that are compatible with these services. If not, then clients desiring to be included in a specific VLAN may never receive critical information.

Other areas of concern for the system administrator include which device will place the priority tag on the frame and how priority will be assigned. Depending on the particular needs of individual networks, it may be advantageous to insert priority/VLAN tags at different physical locations in the network. Assignments may very well be made by end devices, but tagging can occur in the server or in switches (or other devices) anywhere along the transmission pathway. Likewise, the tag must be removed before it reaches its final destination.

Similarly, the system administrator needs to decide *how* frames will be prioritized. There are several options for prioritization. If a particular server is comprised of only latency-sensitive information, such as wire services, video archives, and so on, then every transmission from that particular server could be deemed high priority. In like fashion, it may be more beneficial merely to tag all like frames with the same priority, regardless of origin. For example, all real-time video footage could be tagged with one level of importance, while voice transmissions might receive another. These are all areas of concern that need to be weighed.

6.6   IEEE Standard 802.3x:
Full-Duplex/Flow Control

As noted earlier, the introduction of gigabit Ethernet brings certain concerns relating to network management. Among the most important is how a network will handle the transmission of data at such fast rates. A sensible solution to this concern is flow control, and the IEEE has defined a flow control specification called 802.3x, for use in gigabit Ethernet and other Ethernet-based networks. This spec defines how network devices can communicate with one another concerning the status of their buffers; that is, whether they can receive packets. Here's how the flow control mechanism defined by the IEEE works.

6.6.1 Point-to-point communication

As a network becomes congested, the loss of packets becomes a problem; no matter how much buffer memory a network device possesses, eventually its buffers will become full. Sometimes the only available solution is for the device to drop incoming packets. And as network speeds increase, the risk of packet loss becomes even greater. However, this risk is reduced with flow control, a method of allowing network devices, usually switches, to communicate with each other to avoid dropping packets.

Flow control is ultimately intended to allow network devices to be constructed with limited memory capabilities. In this way, network devices with small buffers are not forced to drop packets during periods of network congestion. Although flow control is not mandated in a network, it makes quite a bit of sense to integrate flow control to ensure a more smooth and reliable delivery system. Whereas asymmetric flow control is only allowed from one device to another but not vice versa (such as in a full-duplex repeater), normal flow control, or point-to-point flow control, can occur with participation from both devices on either end of a packet transmission.

802.3x flow control is a MAC frame-based protocol, meaning that communication is maintained between the MAC control sublayers on devices transmitting to one another. Flow control is initiated when network devices send PAUSE frames to each other. These PAUSE frames are like conversations between the devices, telling each other their status, for example, "I can receive packets," or "Stop sending packets—I'm full." If necessary, the receiving device can send PAUSE frames, which contain PAUSE quanta, for the transmitting device to suspend packet transmission. This time can be extended or canceled with further PAUSE frames. However, if no further PAUSE frames are sent by the time the most recent one expires, the transmitting device will assume the receiving device is ready and will resume sending packets.

In order for a flow-control message to be recognized, the MAC on a network device has to read the right combination of a particular multicast address and the appropriate protocol. This allows the MAC to determine which device is sending the flow-control message. Flow control also stays true to the carrier sense with multiple access/collision detection (CSMA/CD) characteristic of Ethernet, since the interpacket gap is still placed between outgoing packets. Collision detection is utilized to listen to the wire in the event a full-duplex device is connected to a half-duplex one. In addition, designing flow control on network devices is relatively simple, since it accesses the traditional Ethernet MAC protocol.

However, since most gigabit Ethernet devices are operating in full-duplex mode, the need for CSMA/CD is not as crucial as it has been in the past. Obviously, there is no need for collision detection between full-duplex devices, since they can receive and transmit simultaneously. But because these devices are operating in full duplex, which increases the traffic and the speed by which that traffic is traveling, flow control is needed to pace the stress placed on network devices.

Suppose a server fires several requested packets down the wire in rapid succession to a switch (see Fig. 6.10). At the opposite end of the network, the destination port on the switch may be busy, forcing the switch to defer transmission and store the packets in its internal buffers. When a buffer reaches its high-water mark on the receiving port, the switch must send a PAUSE frame to the server in order to avoid dropping frames.

Switches are able to process packets with amazing speed. Since they are not shared devices, they can send packets from many different clients in rapid succession. Assuming that end clients associated with the switch in this example are all requesting large amounts of data from a server or are sending large amounts of information to be stored in the server, if the server is operating at full capacity on one port of the switch, the switch could still soon become inundated with packets even though it is bursting packets as fast as they are being requested by the other ports.

On reception of the PAUSE frame, the server will finish transmitting the current packet and then halt further delivery for the specified amount of time. If the server contains additional buffer space, it can store delayed packets until it has waited the designated period of time

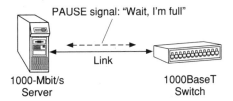

PAUSE signal: "Wait, I'm full"

Link

1000-Mbit/s
Server

1000BaseT
Switch

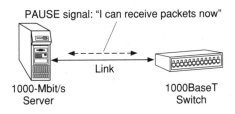

PAUSE signal: "I can receive packets now"

Link

1000-Mbit/s
Server

1000BaseT
Switch

Figure 6.10 Symmetric flow control PAUSE frames can be sent between full-duplex network devices.

or until the switch sends an additional frame indicating that it is ready to receive transmissions again. There are certain situations, however, when action must be taken to keep from overflowing the switch's internal buffers.

6.7 Conclusion

As described in this chapter, associated standards have been adopted to ensure that gigabit Ethernet operates at its full potential. Network concerns such as overall speed, QOS, and latency statistics are eased by the advent of multicast pruning, VLANs, and full-duplex/flow control operation. Alleviating network congestion is crucial to the performance of any network, and the associated standards for gigabit Ethernet will help networks operate even more efficiently.

7

Enhanced CSMA/CD—
Virtual Collision

Moti Weizman*

Director of hardware engineering, Nbase Communications Inc.

7.1 Introduction

The CSMA/CD protocol that lies at the heart of the Ethernet specification was designed at a time when multimedia traffic—such as video and voice—was not present on local area networks. So it's not surprising that CSMA/CD does not come with facilities for handling time-sensitive data. In an effort to rectify this shortfall and outfit Ethernet to carry today's corporate apps, NBase Communications has developed a specification called Virtual Collision. Simply put, this is an enhancement of the Ethernet CSMA/CD algorithm that addresses the needs of time-sensitive applications on shared-media gigabit Ethernet networks.

This chapter discusses the growing need for multimedia support in LANs, the requirements for transmitting multimedia over a network, and the benefits of Virtual Collision over and above the facilities in standard CSMA/CD. It explains how Virtual Collision can guarantee the necessary bandwidth and network characteristics vital to today's and tomorrow's applications while at the same time being very easy to implement.

It's important to note that Virtual Collision will *not* be part of the 802.3z standard. This is largely because of the belief within the IEEE that changes to CSMA/CD should be restricted to those that are neces-

* Moti Weizman is director of hardware engineering at NBase Communications Inc. in Chatsworth, California. He has a BSEE from the Technical Institute in Israel. Moti has 10 years experience in the development of Ethernet products and ASICs, and he is an active participant in the IEEE 802.3z group. He can be reached at motiw@nbase.com.

sary to scale it up to gigabit speed in order to avoid any risk of a delay in finalizing the standard.

7.2 Background

Ethernet is the most widespread local area network. It provides a good solution for a good price and delivers adequate bandwidth for most applications. Installation of an Ethernet network is simple and straightforward. Migrating to switched Ethernet, fast Ethernet, and now gigabit Ethernet to support more aggressive applications is easy. It is based on simple, well-known, and proven CSMA/CD technology.

But Ethernet does have its drawbacks. CSMA/CD results in nondeterministic behavior because its media access rules are based on probability. Furthermore, it lacks a prioritization mechanism, and traffic latency is prone to variance. These weaknesses make it unsuitable for heavy constant-bit-rate (CBR) traffic, including multimedia applications such as video on demand and videoconferencing. NBase has introduced a simple yet innovative modification to CSMA/CD that will make it multimedia capable while retaining all of its advantages. This Virtual Collision mechanism takes advantage of the simple topology of gigabit Ethernet.

7.3 Multimedia: Switch or Share?

The ability to send multimedia over LANs and WANs will drive the next boom in technology and business. Multimedia applications will become a substantial part of our everyday life.

What should network designers use as the infrastructure to carry these applications? Switched gigabit Ethernet can be used to scale a network, allowing it to support more multimedia sessions. Full-duplex switched gigabit Ethernet delivers still more bandwidth. However, the need for large and fast buffers in the switches makes them expensive. From a price perspective, shared media is the more attractive option.

Shared-media gigabit Ethernet will play a part in catalyzing the growth in multimedia traffic—but only if it can simply and cost-effectively support heavy multimedia traffic.

That's where Virtual Collision comes in. Take MPEG-1 and MPEG-2, the most widely used video compression standards. A shared gigabit Ethernet network with Virtual Collision guarantees approximately 200 simultaneous full-screen VGA-quality sessions compressed using the MPEG-1 standard, or 30 high-resolution, high-quality HDTV sessions compressed using the MPEG-2 standard. (See Table 7.1.)

In practice, network designers will still need a mixture of both switched and shared gigabit Ethernet to provide the most efficient and

TABLE 7.1   Video Bandwidth Requirements

Video type	Resolution (pixels)	Uncompressed bit rate (Mbit/s)	Compression standard	Compressed bit rate (Mbit/s)
Video conference	$352 \times 288 \times$ 8×8	6.5	CCITT H2.61	28K
VCR quality (NTSC)	$352 \times 240 \times$ 24×30	60	MPEG-1	1.5
Computer (VGA)	$640 \times 480 \times$ 24×30	220	MPEG-1	4.5
HDTV (high-definition TV)	$1920 \times 1080 \times$ 60×24	3000	MPEG-2	20–35

cost-effective network. A typical network might implement a server farm (Fig. 7.1), where a few servers are connected via a shared gigabit Ethernet hub. This, in turn, is connected to a port on a gigabit switch that distributes the traffic to the workstations. The workstations may be connected via switched/shared 100- or 10-Mbit/s Ethernet. Finally a gigabit switch can be used to interconnect several similar structures.

7.4   Multimedia Requirements

To support multimedia, a network has to comply with a few rules. Today, data, voice, and video traffic are typically carried on separate networks. This is primarily because these types of traffic have different characteristics. Data traffic tends to be bursty—transmitting large,

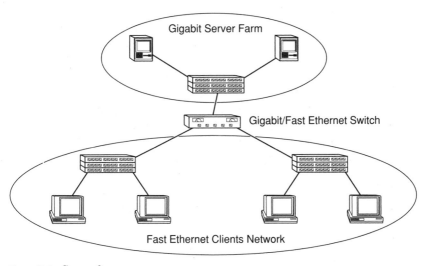

Figure 7.1   Server farm.

discrete chunks of information as fast as possible. Voice and video, on the other hand, tend to be more evenly paced in the rate of transmission but are very sensitive to when and in what order the information arrives. CSMA/CD has proven to be extremely efficient for data transfers in which the transfer rate may vary quite a bit without reducing the reliability of the application. It has also proven to be sufficient for light multimedia usage on an unloaded network. Under a heavy traffic load, however, an Ethernet network may begin to perform poorly, adversely affecting the response time and resulting in unreliable multimedia transfers.

CSMA/CD is not suitable for CBR applications—such as interactive videoconferencing—because it does not satisfy several crucial requirements: maximum delay and delay variability, guaranteed bandwidth, fairness, and prioritization.

Transmit latency (delay) refers to the time between packet queuing at the station and successful transmission onto the wire. This time should be kept very short and as constant as possible. In real-time applications, a long delay will result in a time lag. Constant delay is important to assure smooth video or audio without having to resort to large receive buffer caches, which in turn will dramatically increase the overall latency.

For example, a movie is composed of about 30 frames per second. If the delay difference between two consecutive frames is more than 1/30th of a second, the receiving side must be able to store earlier frames for a time before displaying them. The bigger the variance in the delay, the bigger the buffer size needed, and, consequently, the more expensive the solution.

In order to predict how many real-time sessions a network can support continuously, the network must guarantee a total available bandwidth. To prevent one or more nodes from hogging or capturing the majority of the available network bandwidth for a long period of time, fairness guarantees that all transmitting nodes get to send a minimum amount of data in any given transmission cycle based on the available network bandwidth.

Further, to ensure that critical real-time applications will continue to operate during peaks in network usage, their traffic must be given special priority. For example, on an overloaded network, the requested bandwidth exceeds the available bandwidth; thus, some of the traffic has to be delayed. If the overload period is long, there is a fundamental problem with the network. If it is only for a relatively short period, real-time applications should be given the right of way, since they are the most sensitive to delays. To do this, network bandwidth should first be allocated evenly to all nodes that have high priority. The remaining bandwidth should be divided evenly between all other nodes.

This allocation requires a deterministic calculation of available bandwidth and transmit latency. Ethernet is not deterministic, because the media access rules (CSMA/CD) are based on probability. The available bandwidth and transmit latency for a given station depend on the behavior of all the other stations on the network and cannot be calculated for heavily loaded networks.

While CSMA/CD may guarantee an average bandwidth over a very long period of time, there is no guarantee that one or a few nodes will not acquire the network bandwidth for a relatively long period. During this time, all other stations must wait a random number of slot times (according to the binary exponential algorithm) after colliding. Since the network bandwidth then varies relative to the number of collisions, the delay may vary by hundreds of milliseconds between a frame that suffered many collisions and a frame that was transmitted immediately. The delay may even be in the range of seconds if one of the frames was dropped due to excessive collisions, since higher-layer retransmission must occur.

7.5   Virtual Collision

NBase's Virtual Collision scheme is a minor modification to CSMA/CD that improves CSMA/CD performance. It creates a multimedia-ready, deterministic network that delivers full utilization of the channel, guarantees fairness, provides a priority mechanism, and bounds delay. For simplicity, all explanations of the Virtual Collision algorithm apply to a regular CSMA/CD network without carrier extension.

Note that the Virtual Collision mechanism is restricted to a star topology with a single repeater in the center. However, this limitation is not unique to Virtual Collision nets, but rather is a common restriction on any shared gigabit Ethernet LAN.

The key to Virtual Collision is to bound collision time to guarantee delivery of one good frame, thus ensuring that one of the stations involved in the collision will have a successful transmission. All others will still suffer a collision and will need to retransmit their frames. The station that is allowed to deliver a frame during the collision is the one whose frame arrived first at the repeater. If more than one frame arrives at the same time, an arbitrary choice is made between the stations. To avoid unfairness (i.e., a situation where the closest station to the hub captures the channel repeatedly), the gap between a successful transmission and the next transmission should be bigger than the round-trip delay of the network. This ensures that all stations involved in a collision cycle will have priority over the last station to successfully transmit.

A station should thus comply with the following rules. If it has a frame to transmit, the station must monitor the network for an idle period of at least 76 byte-times (round-trip delay + 12-byte interframe gap, or IFG) and then transmit. If no reception occurs during transmission, the transmission was successful, and the station must again observe the 76-byte idle time before initiating another transmission. If a frame was received during the transmission, a transmit collision has occurred, and the station should stop transmitting, finish receiving the frame, wait the 12-byte IFG, and then immediately retry the transmission.

The simplicity of implementing virtual collision compares extremely favorably to its many tangible benefits. That can be demonstrated by first examining a collision scenario under CSMA/CD and then comparing it to the same collision scenario under the new algorithm. (See Fig. 7.2.)

In the example of an ordinary CSMA/CD network shown in Fig. 7.2, station A is the first to transmit a frame, and the repeater recognizes the frame and distributes it to all other ports. Station B starts a transmission before sensing A's transmission. The actual sequence of events is as follows:

1. Station B's transmission arrives at the repeater, which recognizes this as a collision, since the repeater is still receiving A's transmission. The repeater then begins transmitting the JAM signal to all ports.

2. Station A senses reception (the JAM) during transmission, recognizes this as a collision, and thus stops transmitting. The repeater eventually sees no activity on port A and stops transmitting JAM to the last active port, port B.

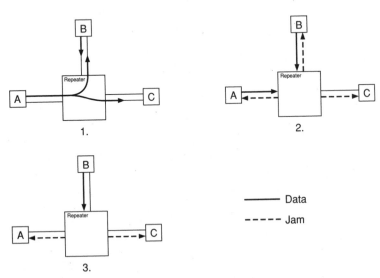

Figure 7.2 Collision scenario in Ethernet CSMA/CD.

3. Station B responds to the reception during transmission, recognizes this as a collision, and stops transmitting. The repeater sees no activity on any port and stops transmitting completely.

Stations A and B must wait a random number of slot times (1 slot time = 64 byte-times) plus 12 more byte-times before retransmitting the packet. The range of the random number depends on how many consecutive collisions the specific frame had just encountered. Thus, the time elapsed leading up to the collision is wasted, and the time until retransmission can be very long for all stations involved (up to 65,547 byte-times).

As noted, it doesn't take much to modify CSMA/CD to make it Virtual Collision-capable. The changes are outlined in Table 7.2.

With the Virtual Collision feature added, the order of events is as follows:

1. Station A is the first to transmit a frame. The repeater recognizes the frame and distributes it to all other ports. Station B starts a transmission before sensing A's transmission.

2. Station B's transmission arrives at the repeater, which ignores the frame because the repeater is still receiving A's transmission.

3. Station B senses reception during transmission, recognizes this as a collision, and thus stops transmitting.

4. Station A completes its transmission, which is successfully received by all other ports, including station B.

Only station B will have to retransmit its frame, and this only 12 bytes after detecting the end of reception from the network. Thus, the

TABLE 7.2   Virtual Collision Modifications

Where	What	CSMA/CD	Virtual collision
MAC	Interframe gap (idle time) after a successful transmission.	12 byte-times.	76 byte-times (12 byte-times + round-trip delay).
MAC	Interframe gap (idle time) after a transmit collision.	12 to 65,547 byte-times total (random number of slot times depending on sequential collision count + 12 byte-times).	12 byte-times.
Repeater	Behavior on collision.	Detect simultaneous reception on more than one port; distribute JAM to all ports; stop JAM when simultaneous reception ends.	Transmit the first frame received to all other ports; arbitrate between ties; all other traffic received is ignored.

time leading up to the collision is fully utilized (by A's frame) and B's retransmission delay is simply the IFG, as opposed to a random (possibly huge) number of slot times under standard CSMA/CD. (See Fig. 7.3.)

7.6   Double Network Diameter!

Usually, the diameter of an Ethernet network is limited by the fundamental rule in CSMA/CD that requires the round-trip delay of the network to be less than the collision window, which is defined as the minimum frame time (64 bytes). That requirement is relaxed when implementing Virtual Collision, since only the end-to-end delay of the network must be less than the collision window of 64 bytes. This is equivalent to a round-trip delay of 128 bytes, or twice the diameter allowable under CSMA/CD.

This is achieved because the worst-case collision recognition time is halved (see Fig. 7.4). The worst-case collision recognition time for existing CSMA/CD occurs when station A starts transmission such that the start of the frame arrives at station B just before station B starts its own transmission. Station B will almost immediately receive the transmission from A and recognize the collision. Station A, however, will recognize the collision only after station B's packet arrives; thus, station A's time from start of transmission to recognizing the collision is the entire round-trip delay. This is the delay from A to B plus the delay from B to A.

With Virtual Collision, that same case will not become a collision at all for station A, since A's frame was the first to arrive at the repeater. Now the worst case occurs when the two stations start transmitting

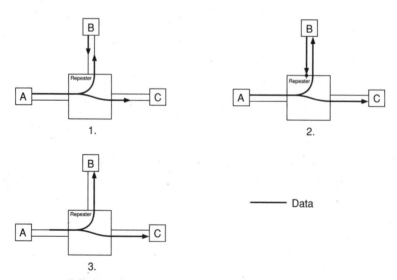

Figure 7.3   Collision scenario in Ethernet CSMA/CD with Virtual Collision.

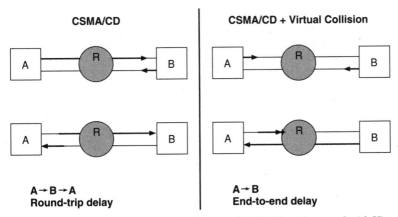

Figure 7.4   Worst collision scenario in Ethernet CSMA/CD without and with Virtual Collision.

simultaneously. Assuming that the repeater picks B to be the first, such that A experiences a collision, the time it takes for A to recognize this collision is merely the time it takes for B's frame to arrive at station A. This is only the end-to-end delay from B to A, which is half of the round-trip delay.

The implication of this for existing CSMA/CD is to allow a doubling of the network diameter. For example, the implementation of the algorithm under fast Ethernet will increase the legal network diameter from 250 to 500 m. The effect of Virtual Collision on gigabit Ethernet with carrier extension is either to double the network diameter or to extend frames shorter than 256 bytes as opposed to 512 bytes, thus significantly decreasing carrier extension's inefficiency.

7.7   Guaranteed Bandwidth and Support for Priority

In existing CSMA/CD, all stations involved in a collision have to retransmit their frames after waiting a random number of slot times. With Virtual Collision, the collision is always utilized to transmit one good frame. In addition, all frames are retransmitted immediately without wasting any time. This increases performance, since the channel can be fully utilized if needed. Finally, no frame is ever dropped due to excessive collisions, since each station is guaranteed to be able to transmit its frame eventually.

In addition to guaranteed bandwidth, a priority mechanism may be implemented by allowing individual stations to burst several frames at a time without an intervening round-trip idle time. If a station that transmitted a frame immediately transmits a new frame after only 12 bytes and does not wait the round-trip time, it is guaranteed to be the first frame at the repeater and is thus guaranteed to be transmitted successfully. This means that a station that had a successful transmission can immediately send a high-priority frame. So, in each collision cycle, a station can transmit one frame, optionally followed by a burst of n high-priority frames. This effectively increases the relative bandwidth dedicated to high-priority frames from that station.

7.8   Fairness

Virtual Collision allocates network bandwidth fairly between all stations that need to transmit on the network. Network bandwidth is divided only between nodes that need to transmit. This can be achieved with no arbiter involved. The network bandwidth available to each node on the network is exactly inversely proportional to the number of stations that need to transmit. To illustrate this, it's helpful to introduce the concept of a *collision cycle* (see Fig. 7.5).

A collision cycle is defined as the period in which exactly one frame was successfully transmitted by each station waiting to transmit a frame at the start of the cycle. Before a collision cycle can start, each station must detect an idle time equal to the round-trip delay plus the 12-byte IFG (a total of 76 byte-times). At this time, each station that wants to transmit a frame will send its frame to the hub. The hub will then broadcast the first frame and ignore the others. After this first frame has passed, the other stations (none of which sent their frames) will wait for the IFG and retry. The next frame in the sequence will then pass the hub. After a station has successfully transmitted its frame, it must wait until the network has been idle for 76 byte-times, so it will not transmit again until the next collision cycle. The cycle ends when the frame from the last station gets successfully transmitted. At this point, all of the stations are waiting for the 76-byte-time idle period, so the entire network will be idle until the start of the next collision cycle, when every station will attempt transmission again. Finally, a station cannot join the collision cycle after it has begun. If the

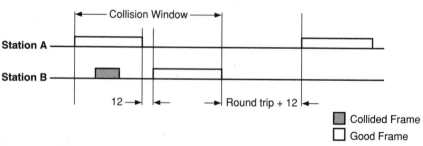

Figure 7.5   Fairness.

station has not collided (i.e., if it was not part of the cycle at the beginning), it must wait for the network to be idle for 76 bytes before it can even attempt a transmission. During the cycle, the network will be idle for only 12 bytes at a time (the IFG).

The fact that the idle time after a successful transmission is equal to the round-trip delay ensures that all stations involved in the last collision cycle will eventually retry successfully. In addition, it ensures that every station that did transmit its frame successfully will not be able to transmit another frame until the next cycle. Furthermore, on each collision, one station will have a good transmission. This creates a deterministic channel where if m stations need to transmit at the same time, the channel will experience $m - 1$ collisions. Each one of the involved stations will have one good transmission in the cycle, one of these occurring on each collision event. The order is also deterministic. The station closest to the repeater will get the first slot, and the station farthest from the repeater will get the last; however, the order of transmission is not important, since each station will get exactly one frame out on every cycle.

7.9 Conclusion

Implementing Virtual Collision addresses the multimedia requirements that are not guaranteed with standard CSMA/CD. The maximum delay will be the number of stations on the network multiplied by the longest allowed burst. The delay is still variable, but since it is bounded to be relatively low, this variability is acceptable. Indeed, the delay should not be longer than several microseconds, even when the network is overloaded, since the priority mechanism will guarantee that the delay-sensitive applications will have higher priority.

For any period of time, the minimum network bandwidth available for use by each node on the network is easily calculable and guaranteed. With CSMA/CD networks, the available bandwidth to each node can only be estimated and is not guaranteed. With Virtual Collision, fairness is guaranteed due to a deterministic utilization of the channel. Finally, a priority mechanism using frame bursts can be easily defined using the existing framework.

This powerful combination allows an affordable, shared gigabit Ethernet network to carry heavy multimedia traffic, thus providing all the benefits of CSMA/CD with the addition of real-time application support.

This chapter demonstrates that the Virtual Collision scheme involves only a few minor modifications to the existing standard. With Virtual Collision, Ethernet can maintain its dominance as the networking foundation of choice for the next decade and beyond.

Quality of Service on Gigabit Ethernet LANs

Nathan M. Walker*
Cisco Systems

8.1 Introduction

For the corporate network, constant growth has become the rule. As the network expands, network managers are seeking new ways to deliver multiple levels of service based on user identification, application, or network bandwidth availability. The notion of *quality of service* has come to embody several levels of service. These levels include a variety of both guaranteed and differentiated (or prioritized) service levels for network traffic. For business operations, these differentiated levels of network service enable separation of mission-critical traffic from less time-sensitive or priority-sensitive traffic.

That separation is proving especially important with the advent of the Internet, followed by the use of Web technology on corporate Intranets, both of which are contributing significantly to increasingly complex network traffic patterns. In fact, Web technology has changed forever the way organizations distribute and access information. Network managers and senior management alike realize this fact, and businesses are quickly adopting and standardizing on Web technology

* Nathan Walker is product line manager for gigabit multilayer switching and gigabit Ethernet at Cisco Systems. Mr. Walker is also vice chairman of the Gigabit Ethernet Alliance and an active participant in this industry initiative. While at Cisco, Mr. Walker has been responsible for fast Ethernet, ATM, and gigabit Ethernet switching product management and product marketing. Prior to Cisco, he held various product marketing and engineering management positions at Hewlett-Packard, SynOptics, and Kalpana, and was director of the computer and data communications business at Ryan, Hankin, Kent, a telecommunications and networking technology consulting firm. Mr. Walker received a bachelor of science degree in electrical engineering and material science from the University of California at Davis.

as the primary transport vehicle for mission-critical information and applications on corporate Intranets.

That makes it imperative for network managers to address and adapt the network infrastructure to this new environment. Traditional shared-media technology in the campus LAN is no longer able to deliver the performance and scalability required to support these new emerging Web applications. Not only is raw bandwidth required to address this problem, but, even more importantly, a set of new differentiated services on the network is needed to deliver information and applications to every desktop, according to business objectives.

Called quality of service (QOS) mechanisms, these differentiated services help to control congestion, prioritize traffic, and, in the end, ensure bandwidth availability. Without these mechanisms available in the Intranet, video, audio, and other interactive applications can block mission-critical applications from communicating or become slow, unpredictable, and even useless. As campus Intranets expand and become more complex, QOS mechanisms will become even more critical.

To a large extent, QOS mechanisms for campus intranets will coincide with the deployment of gigabit Ethernet. In addition, the QOS mechanisms being developed will be applicable to all versions of Ethernet (10, 100, and 1000 Mbit/s). The spread of gigabit Ethernet and QOS can be seen as complementary. Many applications, such as multicast and multimedia, that need the bandwidth of gigabit Ethernet will also require, as congestion grows, the prioritization and bandwidth reservation elements of QOS.

In fact, since Ethernet technology is already widely implemented, gigabit Ethernet promises to be one of the most practical solutions for companies that wish to preserve their installed networking investments while still receiving greater network speed, increased bandwidth, and differentiated QOS. Certainly, a technology like Asynchronous Transfer Mode (ATM) also provides QOS for integrated services across both the local and wide area. However, most organizations today have not implemented ATM technology from end to end in their networks. Because most businesses are unlikely to replace all their existing LAN connections with ATM connections, gigabit Ethernet (and the other forms of Ethernet), along with appropriate QOS mechanisms, is the most cost-effective and pragmatic answer for rapid, reliable data transfer for the vast majority of organizations.

This is not to say that ATM does not have a role in a gigabit Ethernet-based network. Indeed, several other technologies also will play roles as QOS becomes an increasingly essential part of the network paradigm. First, differentiated QOS will need to be offered across multiple transmission technologies (e.g., ATM and all forms of Ethernet). This means that QOS levels will have to be mapped across these technologies in order to be useful in heterogeneous networks and provide end-to-end service. Second, several standards—as well as technical implementations for functions specified in those standards—are required. Examples of these standards include the Resource Reservation Protocol (RSVP), emerging protocols such as 802.1Q and 802.1p, and integrated services over specific link layers (ISSLL). Together, these and other technologies will help provide end-to-end QOS for a wide range of needs in campus Intranets and across WAN interfaces for demanding internetworking environments.

This chapter considers the role of these technologies in providing differentiated QOS mechanisms. It also looks at the components of QOS and how these QOS goals can best be achieved for the network.

8.2   Defining the Basics: COS and QOS

In general, when discussing QOS on the network, there are two basic types of service to consider: class of service (COS) and quality of service (QOS). (See Fig. 8.1.)

A class of service can be defined as a type of traffic that is divided into several differentiated levels of priority. Thus, COS implies differentiated classes of service. This traffic typically includes flows originating from various types of applications or users or network segments. For example, mission-critical applications would be prioritized above best-effort e-mail. Video streaming applications would be prioritized highest during a company-wide broadcast. In a typical network, several, if not all, of these types of applications exist, and network managers must ensure that their networks support differentiated COS for these applications to meet the business needs.

Type	Service elements	Control elements	Enforce control
Quality of service	• Classic definition of QOS • Guarantee specific services • CIR, VBR, ABR, CBR, best effort	• Bandwidth • Latency (delay) • Jitter (delay variation) • Loss of traffic	• Police control elements
Class of service	• Differentiated or prioritized service levels • Best effort	• Priority of traffic	• Police control elements

Figure 8.1   Types of QOS mechanisms and attributes.

Classically, the term QOS has been used to define a set of mechanisms that ensure (or guarantee) specific services such as constant bit rate, variable bit rate and available bit rate (ATM QOS is an example here). These are services to which a contract can be written between a network service provider and the customer. Each of these QOS service types will be assigned specific bandwidth, latency control, loss, and drop characteristics. The guarantees are ensured through policing of parameters to meet the specified service. For instance, constant-bit-rate applications such as audio and video traffic will need a high level of guaranteed bandwidth to run. Therefore, these applications must have priority access to bandwidth and must not be subject to impediments, such as latency and jitter, along the network. To enforce this, traffic is policed to achieve the required specifications. On the other hand, available-bit-rate applications such as traditional data applications can be adequately supported with best-effort QOS guarantees. Best-effort QOS simply means that these applications can function with a wide range of available bandwidth and have greater tolerance for loss. In addition, they are far more tolerant of impediments like latency and jitter.

8.3   The First Steps Toward QOS

Before beginning a deployment of differentiated QOS, users must also consider how well—and at what cost—QOS-enabled devices will operate in their networks. Below are some factors to take into account before deciding on a QOS solution:

- *Cost.*   The cost of implementing differentiated COS and guaranteed QOS is substantial. Cost differences include several elements: equipment cost, management costs, and operation costs. While considering cost may seem obvious, users must look beyond the simple list price of a QOS-enabled Layer 2 or Layer 3 device. More importantly, users must consider the cost of implementing the QOS-enabled devices in their network. Certainly a more expensive device that continues to serve a function in a network has more long-term value than a cheaper item that will become obsolete in a few years.

- *Migration path.*   Closely related to long-term value, the migration path from current network equipment to equipment with improved network services ensures that QOS products and technologies are backward compatible and progressively more robust. Consideration of how the existing network equipment will impact the delivery of differentiated QOS services should be given. When purchasing a product, users should review a vendor's history of providing a clear migration strategy and make certain that the vendor intends to keep delivering backward compatibility and state-of-the-art technologies.

- *Mapping across multiple technologies.*   Most internetworking environments, including gigabit Ethernet-based environments, will be built on a multitude of technologies, or a heterogeneous environment. So, QOS must be able to be mapped across that environment, regardless of the technology mix. Users should therefore take care to seek out vendors that offer products with QOS features that are translatable across various network layers and technologies.

- *Applications usage.*   Users certainly should evaluate the applications that must be supported in their environments before considering the necessary level of QOS support. As mentioned above, constant-bit-rate applications such as multimedia applications typically require a high level of QOS. However, if a network is designed principally to send and receive low-bandwidth data applications, best-effort QOS should suffice.

- *Network management.*   In particularly complex networks with many QOS elements, robust network management is essential. Management services should allow network managers to monitor network operations efficiently, set and police policy, drill down to diagnose troubles, and provide a means to resolve any problems. In addition, management may include a flow-based paradigm to visualize traffic flows by service level in the network, including aggregate traffic flows, flows by application, and flows by class of service. Enhanced network monitoring will also provide the capability to observe bandwidth allocation, excess usage, packet loss, and all policed activity.

8.4   An Enterprise-Wide Solution

In order to work efficiently, QOS functionality must be installed throughout the entire network, providing an end-to-end solution. Also, it should be remembered that QOS features constitute only one piece of the enterprise solution for network services. A variety of other hardware and software elements are also required to deliver a complete set of enterprise-wide network services, connectivity, and reliability. Both these factors make it important to consider requirements for building a robust enterprise-wide solution.

- *Standards compatibility.*   Because gigabit Ethernet is compatible with existing IEEE 802.3 standards, it is the ideal solution for achieving greater speeds and increased bandwidth in installed Ethernet networks. Compatibility with all transmission technologies—including all types of Ethernet, ATM, and WAN standards—is important. As for the campus, gigabit Ethernet equipment will help ensure high speed in the campus core. Furthermore, with QOS features, gigabit Ethernet devices will give users both the standards compatibility and the services they need on their networks.

- *Multiprotocol support.*   QOS-enabled Layer 2 and Layer 3 devices must support a variety of existing and emerging protocols and standards. These include 802.1Q/p, RSVP, ISSLL, popular routing protocols (RIP, OSPF, BGRP, etc.), and popular network protocols (IP, IPX, etc.) in addition to popular Internet and management protocols. This multiprotocol support will ensure compliance with installed devices in existing environments while helping companies to protect their technology investments now and in the future.

- *Coexistence with VLANs and Layer 3 switching.*   VLANs are now a central element for solving specific problems within the enterprise network. Layer 3 switching is emerging to address requirements for acceleration of network throughput in the enterprise core. Therefore, network elements should support VLANs and high-performance Layer 3 switching as well as QOS.

- *Seamless migration.*   A network where QOS or QOS-capable hardware is installed should not be subject to a disruption in service or a redesign. Instead, old equipment must operate seamlessly with the newer QOS-enabled network elements. The added functionality should come online seamlessly, without notice by network users, while enabling the necessary differentiated COS or guaranteed QOS functionality. The service information must pass through the existing installed equipment. To ensure seamless migration, users should consider vendors that offer a comprehensive line of internetworking products with a variety of QOS options.

- *Expandable hardware and software options.*   The company network should be able to gain richer functionality through hardware and software additions. Users should pay special attention to how easily various network architectures allow QOS to be added and enhanced through both software and hardware. Again, users should assess the offerings of major internetworking companies with complete product lines.

- *Support for both cell and frame technologies.*   QOS support in high-bandwidth technologies like gigabit Ethernet is important and should include support for other popular, high-speed switching technologies such as ATM and frame relay. In addition, as high-end environments become increasingly heterogeneous, QOS elements should be interpretable across technologies. Thus, hardware and software will map service levels across different technologies, for example, by mapping frame-based QOS to cell-based QOS.

8.5   QOS Components and Enablers

Today, capabilities are becoming available to provide guaranteed QOS over cell-based ATM networks, and more should be available in the near future. Differentiated COS will emerge as a lower-cost and simpler approach for campus-based networks. For both types of service, there are a number of functions and technologies required to deliver a range of differentiated QOS capabilities. (See Fig. 8.2.) These include:

- Scalable bandwidth with a consistent data format

- Packet or flow priorities and resource reservation

- Queuing methods

- Policing methods

- Discard methods

- Mapping service levels across technologies

- Low-cost silicon for implementation

8.6   Scalable Bandwidth
with a Consistent Data Format

Gigabit Ethernet, being an extension to 100-Mbit/s fast Ethernet and 10-Mbit/s Ethernet, provides a consistent data format (e.g., the 802.3

QOS component	Function	Examples
Scalable bandwidth with consistent data format	• Scale speed with easy handling	• Ethernet, ATM
Packet or flow priorities with resource reservation	• ID and tag information • Signaling for resources	• IEEE 802.1Q/p, IETF—RSVP, IP TOS field
Queuing methods	• Prioritize traffic	• FIFO, priority queuing, custom queuing, weighted fair queuing
Policing methods	• Detect and discard or modify traffic violations	• Reservation policing, port policing
Discard methods	• Purposely drop data	• Tail drop, random early discard
Mapping service levels across technologies	• Apply QOS features across LAN/WAN technologies	• IETF—ISSLL

Figure 8.2   QOS component elements.

frame format) and is suited for use within buildings, on campus backbones, or even in data centers. ATM technology also provides scalable speed and a consistent format. However, ATM delivers this at a higher cost. In fact, Ethernet, including gigabit Ethernet, is expected to deliver differentiated COS functionality at relatively low cost and be more widely deployed in campus networks.

First of all, gigabit Ethernet is a robust enough solution to provide the bandwidth necessary to divide traffic into various classes, thereby providing COS support. As mentioned previously, this is a necessary condition for support of applications with differing bandwidth and QOS needs. Second, gigabit Ethernet, because of its widespread deployment, is a highly cost-effective and scalable format for delivering QOS. It creates the opportunity for a consistent Ethernet infrastructure, thereby preserving the value of the installed Ethernet base.

8.7   Packet or Flow Priorities and Resource Reservation

Differentiated QOS requires the identification and tagging of information on the network and signaling to reserve network resources. These types of functions will be delivered by a combination of IEEE-802.1Q/p, Resource Reservation Protocol (RSVP), and IP type of service (TOS) field information. IEEE 802.1Q and IEEE 802.1p will enhance Ethernet's ability to support differentiated COS at Layer 2. In contrast, RSVP and IP TOS (precedence subfield) provide the tools for Layer 3 support.

The IEEE 802.1Q/p work focuses on two areas: priorities and multicast groups. IEEE 802.1Q specifies explicit priorities in the virtual LAN (VLAN) packet header for identification of traffic priority. IEEE 802.1p specifies multicast groups within a switched IEEE 802 network to reduce the scope of flooding when traffic such as IP multicast is used for multimedia applications. Further, IEEE 802.1p specifies the use of priority queuing mechanisms to support traffic that may need lower jitter or higher priority than normal, best-effort traffic.

With 802.1Q/p, priority would be maintained in one of two ways. In the first case, the host would ask the network for priority allocation. The network, in turn, would respond that a particular flow should receive a particular priority. Alternatively, priority would be assigned on a per-port basis. In this case, there would be a maximum priority that any one port could use.

In order to use these mechanisms, a bridge or switch, for example, will need to have access to multiple queues. Since much installed equipment has only a single queue on which frames are processed, older equipment will not be able to use 802.1Q/p unless hardware is upgraded. Instead, users may have no choice but to throw bandwidth at the bottleneck, especially at the usual point of congestion: the backbone.

Available now, RSVP is used to reserve network resources by allowing applications to request a specific QOS for a data stream. Hosts and routers, for instance, use RSVP to deliver these requests to the routers along the paths of the data stream and to maintain the router and host state to provide the requested service, usually bandwidth and latency. In this case, RSVP specifies that each router between network endpoints will participate in the RSVP signaling sessions to reserve, tear down, and manage appropriate resources. This model provides end-to-end, fine-grained resource control but can lead to scalability difficulties in large, high-speed networks. Without sufficient network bandwidth, if more than a small subset of traffic is subject to RSVP, traffic may slow down unacceptably and additional QOS features may have to be implemented. Also, the reservation requests need to be translated into network policing to enforce packet classification and policing throughout the network.

IP type of service (TOS), a Layer 3 approach, provides the necessary information for packet flows by informing the network as to the type of service requested. Briefly, IP TOS allows a network administrator to sort network traffic into various classes of service at the perimeter of the network and implement those classes of service in the core of the network using priority, custom, or weighted fair queuing (see Sec. 8.8). Therefore, IP traffic with higher-priority service receives precedence over traffic with lower-priority service. IP TOS can be broken into two separate subfields: TOS and precedence. Because the TOS subfield is rarely used, only the precedence subfield should be considered a viable solution by users. With the precedence subfield, a network operator may define as many as six classes of service and then use the extended access control lists to define network policies in terms of congestion handling and bandwidth allocation for each class.

8.8 Queuing Methods

Queuing is simply a method used by a network to prioritize traffic based on predetermined criteria. In general, there are four types of queuing: first in-first out (FIFO), priority queuing, custom queuing, and weighted fair queuing.

FIFO is the basic method of queuing used by routers or switches with no QOS qualities. With FIFO, there is no sorting of traffic. Traffic is simply serviced as it arrives.

Priority queuing, on the other hand, ensures timely delivery of a specific protocol or type of traffic because that traffic is always transmitted ahead of other types of traffic. Priority queuing works by applying a set of filters or access list entries to each message forwarded by the router or switch. These filters inspect attributes of the packet, such as the source or destination identity, the transport protocol, or the application, and then prioritize the message according to the predetermined network requirements (network policies). The queuing algorithm places the packet in a queue based on the priority and, in transmission, gives the higher-priority queues preferential treatment over low-priority queues.

Custom queuing handles traffic by assigning different amounts of queue space to the various classes of packets and then servicing the queues in a round-robin fashion. Therefore, a particular protocol, user, or application can be assigned more queue space, although this entity can never monopolize the entire bandwidth. Custom queuing applies a set of access-list entries to each message forwarded. These access-list entries inspect attributes such as the identity of the source or destination system, the transport protocol, or the application to classify the message. The queuing algorithm then places them in selected queues. The router or switch services queues in round-robin order. The amount removed from a queue on each pass varies by configuration. This ensures that no class of packets achieves more than a predetermined proportion of the capacity when the line is under stress.

Weighted fair queuing ensures that queues do not starve for bandwidth and that traffic gets predictable service. Low-volume traffic streams receive preferential service, transmitting their entire offered loads in a timely fashion. High-volume traffic streams share the remaining capacity, obtaining equal or proportional bandwidth. Weighted fair queuing also ensures that the major causes of inconsistent response time—trains of packets, one immediately after another—are sorted into separate streams or conversations and forced to interleave. The algorithm also addresses the problem of round-trip delay variability. If multiple high-volume conversations are active, their transfer rates and interarrival periods are made quite predictable.

For the gigabit Ethernet-based campus Intranet, priority queuing is arguably the best choice. On the network edge, weighted fair queuing and custom queuing can both be accomplished through software, giving robust functionality at a reasonable cost for the WAN. However, neither can be implemented with software at the network core, which requires a hardware solution for 1000-Mbit/s links. Weighted fair queuing can be implemented with hardware but probably is too expensive an option for most Ethernet networks. Therefore, priority queuing offers the best of both worlds: a hardware implementation at a reasonable cost. It provides the functionality needed by most campus Intranets and does so with a simple design, making the solution relatively easy to configure and manage.

8.9   Policing Methods

Traffic policing is a mechanism used to detect and discard or modify traffic that violates the traffic contract agreed to at connection setup. The objective of policing is to protect the network and its users from either malicious or unintentional misbehavior by another network user. For example, misbehavior may be caused by malfunctioning hardware or software. The policing function checks the compliance between the monitored or measured parameter values and the negotiated parameter values, then takes appropriate actions in the case of violation. Generally there are two types of policing: reservation policing and port policing.

In reservation policing, the network device checks parameters such as flow rates and frame sizes against the reserved parameters. If a violation is detected, specified actions can be taken. Typically, this type of policing is confined to the edges of the network.

In port policing, a port uses a particular value, such as a TOS value, to ensure that a transmission gets assigned an appropriate priority. Generally, this type of policing is more appropriate to the center of the network, as it is less expensive and easier to configure and manage than reservation policing. In addition, reservation policing suffers from the same shortcoming as many other reservation-based techniques: a lack of scalability. Because a particular strength of Ethernet is scalability, port policing is a better option for gigabit Ethernet-based networks.

8.10   Discard Methods

In discard, the network purposely drops data as it receives more traffic than can be handled by buffering. Discard accomplishes two major goals. First, by dropping packets, it informs high-volume senders that their transmissions are flooding the network, thereby prompting them to cut their bandwidth needs. Second, it helps to allocate bandwidth to other users until the network recovers from its congested condition. For gigabit Ethernet-based networks, there are generally two types of discard: tail drop and random early discard (RED).

Tail drop mandates the discard of packets trying to oversubscribe a queue. In other words, if a packet arrives into a congested queue, the tail drop mechanism will simply discard the arriving packet.

RED, on the other hand, is a discard mechanism that does not wait until a queue is completely full before beginning to drop packets. Instead, once a queue reaches a certain threshold of congestion, it will begin to discard a predetermined percentage of arriving packets. Through this random discard, RED increases the probability that discarded packets will have originated from high-volume senders. These senders, therefore, are encouraged to cut their bandwidth needs until the backup clears. It is this ability to target senders most responsible for network congestion that makes RED superior to tail drop discarding.

8.11 Mapping Service Levels
Across Technologies

Mapping service levels across technologies simply means applying QOS features such as reservations and queuing from one network technology to another; for example, mapping QOS service levels from Ethernet to ATM or Ethernet to frame relay. This may involve shifting traffic from one service level to another at the interface point between the technologies (e.g., shifting Ethernet at priority 1 to ATM constant-bit-rate service). This mapping will also occur within Ethernet networks where Layer 2 priorities in switched networks will need to be shifted based on Layer 3 priority information. Currently, there is no standard for this mapping task. Mapping schemes tend to be quite vendor specific, and users should check with their internetworking vendors to ensure the best method of mapping across gigabit Ethernet and ATM, for example. This applies whether a user is trying to map services within the LAN or from LAN to WAN.

Fortunately, greater industry-wide mapping guidance is in the offing. Currently, major internetworking vendors are hammering out specifications for a future integrated services over specific link layers (ISSLL) standard. ISSLL specifies extensions to the IP architecture that allow applications to request and receive a specific level of service from the internetwork as alternatives to the current IP best-effort service class. As all forms of Ethernet, including gigabit Ethernet, and ATM continue to emerge on the network, such service mapping will become increasingly vital in offering complete end-to-end services. (See Fig. 8.3.)

In the meantime, users can map service levels in a number of different ways. For example, when mapping RSVP to a lower-level service, users probably should turn to queuing. Queuing, for instance, may be implemented in a virtual network or channel to map RSVP to that virtual circuit or channel. In the case of a multiplex service, queuing may not be the best solution. Users now may want to consider setting up a separate channel for the flow that needs QOS. This will provide an even finer measure of control.

8.12 Low-Cost Silicon:
An Important Enabler

For many networks, end-to-end low-cost, robust differentiated QOS will not be possible without the existence of low-cost silicon, which

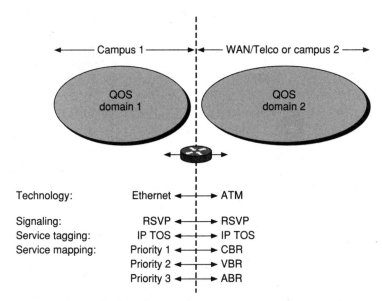

Figure 8.3   Example of service mapping.

increasingly is enabling services at lower prices. As the cost of silicon has dropped, the ability to afford increased functionality, such as the addition of extra queues, policing, and discard techniques, has risen. As with most other technologies, this trend toward greater performance at a lower price undoubtedly will continue.

Despite the falling cost of silicon, users still must decide which type of QOS service will best serve the network application and users for the least cost. For the campus Intranet, users essentially have two options: differentiated COS versus guaranteed QOS or varying degrees between these. In both cases, functionality embedded in silicon ASICs improves both speed and ability to handle increasing functionality.

Processor-based solutions that rely on software to provide the network with QOS will be more flexible. However, because they gain their functionality from software, processor-based solutions will also be inherently slower in the application of QOS functionality. Nevertheless, some companies may still choose to deploy this type of solution at the WAN edge, where traffic tends to be far less busy and QOS functionality is high.

Hardware-based solutions will deliver higher speed and application of QOS functionality. At the network center, peak performance is critical, and hardware-based solutions will meet this need.

8.13 Choosing the Right QOS Combination

To meet speed requirements, QOS functionality elements and implementation must be carefully matched to ensure low cost and high performance. For network managers, the challenge is to choose the best combination of hardware and software that has the best QOS functions for their needs; providing enough QOS functionality without incurring excessive costs. Here are the options open to them:

IEEE 802.1Q/p on its own.   When combined with a queuing mechanism, IEEE 802.1Q/p allows switches to prioritize and direct traffic. Because it can be implemented in inexpensive hardware, 802.1 Q/p is typically the first step in adding QOS functionality to a network. However, the solution does have a downside. There is no admission control or policing mechanism to 802.1Q/p. Therefore, the network can become oversubscribed if flows with the same priority are sent at the same time. In addition, if high-priority traffic—particularly if it also is high-bandwidth—floods the network, it will lock out all other traffic. So, network managers must be very careful to design prioritization schemes that, in all probability, will not overwhelm the network.

RSVP on its own.   RSVP provides signaling functionality for Layer 3 devices and gives a finer grain of control over the network. As mentioned previously, RSVP allows network managers to create access lists so that bandwidth can be reserved for traffic passing through Layer 3 devices. While it does reserve needed bandwidth for high-priority traffic, RSVP alone also has limitations. First of all, it cannot communicate with Layer 2 devices. Second, it must be able to talk with end station applications to be truly effective. Third, using RSVP to make lots of reservation messages will cause a flow to become unacceptably sluggish and thus reduce the network performance. RSVP also requests acknowledgments for the reservations. Another issue is that network devices need to be able to reserve, to the best of their ability, the resources needed for each RSVP request. The ability to reserve resources will vary from device to device.

Using both IEEE 802.1Q/p and RSVP.   Installing both of these mechanisms definitely is preferable to installing just one. In short, the combination of 802.1Q/p and RSVP allows users to draw on the strengths of each. With these mechanisms, information from an RSVP-enabled Layer 3 device can be translated for the Layer 2 switch. In other words, the Layer 3 device can tell the Layer 2 switch which packets receive precedence. The switch then can forward those priority packets. The chief shortcoming of this solution is that it adds complexity to the network structure. As more control is added, more equipment (and software) will need to be configured and monitored.

Still, as QOS becomes increasingly significant in the campus, the combination of 802.1Q/p and RSVP will probably become one of the most popular installations used to deliver differentiated QOS or COS.

Using IP TOS (precedence field) and RSVP and 802.1Q/p. This combination of QOS mechanisms has all the benefits of the previously discussed RSVP-802.1Q/p combination, with increased functionality for IP implementations, but adds even greater complexity. The precedence field sorts traffic into various classes of service at the perimeter of the network and implements those classes of service in the core of the network using priority, custom, or weighted fair queuing. This eliminates the need to classify traffic explicitly at each interface in the core network. Certainly a network with this type of QOS can be quite difficult to configure and maintain. However, this complexity will be needed for particularly robust networks; for example, Internet service provider (ISP) networks or campus networks with high-bandwidth multimedia. (See Fig. 8.4.)

8.14   Implementation

Network designers need to consider *how* QOS should be implemented as well as *where* QOS functionality is needed. The *where* will indicate a starting point for deployment in the early stages of Ethernet QOS deployment. For instance, network administrators will have to judge carefully the value of implementing QOS in congested backbones. In

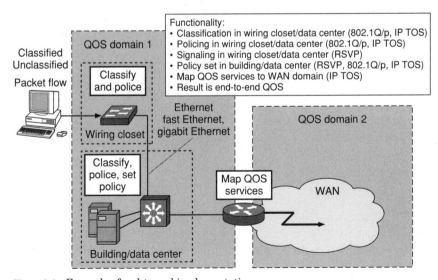

Figure 8.4   Example of end-to-end implementation.

some cases, they will find that QOS is an effective, cost-sensitive way to provide greater service to users, especially to high-bandwidth users. But as campus Intranet traffic grows even more, administrators may have to turn to both higher bandwidth—such as gigabit Ethernet—and QOS. For most companies, such an implementation remains in the future. But, with the explosive growth of large, high-priority applications, that day may prove to be sooner rather than later.

Generally speaking, QOS should be implemented where it is needed most. Therefore, users should implement QOS where there is most likely to be congestion. A selective implementation will not only provide services at congestion points but will also be cost-effective. However, an end-to-end implementation—applying QOS at the desktop, throughout the network, and at server locations—is a required solution for users who constantly run mission-critical applications or high-bandwidth applications such as video and other multimedia applications. These applications require differentiated (or prioritized) service levels to prevent the blocking of best-effort traffic. In addition, end-to-end QOS provides a high degree of network flexibility such as mechanisms for implementing user- and application-based network policies.

8.15 Policy

As the network increases in complexity, the application of policy management to applications and user groups will become ever more important. While policy can be implemented for nonreserved traffic, it will apply particularly to a mechanism like RSVP, which reserves bandwidth for certain types of traffic, users, or applications. For instance, a network manager may want to give precedence to high-priority traffic coming off a particular server. This policy will help ensure that this traffic doesn't get dropped due to congestion on the network. In addition, policies will help ensure that certain user groups are not arbitrarily starved of bandwidth because other user groups believe their traffic is more important, even if it is not. Policy will help mediate such intracompany disputes.

8.16 Conclusion: QOS as a Means for Network Control

With an emerging group of Web-based applications, gigabit Ethernet in combination with QOS will provide the service levels and high bandwidth needed by users while limiting such network impediments as latency and jitter. Moreover, by offering such mechanisms as reservation, prioritization, and policy, QOS will enable network managers to exercise control over bandwidth allocation. This will help ensure that high-priority traffic does not get dropped in favor of lower-priority traffic. In a network with multiple classes of traffic and many, often competing user groups, QOS will provide not only the bandwidth where it is needed most. It also will help keep the peace.

Implementing VLANs in a Gigabit Ethernet Network

Earl Ferguson*
Cofounder and chief technical officer, Foundry Networks, Inc.

9.1 Introduction

The ability to create virtual LANs (VLANs) on switched networks has existed for several years. Pioneered by Ethernet and ATM switch vendors in the early 1990s, VLAN functionality has become a standard feature on many LAN switch products today.

The technology has been slow to catch on initially in spite of its widespread availability. Users have tended to perceive the cost of implementing VLANs—primarily the cost of increased network management—as outweighing the benefits these networks offer. However, as VLAN implementations have matured, the cost of the network management to support them has decreased, and better and more automated methods for identifying host membership have been developed. Also, as users begin to deploy large switched 10/100-Mbit/s networks with gigabit Ethernet backbones and server farms, efficient broadcast management will become increasingly important, and virtual LAN technology is one of the least costly tools to accomplish this task.

The IEEE has recognized the importance of VLANs in the future of networking by forming the 802.1Q committee to further interoperabil-

---

* Earl Ferguson is the cofounder and chief technical officer for Foundry Networks, Inc. Before launching Foundry, Earl held a number of significant technical and management positions, including vice president of engineering at Centillion Networks and Tri-Data and director of internetworking hardware at Network Equipment Technologies. Earl holds six patents in internetworking technologies. He has been working in the computing and network industry since 1961. He received a master's degree in mathematics from the University of Michigan and a bachelor's degree in mathematics from the University of Washington.

ity of multiple vendors' products. Draft standards have been developed to propose multivendor ways of identifying and controlling VLANs. The method for doing this is known as 802.1Q tagging.

This chapter explores the often unique considerations for implementing VLANs on networks that combine 10- and 100-Mbit/s switched technologies with gigabit Ethernet switching.

9.2   VLANs Defined

VLANs are groups of networked end stations that communicate as if they were on the same physical LAN segment, even though they may be located on different physical segments throughout a switched network. The VLAN functionality is embedded in the switches, and the virtual LAN topology may be defined across one or more switches.

A major benefit of VLANs is that members of a common workgroup who share the same resources can be located on different physical segments of the network. An example might be a user in building 5 on a campus of five buildings who can access a server in building 1 without having to go through any costly router hops because the user and server are on the same VLAN. The packets are switched at or close to wire speed between client and server. At the same time, this particular user's broadcast/multicast traffic (e.g., IP ARP) does not interfere with that of other users in the network.

Another way to characterize a VLAN is as a *logical broadcast domain*. Switches are designed to forward packets from the source station's port to the destination station's port. Since switches, like bridges, make a simple forwarding decision based on MAC address information in the packet, they are required to forward broadcast packets to all switch ports in order that all stations on the network that need to see the broadcast will be guaranteed to receive it. However, once VLANs are defined on a switched network, broadcast/multicast traffic from one member station is only forwarded to switch ports to which other member stations of the same VLAN are attached.

Traditionally, routers have been used to provide broadcast control. In the past few years, switching has become a popular replacement for routing in many areas of the network—including workgroup, intermediate distribution, and even backbone levels—because it is both faster and much less expensive than routing. Without broadcast control, however, large switched networks are at risk of broadcast congestion problems. Adding VLAN capability to switches allows users to create large switched networks while still maintaining the necessary broadcast controls.

As this chapter demonstrates, broadcast control in a switched network will become more important as the network transmission rates are increased from 10 to 100 Mbit/s and ultimately to 1 Gbit/s. Gigabit Ethernet changes the requirements for designing switched networks. It also changes the requirements for designing the switches themselves.

This chapter will explore what these changes mean and the role VLAN technology will play in gigabit Ethernet LANs. It will look at the ways in which gigabit Ethernet changes the way users want to design their networks and how vendors must change the way they design products. Most importantly, it will discuss how VLANs can be used in combination with gigabit Ethernet to create efficient, cost-effective networks.

9.3   Gigabit Ethernet Changes the Rules

Gigabit Ethernet offers network designers and managers a very powerful bandwidth tool with major benefits for today's network environments. Gigabit Ethernet removes backbone bottlenecks. It provides the bandwidth needed to support quality of service (QOS) for many applications and thereby eliminates the need to make basic infrastructure changes. And, of course, gigabit Ethernet allows users to continue in the successful and well-understood Ethernet network paradigm.

This new tool also changes some of the old rules, both for network design and for switch hardware design. Three of the most important changes brought about by the advent of gigabit Ethernet are:

- A need for greater control of bandwidth usage in networks—particularly to avoid flooding traditional LAN segments with broadcast/multicast traffic

- A requirement for hardware-based switching services in the design of gigabit switches themselves, to take advantage of gigabit performance

- The ability to design networks that are cost-effective *and* supportive of applications that require quality-of-service (QOS) guarantees

Gigabit Ethernet poses new challenges for network equipment designers and users alike. Users must deploy more stringent broadcast/multicast traffic control in their networks to prevent broadcast packets from becoming an overwhelming problem. Network equipment designers and manufacturers, for their part, have to provide users with the tools necessary to implement effective broadcast traffic control.

9.4   Broadcast/Multicast Control

A major benefit provided by gigabit Ethernet is the ability to consolidate traffic from many separate 10- and 100-Mbit/s network segments onto a single gigabit building or campus backbone. However, implementing such a backbone can pose a problem: broadcast traffic from the gigabit backbone may overwhelm the lower-speed segments. This problem stems from the fact that the effect of flooding the network with broadcast/multicast traffic is one or two orders of magnitude greater at gigabit data rates than it is with 10- or 100-Mbit/s networks.

The seriousness of the problem depends on some key factors: the number of users on the network, the network protocols used, and the number of servers connected to the gigabit backbone. A fourth key factor is the rise in the number of multicast applications such as one-way video multicast.

Clearly, the number of users on the network has a major impact on broadcast traffic load. A general rule is that, depending on the other two key factors, a flat switched network can effectively support between 1000 and 2000 users before broadcast control of some kind must be introduced (in some cases, this number may be much smaller). So-called chatty protocols such as AppleTalk, older NetWare (3.11 and earlier), and NetBIOS create more broadcast traffic and therefore may also contribute to the problem when deployed on faster network infrastructures with centralized servers. Finally, more servers centralized within the organization will create a greater load of client-server broadcasts that will be propagated throughout the network in the absence of broadcast control. The solution to the problem is to implement some form of broadcast control so that broadcast/multicast traffic doesn't overcommit lower-speed segments.

Clearly, an uncontrolled broadcast storm on a gigabit Ethernet backbone with downlinks to lower-speed segments would bring the entire switched building or campus network down. It is crucial that network designers ensure that their networks are protected against such occurrences before gigabit Ethernet is introduced.

But a broadcast storm is a worst-case scenario. It is equally important to understand what happens under normal network conditions. Figure 9.1 shows a likely network topology employing gigabit Ethernet. This topology includes a combination of shared and switched 10- and 100-Mbit/s connections to desktop workstations. The desktop connections are aggregated from wiring closet switches onto gigabit links in the risers and connected to a backbone composed of gigabit Ethernet switches. The gigabit Ethernet switches are connected by trunk links at the core of the network and provide gigabit connections to centralized servers.

In a traditional network design, servers and clients are colocated and the 80:20 rule holds (i.e., 80 percent of the traffic is local to the workgroup, and 20 percent travels over the backbone). Broadcast traffic between servers and clients is contained in the workgroup segment, and the bandwidth of the workgroup segment is usually adequate to support local traffic.

However, in Fig. 9.1, the servers have been moved to centralized server farms and consolidated to use larger superservers of multiple-user workgroups. It is desirable to keep this as a switched Layer 2 network in order to preserve the performance of the client-server traffic (avoiding slower router hops), but then the broadcast traffic from the servers and their associated clients is forwarded to all segments. This can result in a manifold increase in broadcast traffic on all of the individual segments. Using a gigabit backbone removes limitations in the backbone but only distributes the problem to the workgroup segments. This may not initially be a serious problem on 100-Mbit/s segments, but it will likely steal enough bandwidth on shared 10-Mbit/s segments to slow response times or make the segments virtually unusable. The problem is compounded at peak traffic times during the day, most notably first thing in the morning when the majority of users first turn on their client stations and a flood of broadcast packets hits the network.

Broadcast control is necessary to prevent debilitating broadcast storms and ensure efficient use of bandwidth on lower-speed portions of the network. VLAN technology is the primary broadcast control tool for Layer 2 switched networks today.

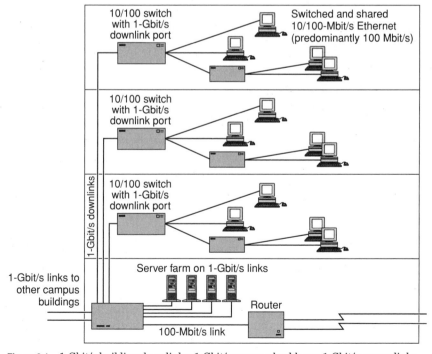

Figure 9.1   1-Gbit/s building downlinks, 1-Gbit/s campus backbone, 1-Gbit/s server links.

9.5 Implementing VLAN Technology in Gigabit Switches

If VLANs are the primary tool for broadcast/multicast control on networks with gigabit Ethernet backbones, it is incumbent on switch vendors to deliver products that properly support VLANs. Here again, gigabit speed changes the rules.

Many manufacturers of 10-Mbit/s Ethernet switches initially used standard processors running switching software as the frame-forwarding engine at the core of their products. Some vendors have since moved to hardware-based forwarding to gain performance and/or reduce costs, designing application-specific integrated circuit (ASIC) chips to provide frame forwarding in their switches.

With the advent of 100-Mbit/s Ethernet switches, hardware-based forwarding engines have become much more common. Frame-forwarding latency can be reduced by as much as 90 percent over processor-based switches, giving 100-Mbit/s switches a necessary performance boost.

At gigabit speeds, switches require hardware-based frame-forwarding engines. This, in turn, means that gigabit Ethernet switches will require hardware-assisted VLAN support. One way to implement the VLAN forwarding is to use a mask for all ports belonging to a VLAN. Each port then is assigned the mask of the VLAN of the port so that hardware can directly flood broadcast/multicast and unknown unicast traffic only to the ports of the VLAN. For ports supporting multiple VLANs, a lookup of the VLAN ID found in the 802.1Q tag is performed to obtain the mask. The VLAN ID lookup is included in the general hardware lookup process that takes place each time a new frame arrives at the switch port.

It will be important for users to query vendors about the components they use in their switches and to understand the impact of these design choices. Users will be well advised to select switches that utilize designs that can meet the requirements of increasingly heavy backbone loads.

9.6 Gigabit Ethernet and QOS

One of the more pressing concerns for network planners and managers in recent years has been how to support applications that require more stringent quality-of-service guarantees from the network. The debate has focused on two different approaches. One approach is to build into the network itself the ability to provide detailed—and complex—bandwidth management. When differing QOS levels are called for, the network employs mechanisms to manage the available bandwidth in such a way as to ensure that high-QOS traffic gets the bandwidth it needs. This is the approach taken by ATM.

The other approach uses a combination of shared and switched 10-, 100-, and now 1000-Mbit/s (gigabit) Ethernet to ensure that there is enough bandwidth to obviate the need for embedded traffic management and QOS guarantee mechanisms in the network itself. (Essentially, this approach is like assuming that there isn't a need for commuter lanes if the highway is wide enough to begin with). The addition of gigabit Ethernet technology provides sufficient bandwidth to defer implementation of full QOS guarantees. With the addition of simple priority queuing mechanisms in the network switches, the need for complex embedded traffic management may be deferred indefinitely. This includes applications such as real-time video and voice that demand stringent QOS guarantees regarding delay and delay variation. VLANs provide an ideal point of priority control.

However, an Ethernet network still requires that broadcast/multicast traffic be managed to prevent it from unduly diminishing the bandwidth supply, even when gigabit technology is used. It also means that networks must be designed to maximize bandwidth at critical points in the topology (an issue explored in Sec. 9.7). But overall, judicious use of switched 10/100-Mbit/s and gigabit Ethernet, combined with VLANs for broadcast and QOS priority control, can eliminate the need to embed complex bandwidth management and QOS guarantees in the network itself.

The IEEE 802.1Q and 802.1p committees are addressing these issues. The proposed mechanism for carrying QOS priorities is the VLAN tag of 802.1Q.

9.7 Designing Gigabit VLANs

This section looks at how VLANs can be implemented on primary and secondary gigabit Ethernet backbones as users migrate toward higher-bandwidth networking. Specifically, it explores the use of VLANs to control traffic flows on 100-Mbit/s and gigabit links as they might be implemented on building/campus backbones and in the building risers.

The migration path that most users will take as they deploy gigabit Ethernet will start in the primary campus or building backbone. The backbone is where today's networks tend to run out of bandwidth first. The key factors contributing to this backbone bandwidth crunch are centralizing servers in server farms (which turns the 80:20 rule upside down) and providing users access to company Intranets and to the Internet, and may eventually include the support of voice and video services. This high-speed primary backbone will most often consist of one or more gigabit Ethernet switches aggregating multiple 100-Mbit/s connections from wiring closets on one end and centralized servers on the other. If the backbone comprises two or more switches, a considerable amount of traffic will be carried across gigabit Ethernet trunks.

From there, the next logical step to increase performance, if necessary, will be to aggregate traffic onto gigabit switch ports in the wiring closets and run gigabit connections through the risers to the backbone switches. This will usually have to be accompanied by gigabit connections to those servers in centralized farms, depending on the flow of traffic in and out of the servers. (If there are high traffic volumes into a few centralized servers, gigabit links are needed to those machines in order to avoid a bottleneck caused by a stepdown to a lower-speed link at the server.)

Finally, for those users with an insatiable need for speed, gigabit links can be distributed to workgroup users. While this won't be in most users' plans for quite some time to come, it is still important to keep it in mind as the primary and secondary backbones are being put in place.

VLAN technology provides users with an excellent method for controlling broadcast and multicast traffic on today's switched 10/100 networks, and it will become even more valuable once gigabit switching is added. But before looking specifically at how VLANs can be implemented as users migrate toward gigabit LANs, it is necessary to discuss the emerging VLAN standards and what they will mean to users. This chapter will also look briefly at how inter-VLAN connections can be made using routers or Layer 3 switches.

9.8   VLAN Standards and MultiVendor Networks

While most major switch vendors offer VLAN support on their products, as of this writing there are no standards governing implementation of virtual LANs over switched Ethernet networks. However, the IEEE is near completion of two proposed standards that will go a long way toward providing for support of VLANs on multivendor switched networks.

One of these is 802.1Q, which, among other things, provides for a standard method of tagging packets with VLAN membership information. VLAN tagging will allow switches from different vendors to recognize VLAN membership and process the packets accordingly. The 802.1Q specification also provides for standardization of two other important VLAN functions: prioritization and administration. To this end, 802.1Q provides a mechanism by which 802.1p priorities can be carried over the VLANs. It also provides for standard VLAN administration, which is key to interoperability.

The 802.1p specification provides a means for establishing priorities for traffic that can then be carried in the 802.1Q tag. It provides protocols for group address (multicast) assignment and resolution between multiple applications. 802.1p provides the equivalent of IP multicast for Layer 2 traffic.

Support for these standards is key to the widespread deployment of gigabit Ethernet. As discussed, VLANs will offer a valuable tool for controlling broadcast traffic in switched gigabit networks. At present, many users already have VLAN-capable switching equipment in place from one or more vendors. Not all of those vendors will offer gigabit products, and even those that do may not have the best gigabit switches to meet each user's unique needs. Therefore, many users may find they have to mix and match products from different vendors. In these situations, standards-based interoperability will be crucial.

The 802.1Q and 802.1p specifications should be complete and approved by mid-1998. In the interim, however, some vendors are offering products based on a stable, prestandard draft of the specifications, especially 802.1Q.

Users should look for hardware-assisted VLAN functionality on gigabit Ethernet switches at the very least, and they may want to extend that criterion to the selection of 100-Mbit/s switches as well. It is clear that whatever VLAN packet tagging and prioritization is done in the switches will be improved if it is implemented in hardware to ensure that the performance is adequate.

9.9   VLANs and Routing in High-Speed Networks

An OSI Layer 3 connection is required to forward traffic between VLANs. In other words, when a station on one VLAN sends a packet to a station on a different VLAN, it must be *routed* between the VLANs. This means that the packets must pass through a router or a Layer 3 switch. This does not mean that users will have to find routers or Layer 3 switches that have gigabit Ethernet interfaces in all cases, however. There are different options for providing inter-VLAN packet forwarding.

9.9.1   Internal routing/Layer 3 switching

Some vendors have built wire-speed Layer 2/Layer 3 switching and/or routing into their gigabit Ethernet switch products. The switching component provides very low-latency packet forwarding at gigabit speeds. Wire-speed Layer 3 switching can decrease latency considerably over conventional routing, offering up to 10 times the performance of traditional routers. Because it is internal, the Layer 3 switching function forwards packets between VLANs (or subnets, for that matter) without having to send the packets on a round-trip to an external device. Just eliminating a trip to an external router can save between 50 and 150 μs.

A traditional router can also be built into a switch, providing full wire-speed routing. The combination of router management (route determination, error handling, broadcast firewalls, etc.) and wire-speed Layer 3 switching performance provides significant price/performance advantages.

9.9.2   External routing

If an external router is used, it does not necessarily have to be connected to a gigabit link. Depending on how much inter-VLAN traffic is expected, it may be sufficient to connect the external router to a 100-Mbit/s port on one of the backbone switches. If heavier traffic is expected, a gigabit port connection may be required.

A well-defined VLAN architecture can also reduce traffic flowing through routers and Layer 3 switches by ensuring that the largest amount of multicast traffic stays within the individual VLANs.

9.10   Using Gigabit Ethernet and VLANs to Relieve Backbone Congestion

As mentioned previously, the first phase of gigabit Ethernet deployment is being used to provide congestion relief in the backbone. Figure 9.2 shows a network in this first phase, with gigabit Ethernet only in

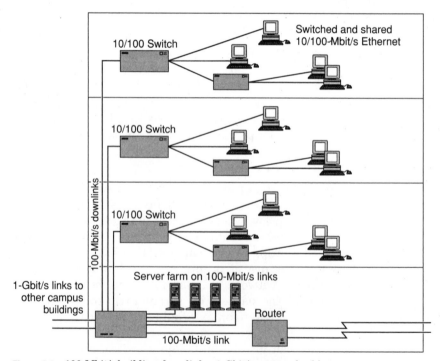

Figure 9.2   100-Mbit/s building downlinks, 1-Gbit/s campus backbone.

the backbone and 100-Mbit/s links coming from the wiring closets. Switched 100-Mbit/s links are also used to support centralized servers.

Gigabit Ethernet in the backbone provides enough bandwidth to accommodate even the most demanding enterprise applications. VLANs are used to control broadcast/multicast traffic on the backbone and riser links and to interconnect members of distributed workgroups. VLAN technology prevents the lower-speed downlinks from being flooded by unnecessary multicast traffic. Since the VLAN topology is implemented over trunk links, support for standard VLAN definition and administration is imperative.

If there is a high enough level of traffic between VLANs on an individual floor (e.g., to and from servers located on the individual floor), Layer 3 switching can be used to handle inter-VLAN traffic at that level rather than have it all go to the backbone-attached router or Layer 3 switch.

If, on the other hand, most or all servers are centralized, there will be little benefit to Layer 3 switching on the floors, since most packets will travel across the backbone to and from servers. In the case of the network shown in Fig. 9.2, all of the servers are located in the centralized server farm. If these machines are mostly dedicated to serving users in single workgroups or departments (that is, one server to one workgroup or department), there should be minimal inter-VLAN traffic. Each server can be a member of the VLAN associated with the workgroup or department it serves, and few users will need to cross VLAN boundaries to make connections to other servers.

Some servers, however, will need to be available to users from multiple VLANs. These include e-mail servers, Intranet servers, certain database servers, and other company-wide resources. There are two ways to accommodate this: applying inter-VLAN routing or making the servers members of multiple VLANs. The first approach—putting a router between all users and a company-wide resource—could create a serious bottleneck in the network topology. The second approach—making individual servers members of multiple VLANs—is a good solution in most cases.

9.11   Using VLANs to Reduce Congestion in the Secondary Backbone

Once the data rate on the primary backbone is increased to 1 Gbit, a combination of switched 100-Mbit/s downlinks and VLANs to control broadcast traffic may be used to effectively stave off increasing the bandwidth in the risers for quite some time. Depending on the level of broadcast traffic on the network, VLANs can significantly improve performance in networks that combine 100-Mbit/s switching at the departmental level with 10-Mbit/s workgroup switches.

Just how long users can expect 100-Mbit/s downlinks to be adequate in the risers, of course, will depend on both the network topology and the applications being supported. Large file transfers with low response time requirements, such as document images, heavy multimedia traffic, desktop video conferencing, or voice traffic, are examples of applications that may require a faster migration to gigabit in the secondary backbone. Centralizing servers also contributes to the need for more bandwidth sooner in the risers.

Once overall network load or QOS requirements for certain applications necessitate it, however, secondary backbone switches can easily be deployed in the wiring closets to aggregate 100-Mbit/s traffic onto gigabit downlinks. In this topology, illustrated in Fig. 9.1, VLANs play the crucial role of providing multicast traffic control to prevent flooding the 100- and 10-Mbit/s portions of the network with an overload of unnecessary multicast packets.

Again, for those networks where a significant amount of the inter-VLAN traffic is on the same floor, a Layer 3 switch is called for in the wiring closet to keep large amounts of traffic from going across the backbone to the router.

9.12 Gigabit-Enabled Server Farms and Workgroups

Once gigabit downlinks have been put in place on the secondary backbone to aggregate 100-Mbit/s traffic from the wiring closets, many users will find that gigabit links may also be required to some of their servers. Large superserver connections, especially, may become bottlenecks as packets from the entire enterprise network flood in at gigabit speeds only to be choked down to 100 Mbit/s on the server links.

Here again, VLANs can help manage broadcast traffic between servers and their associated client stations. At present, VLAN identification is provided using vendor proprietary information. With the adoption of the 802.1Q standard of VLAN tagging, server packets can carry standard information that can be interpreted by any vendor's switch that is capable of supporting the standard.

There are three different methods of performing the 802.1Q packet tagging process. One is to use the switch itself to provide the separation of packets and append the VLAN ID. This works well when one server is a member of one or a small number of VLANs. If large superservers are used that must be members of several VLANs, it may be better to outfit them with network interface cards (NICs) that are capable of providing VLAN packet tagging. A third method for dealing with servers that must be members of multiple VLANs is to use multiple NICs, each associated with a different VLAN.

While the vast majority of desktops will remain on a combination of 10- and 100-Mbit/s shared and switched network connections, gigabit Ethernet can easily provide a cost-effective means to support isolated high-performance workgroups within the overall network. Again, VLAN technology will be critical in preventing small workgroups of gigabit-attached end stations from overwhelming lower-speed network segments with multicast traffic.

9.13   Planning for VLANs and Gigabit Ethernet

VLANs will, in many cases, be key to the successful implementation of gigabit Ethernet in production networks. But designing and implementing virtual LANs is a major project in its own right, one that warrants thorough, careful planning. It requires a complete understanding of current traffic flows on the network as well as fluency in the VLAN administration tools that will be used. VLAN implementation usually entails a certain amount of iterative learning as well. It is not surprising, then, that users who have experience with VLANs prior to the introduction of gigabit Ethernet will generally find that the deployment of both technologies goes more smoothly than will those who try to do everything in one big jump.

A complete guide to implementing VLANs is beyond the scope of this book. However, there are a few things that may help readers plan for the introduction of VLANs, especially when gigabit switching is also projected:

Understand the current traffic flows in the network. The importance of this can't be overemphasized. In most cases, the network manager or designer will want to group users into VLANs based on the heaviest traffic flows. This information is also critical when planning where to deploy servers, Layer 3 switches, and routers. For instance, when routers are used to interconnect different VLANs, it's a good idea to minimize the number of router hops between them. End users who most commonly access servers on their own VLANs but also often communicate with resources on other VLANs shouldn't have to suffer the consequences of too many router hops. Earlier in this chapter, it was noted that, in the context of a switched network, VLANs offer the advantage of allowing workgroup users in different network locations who would otherwise be on different subnets to communicate with each other and their shared resources without suffering the delay associated with router hops. Wire-speed Layer 3 switches or the new wire-speed routers, however, can eliminate this concern. Because these devices forward packets at or near wire speeds, even in a heavily subnetted network, users pay only a very small penalty for being on different subnets. However, it is still important to manage broadcast traffic to ensure that the maximum amount of bandwidth is available, especially on the lower-speed segments of the network. Here, Layer 3 switching plays an equally important role: it allows network administrators to define relatively small VLANs (i.e., broadcast domains) without being overly concerned about the delays associated with router hops.

Plan a viable switch topology.   It is useful to plan for as close to a fully switched network as possible, given budget considerations. This is not to say that VLANs don't provide any benefits when there are multiple end stations attached to a single switch port. However, it may be more difficult to administer and manage such a network, and the fully switched topology will be able to withstand the increased bandwidth demands of modern applications longer than shared segments.

Identify the VLAN requirements.

- Classify the workstations and servers into logical groups to be included in a single VLAN.
- Classify by server access requirement, type of work performed, common activities, or functional relationships.
- If there are groups with security requirements (e.g., personnel, corporate finance), then separate them into a VLAN per security interest.
- If the size of a workgroup starts to exceed a few hundred, consider using multiple VLANs with a Layer 3 switch or wire-speed router to connect them.

Identify the best VLAN implementation.

- *Number of VLANs.*   Can the net manager define a large enough number of VLANs to meet needs?
- *Interswitch VLANs.*   Does the vendor's product allow the net manager to extend VLANs across multiple switches?
- *VLAN definition.*   Can the net manager define VLAN membership in the way that works best for the environment?
- *VLAN membership.*   There are several ways to define VLAN membership: by switch port, by MAC address, and by Layer 3 information such as IP subnet, IPX network address, and Layer 3 protocol. Port-based VLANs are the simplest to administer and work very well in a fully switched network.
- *VLAN administration.*   Is the vendor's tool set for managing and administering VLANs comprehensive and easy to use? For exam-

ple, do the tools allow the net manager to conveniently administer VLANs that span multiple switches?

Identify the best VLAN-enabled switches for the network's needs.

■ *Scalability.*   Does the vendor's product line easily scale from 10-Mbit/s to 100-Mbit/s to gigabit speeds?

■ *Standardization and interoperability.*   Is there support for the 802.1Q VLAN standard?

■ *Hardware assistance.*   Do the switches support hardware-assisted VLAN functionality, at least on the higher-speed ports?

9.14   Summary

VLAN technologies provide a reasonable and effective alternative for managing high bandwidth requirements. Gigabit Ethernet will provide network planners and managers with a powerful new bandwidth tool for supporting more efficient network topologies, such as server farms, as well as the next generation of high-performance network applications. VLANs, together with VLAN-capable Layer 2 and Layer 3 switches and wire-speed routers, will help in managing these high bandwidth requirements.

Gigabit Ethernet will initially be deployed in the primary campus or building backbone, supporting downlinks from 100- and 10-Mbit/s switches in the wiring closets. It will then migrate to the secondary backbone to aggregate multiple 100-Mbit/s links in the risers. At the same time, users will begin to use gigabit links to support centralized superservers. Soon, gigabit Ethernet may even find its way to some high-performance workgroups.

In many cases, successful implementation of gigabit Ethernet in such mixed 10/100/1000-Mbit/s networks will benefit from the use of VLAN technology to control broadcast/multicast traffic. In order to ensure adequate performance, all gigabit switches, and possibly 100-Mbit/s switches as well, must support hardware assist for the VLAN functionality. Support for VLAN standards is also recommended for use in multivendor networks.

3

Gigabit Ethernet and Routing

Purpose

To explain the technologies available for implementing routing on a gigabit Ethernet network.

What Is Covered

Gigabit Ethernet throughput solves plenty of problems for network managers. But it's a Layer 2 technology. To control traffic, network managers still need to deploy Layer 3 routing intelligence in the LAN. This section of the book describes the different routing techniques available to network managers for use on gigabit Ethernet networks.

Chapter 10 discusses multilayer switching, a recent and exciting approach to deploying routing in high-performance switched networks. Multilayer switches combine the performance of a conventional Layer 2 switch with the smarts of a Layer 3 router.

Chapter 11 focuses on IP switching, a technique developed by Ipsilon Networks before it subsequently was bought up by Nokia Telecommunications Inc. (Helsinki, Finland). Simply, this widely licensed technology allows full-function IP routing to be supported directly by gigabit Ethernet switches.

Chapter 12 describes the Cells in Frames (CIF) protocol. CIF allows network managers to use the QOS and flow-control techniques already developed for ATM, but implemented on less expensive and easier-to-use gigabit Ethernet equipment.

CIF is the brainchild of Larry Roberts, who is widely credited as one of the founding fathers of the Internet.

Chapter 13 deals with the subject of NT switching. This new class of network device is being pioneered by the start-up Berkeley Networks Inc. (San Jose, California). NT switches use high-performance hardware to forward frames, and the Microsoft OS for the clever stuff—like routing, management, security, and directory services.

Contributors

Cisco Systems Inc.; Nokia Telecommunications Inc.; ATM Systems (an AMP company); Berkeley Networks Inc.

Gigabit Ethernet and Multilayer Switching/Routing

Jayshree Ullal*
Vice president, marketing, Enterprise Line of Business, Cisco Systems Inc.

10.1 Introduction

Gigabit Ethernet has already cleared a lot of the hurdles that face any emerging technology. But challenges remain. For all its speed, gigabit Ethernet is not going to do away with the need for Layer 3 routing. Network managers will require their gigabit Ethernet-based networks to support a high degree of intelligence while still providing the speed and bandwidth advantages of switching. In other words, they are looking for reliable, multilayer combined switching/routing solutions that also add value through features that enhance scalability, manageability, and security. If leading internetworking vendors do not offer such robust features with their gigabit Ethernet solutions, users will look to other high-speed technologies to support their increasingly complex networks.

This chapter contains a discussion of key areas to examine carefully when implementing multilayer gigabit technologies at the foundation

* Jayshree Ullal is responsible for the overall marketing of the Enterprise Line of Business—including strategy, product management, marketing positioning, programs, and solutions. Ullal is responsible for growing Cisco's business in key enterprise markets including wiring closet switching, data center and backbone technologies, Layer 2–3 convergence, and customer premise equipment for remote WAN access with end-to-end network services. She continues to drive the marketing and product strategy for emerging technologies such as multilayer switching, gigabit Ethernet, ATM, and policy-based services. She directed Cisco's strategy in the LAN switching arena to become the market share leader, growing this business from $10 million in 1993 to a $2 billion run rate in 1997. Prior to joining Cisco, Ullal was the vice president of marketing at Crescendo and general manager and director of Internetworking at Ungermann-Bass. She holds a B.S.E.E. and an M.S. in Engineering Management from Santa Clara University.

level of a network. Because the term *network scalability* has been overused and underexplained in this context, the first part of this chapter defines scalability and discusses its various characteristics. Then, because network managers will need to make intelligent choices on multilayer equipment, this chapter takes a close look at the essential features and functions that these implementations should accomplish. A list of criteria by which to measure current and future gigabit implementations is also provided. Finally, the leading proposals for multilayer networking are discussed and perspectives are shared on switched gigabit Ethernet in current networks and within the networks of the future.

10.2   Scalability: An Extended Definition

If users are asked what elements contribute to scalability, they will probably respond "bandwidth" and "speed"—the classic ingredients of throughput. However, in today's world of complex networks, scalability should take on a broader meaning, which should also include the following elements:

- *Network design scalability.*   MIS professionals should protect their companies' existing investments by effectively designing networks with scalability as a key option. The goal is to gain the benefits of advanced services and capabilities while leveraging the network's installed base and avoiding forklift upgrades.

- *Network performance scalability.*   By upgrading the network, MIS professionals want to not only increase raw throughput but provide reliable, superior performance and better services to more users running more complex applications.

- *Migration of the installed base.*   Users should be able to take advantage of the added speed and bandwidth quickly and easily with few adjustments to their work processes. In addition, new users and even new locations should be able to link onto the network without difficulty and leverage all existing and future network services.

- *Network services scalability.*   Not only should the network maintain high packet throughput and availability; it should also be possible to activate other network capabilities and services—including virtual LANs (VLANs), more advanced management services, and network controls—to handle increased network complexity and multimedia capabilities.

For most companies, achieving this rigorous level of scalability means, in part, installing Ethernet-compatible equipment. Ethernet is the most widely installed networking technology in the world, making it most likely to scale effectively for companies that need to ensure scalability of network design, network performance, user migration, and network services. With the addition of gigabit Ethernet on the backbone, Ethernet also provides very high levels of bandwidth availability and speed, thereby also conforming to the more traditional definition of scalability.

However, even with gigabit Ethernet installed in an Ethernet-based network, maximum scalability is still not ensured. In order to achieve true scalability—and the greatest benefits for both users and MIS professionals—an integrated system of switching and routing hardware and software is required.

10.3   Layer 3 Switching: Combining Intelligence and Speed

Traditionally, switching and routing technologies have been disparate and distinct. In the past, switches have acted much like bridges. They delivered fast throughput and low latency, thereby helping to speed data through the network for a relatively low cost. However, they contained little embedded intelligence. Therefore, these switches—called Layer 2 switches—exposed the network to numerous deficiencies, including an inability to scale or segment the network. Additionally, networks were open to such problems as address limitations, spanning tree loops, and broadcast storms.

Routers have been providing a solution to these problems since the late 1980s. Routers allow for the interconnection of disparate LAN and WAN technologies, while also implementing broadcast filters and logical firewalls. In other words, routers make the network more stable and robust, thereby permitting internetworks to scale globally. However, routers have a downside: they tend to slow down the network, as they examine each packet before forwarding it to its destination.

Now, as networks continue to evolve, Layer 3 switches are integrating routing technology with switching for the best of both worlds: speed *and* intelligence. This approach yields very high routing throughput rates and adds stability and scalability to the network. Even in highly segmented networks, communications must reach their destinations and do so quickly. In other words, routing and switching are working together as integrated elements, which, if the Layer 3 gigabit switch is doing its job, should provide three basic—and essential—functions:

- *Route processing.* Layer 3 switches should coordinate networkwide connectivity and furnish the flow and topology information needed to provide value-added network services to communications between network users.

- *Network switching.*   Layer 3 switches should provide data forwarding based on network/link layer information and handle movement of packet traffic in and out of each device and across a switched fabric at wire speed.

- *Network services.*   Layer 3 switches should provide network layer processing and value-added services such as multicast, security, policy management, and multimedia.

In addition to integrating switching and routing, and because corporate networks are often not homogenous, Layer 3 gigabit switches also should integrate cell and frame technology to provide support for any LAN media. Therefore, Layer 3 switches should be able to be implemented throughout the network—wherever the demand exists for greater bandwidth, access to services, or manageability. (See Fig. 10.1.)

10.4   What to Look for in a Layer 3 Solution

For any company, a Layer 3 gigabit switching solution is not only a way to provide both high throughput and intelligence—it is also is an investment. Therefore, users should require a certain set of features and functionality before they commit to any solution. The following is a list of features that users should expect from their Layer 3 investments:

- *Standards compatibility.*   Any Layer 3 solution should be built on widely accepted IEEE industry standards. In the case of Ethernet technologies, users should look to vendors with a history of providing a robust line of reliable products that comply with IEEE 802.3 mech-

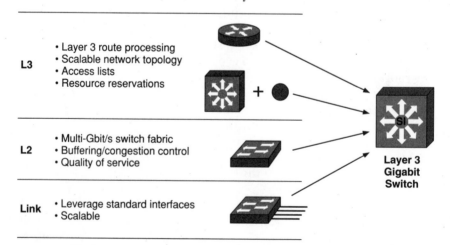

Figure 10.1   The Layer 3 gigabit switch.

anisms. This will help ensure both backward and forward compatibility for installed networks.

- *Multiprotocol support.* Switch architecture should allow for support of popular existing routing protocols, including Open Shortest Path First (OSPF), Border Gateway Protocol (BGP), and Interior Gateway Routing Protocol (IGRP). This multiprotocol support will ease installation of switches into an existing environment and will help companies protect their investments in their current technology infrastructures.

- *Adaptability with virtual LANs.* As networks rapidly adopt Layer 2 switching, it becomes increasingly important to offer hardware- and software-based functionality for controlling broadcasts and multicasts. VLANs provide these capabilities and are effectively used within Layer 2 switching domains. VLANs operate transparently to Layer 3 protocols, establish contained broadcast boundaries, and enhance the design of switched internetworks with these broadcast control functions. True Layer 3 switching is compatible with VLANs and can be used to extends them across the campus.

- *Seamless migration.* The integration of Layer 3 switching should not entail a redesign of the network or force any disruption in service. Instead, it should allow for the rapid evolution of a network while preserving the installed base of hardware. In fact, with the proper solution, many users may find that a simple software upgrade will provide Layer 3 switching functionality to existing high-end routers—regardless of whether the network is strictly router-based or is configured as a multiservice network.

- *Expandable hardware and software options.* Layer 3 switching architectures should enable the network to expand easily and gain enhanced functionality with currently available hardware and software add-ons. Such enhanced functionality may include greater mobility services, multimedia services, or broader security.

- *Support for both cell and frame technologies.* Layer 3 switching technologies should include support for other popular, high-speed switching technologies such as ATM and frame relay. It should be possible to incorporate Layer 3 switching into high-end heterogeneous environments easily.

- *Support for enhanced security.* Campus Intranet security should be delivered by two methods: port security and access security. Controlled by keeping a record of the MAC addresses active on all ports of a switch in a table, port security provides the ability to limit which workstations are permitted to connect to a specific port or switch. Access security, which is controlled by Layer 3 and VLAN membership policies within the Intranet, allows users to gain access to specific areas of the network or subnets. With access security, each workstation in the network is identified by a unique Layer 3 address, placing limits on which specific addresses can access or communicate with resources on specific subnets within the campus Intranet or the commercial Internet.

- *Management options.* Layer 3 switching solutions must be supported by advanced management services. With switched Intranets, campus networks have become increasingly complex and subject to failures. Therefore, comprehensive management solutions must be available to manage the whole switched Intranet, not just the network devices. These management solutions must ensure high levels of availability, performance, and security.

10.5   Roll Call: The Leading Layer 3 Gigabit Solutions

Among top internetworking vendors, there are currently two distinct approaches to Layer 3 gigabit switching technology:

- *Packet-by-packet technology.* Packet-based Layer 3 switches act very much like traditional routers. They examine every packet before forwarding it on to its final destination, and many support major routing protocols such as OSPF. The main difference from a traditional router is that the high-performance characteristics are primarily due to the implementation of a hardware switching path instead of software-based switching. Basically, packet-by-packet technology accelerates network performance by improving the Layer 3 forwarding of traditional routers. As a result, packet-based technologies have the same scalability attributes as a router-based network and can grow very large without becoming unstable.

- *Cut-through technology.* Cut-through switches attempt to gain the advantages of both Layer 2 and Layer 3 technologies by examining the first few packets of a transmission and, once a destination is determined, transmitting all remaining data directly using Layer 2 (cut-through). These switches offer the benefit of an overall network speed increase and have the same scalability attributes as bridged networks. Essentially, the increased performance comes from avoiding the examination and rewrite of every packet being transmitted that would normally be performed by a traditional router.

Within the context of packet-based and cut-through Layer 3 switch technologies, leading internetworking vendors have all taken different approaches to providing what they believe are the best, most robust

Layer 3 switching solutions. As such, it is worthwhile to review briefly each vendor's approach.

SecureFast Virtual Network (SFVN). SFVN is Cabletron's Layer 3 scheme. The aim of this primarily cut-through strategy is to move traffic quickly between VLANs by minimizing the Layer 3 routing functions of optimized switches. With SFVN, Cabletron embeds Layer 3 virtual routing services directly into its high-end switches. Equipped with SFVN technology, Cabletron switches have the ability to route the first packet and switch all successive packets of communications between VLANs. Because only the first packet is routed, data is transmitted quickly through the network and latency problems are avoided. This method of transmission closely resembles several other high-end switching schemes, including ATM and frame relay. In addition, it supports both packet-based and cut-through switching, earning the technology high marks for possible incorporation into heterogeneous networking environments. One consideration for MIS managers is that SFVN can only provide full functionality in a complete SFVN environment. This means that MIS professionals must understand that they are buying into a proprietary networking methodology.

Fast IP. 3Com's Layer 3 solution, Fast IP, adds intelligence at the client level to speed traffic through enterprise networks. With client software installed, desktops and servers supported end-to-end with Fast IP determine the fastest communications path between two points. The idea is to cut latency by reducing the number of router hops a packet must make while traversing subnets. Here's how it works. Imagine, for instance, that one Fast IP-enabled client sends data to another Fast IP-enabled client in an interconnected building. Once the Layer 3 communication is established, the desktop and servers determine whether a lower-latency Layer 2 path is available. If so, the communication—supported by a distributed next-hop routing protocol—is sent along the Layer 2 path, thereby providing more rapid data transmission. If not, the communication reverts to the controlled Layer 3 path for transmission. The good news is that Fast IP, because it works on the desktop and server levels, requires no changes to a network's routers and switches. The solution, in fact, appears to be very amenable to application in heterogeneous environments that contain the internetworking hardware of numerous vendors. The downside is that the solution is not as open as it first appears. As currently configured, Fast IP will only suit PCs, workstations, and servers that have 3Com network interface cards (NICs) installed. This leaves many networks out of the running for the Fast IP solution.

IP Switching.   Ipsilon's IP Switching method is rather similar to Cabletron's SFVN, as it relies on the cut-through method of speeding data through the network and reducing latency. Since developing the spec, Ipsilon has been acquired by Nokia. With IP Switching, an IP-enabled switch examines the first or first few packets of an incoming communication. Once a virtual channel is established, all subsequent data on that channel can flow through an ATM switch. Because each further packet is not examined—that is, flow labeling is performed only at a single location within the IP switch network—the communication is forwarded more quickly through the network than it would be through a network built strictly with routers. In addition, IP switching relegates much of the packet labeling and cache lookup to the edge of the network. Supporters regard this as an advantage, as the network edge typically receives fewer packets and therefore is less crowded than the center of the network. On the other hand, some networks that experience a lot of traffic at the edge may not benefit from offloading the per-packet work.

Switch Node (IP AutoLearn).   Bay Networks' Switch Node method is essentially a packet-based Layer 3 switching solution. Called a routing switch by its producer, it borrows heavily from router technology but adds several innovations to speed data through a network. First, with Switch Node, Bay Networks distributes the Layer 3 intelligence across switching interface cards. Because these cards carry out switching decisions locally, the overall performance of the central switch is maintained and latency is cut. With the addition of new switching interfaces, the network also is able to scale. Second, the Switch Node uses a technology that Bay Networks calls IP AutoLearn. According to Bay, IP AutoLearn automatically builds IP forwarding address tables. This means that switches can be installed into a network with no configuration, much like a Layer 2 switch. Switch Node then operates as a Layer 2 switch while it listens to IP traffic and learns the locations of IP devices. After it builds a database of Layer 3 topologies, it forwards all IP traffic at wire speed between the locally attached subnets or VLANs at Layer 3. However, some configuration must occur, or the switch would act purely as a Layer 2 switch with no Layer 3 intelligence. Essentially, IP Auto-Learn permits the snooping of IP ARP packets as they pass from client to server and back. This allows the switch node to switch all other packets within that flow in hardware. However, if a network administrator is utilizing any type of security traditionally found on routers—like access lists—this information will be ignored as flows are learned and switched locally on the Switch Node modules. Bypassing these key router functions opens the potential for security breaches between VLANs. Another key issue is that since the IP

AutoLearn mode must rely on an external router, there must be as many connections to the router as there are subnets being serviced by the Switch Node. And even though the Switch Node supports tagging or trunking multiple VLANs or subnets from Switch Node to Switch Node with a draft release of the 802.1Q standard, this support is currently only available for Switch Node–to–Switch Node connectivity, and the IEEE standard has not yet been finalized. Also, when IP AutoLearn mode is enabled in a mesh of Switch Node products, Spanning Tree will determine a single forwarding path, and no load balancing can occur as it would in a traditional router-based network. Further, when a link failure occurs, Spanning Tree convergence time is lengthier than the time taken by routing protocols to provide a backup path.

NetFlow/Tag Switching. Cisco Systems has developed two technologies to address the integration of Layer 2 and Layer 3 switching, with each serving a distinct internetworking function. These technologies are NetFlow LAN Switching and Tag Switching. NetFlow LAN Switching (NFLS) is implemented in Cisco's Enterprise-class LAN switches, enabling them to perform packet-by-packet Layer 3 switching at speeds equivalent to current Layer 2 switching performance. The two hardware components of NFLS are a Cisco IOS-based Route Switch Module, which is a fully functional RISC-based routing engine, and a NetFlow Feature Card, which is an ASIC-based multilayer switching engine capable of wire-speed packet-by-packet Layer 3 switching. NFLS allows multilayered switches to combine high-performance Layer 3 switching while applying advanced network services on a per-flow basis—thereby providing enhanced security, QOS, and traffic accounting information. Cisco's Route Switch Module, using Cisco's IOS software, performs all necessary route processing functions at the same time it applies relevant services. The NetFlow Feature Card automatically identifies traffic flows between end points and then switches packets in these streams at wire speed. The NetFlow Feature Card automatically populates its forwarding cache with information from the Route Switch Module so that all packets corresponding to a flow can be forwarded by the multilayer ASIC switching hardware. By identifying flows using both network-layer and transport-layer information, NetFlow LAN Switching allows Cisco IOS services to be applied on a per-user, per-application basis. *[Editor's note: In the interests of a balanced discussion, it is only fair to point out that the Route Switch Module is hardly an inexpensive addition to Cisco's portfolio. At press time it cost $20,000.]* On another front, Cisco's Tag Switching is aimed at scaling large router networks and wide-area backbones where there is a combination of routers surrounding Layer 2 core switches, such as ATM switches. In the Tag Switching architecture, ATM core switches run Layer 3 routing protocols such as OSPF or EIGRP, creating a peer relationship between the routers and the ATM core switches. This alleviates the scalability concerns including router convergence time, virtual circuit consumption, and signaling load issues, that can arise when routers have too many adjacent peers. Tag Switching integrates traffic engineering and QOS enforcement features across packets over SONET and HSSI networks as well as ATM by assigning tags (fixed size labels) to multiprotocol frames.

Naturally, each vendor's solution has a more complete list of strengths and weaknesses than presented above. For more complete information, readers should contact individual companies.

10.6   Configuring the Network for Switched Gigabit Ethernet

Now that the leading Layer 3 switch technologies have been discussed, the logical thing to do is examine where they can function within a network.

Currently, the most frequent application of switched gigabit Ethernet—and the one most often installed to provide the needed performance and scalability for the company Intranet—is on the building backbone. For this application, gigabit Ethernet is deployed for backbone links in the building riser connecting a centrally located switch with each wiring closet. Each wiring closet switch has a gigabit Ethernet uplink. Multimode or single-mode fiber cable are used to achieve the required distance. A gigabit Ethernet switch is centrally located in the building data center with connection to servers, routers, and ATM switches where needed. The server connections may use copper or short-distance fiber for lower cost. Routing and ATM services are provided as needed for high-speed connection to the wide area network.

A second popular application for switched gigabit Ethernet is at the campus backbone. Here, gigabit Ethernet links are used to connect switches in each building with a central campus switch. Full-duplex operation is used to achieve maximum throughput and distance with fiber media. Either single-mode or multimode fiber can be used. A gigabit Ethernet switch is located in a central location with connection to servers, routers, and ATM switches where needed. The server connections within the campus data center may use copper or short-distance fiber for lower cost. Routing and ATM services are provided as needed from the campus data center for high-speed connection to the wide area network.

Gigabit Ethernet also is suitable for connecting high-performance servers to the network and can even be deployed to power users at the desktop. At the server level, many environments are already being overrun by the march of technology. As processing power doubles every 18 months or so, servers are growing quickly in power and throughput. In addition, servers are increasingly being centralized within large enterprises. The result: high-performance UNIX servers are able to flood three to four fast Ethernet (100 Mbit/s) connections simultaneously. Therefore, more and more, these systems require a faster network and a switched environment.

Gigabit Ethernet is even being considered for power desktop users. Granted, this application is only just emerging. However, specific, bandwidth-hungry applications such as film postproduction and image processing, which require the transfer of very large files between desktops and servers, soon will require the availability and speed of gigabit Ethernet. Inevitably, gigabit Ethernet will migrate from the backbone and server locations as these applications become constantly larger and more demanding. At first, users undoubtedly will share the 1000-Mbit/s Ethernet. However, switched gigabit Ethernet is not even out of the question at the desktop for power users a few years down the road.

10.7   Conclusion: The Future Is in the Network

Without a doubt, gigabit Ethernet switches are in the near future of many, if not most, companies. With the coming of Internet- and intranet-based applications, companies will need the speed, bandwidth, and scalability these switches offer both on the backbone and at the server level. The majority of MIS professionals have come to realize that they also will continue to need the intelligence traditionally provided by routers. (See Fig. 10.2.)

Increasingly, Layer 3 gigabit switches will be the answer for networks overwhelmed with users and the demand for more services. Layer 3 switches provide the speed and bandwidth inherent in the Layer 2 switch along with the services and scalability of routers. In addition, many Layer 3 switching solutions provide other vital features, such as support for widely accepted standards, support for popular protocols, adaptability with VLANs, expandable hardware and software options, seamless migration, support for both cell and frame technologies, advanced Layer 4 security, and options for advanced management services.

However, users should be wary. Leading internetworking vendors all market very different Layer 3 solutions. Some are cut-through–based solutions; some are packet-based. In addition, products vary in quality.

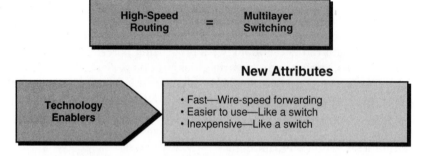

New Attributes

• Scalability—Establish and manage network hierarchy and topology
• Packet forwarding—Via network address
• Controlled access—Security
• Controlled forwarding—Multicast, QOS
• Troubleshooting

Figure 10.2   The evolution of routing.

While all will probably provide greater network speed and relieve bottlenecks, only some will support the full list of vital features mentioned previously. Before making a decision, it is truly incumbent on users to examine thoroughly the options available so that the most robust solution available is purchased. After all, the network has become central to the health and profitability of the corporation, and this reliance on the company network will certainly only increase in the future.

IP Switching and Gigabit Ethernet

Allen Beasley*

11.1 Introduction

Gigabit Ethernet exploded onto the networking landscape in the spring of 1996 with a dozen start-up companies announcing their intent to develop gigabit Ethernet products, the established networking vendors quickly announcing how gigabit Ethernet fits into their product strategies, and the $220 million acquisition of a 50-person gigabit Ethernet start-up that had yet to ship a single product. Within a few months of these announcements and events, Dataquest Inc. (San Jose, California), a notable market research firm, published a gigabit Ethernet market forecast that predicted product shipments growing from zero in 1996 to $73 million in 1997 to $2.9 billion by the year 2000, for an annual market growth rate of almost 250 percent.

A similar phenomenon had occurred a few years earlier in the networking industry when early ATM products were announced and industry analysts predicted that the market for ATM networking equipment would grow from an estimated $100 million in 1994 to over $2 billion in 1997. Only a couple of years later, however, analysts had a much less sanguine view of the ATM marketplace, downgrading their projections to a mere one-quarter of what was predicted during the initial announcement phase for ATM products. Interestingly, during the

---

* Allen Beasley was product marketing manager for Ipsilon Networks until December 1997.

Beasley began his career as an investment banker in the Technology Group for Alex. Brown & Sons in San Francisco, California. While at Alex. Brown & Sons, Beasley worked on the initial public offerings of several networking companies and was also engaged in a number of merger and acquisition assignments in the networking industry.

In 1990, Beasley graduated Phi Beta Kappa from Stanford University with a BA in economics with Honors and Distinction. He received an MBA from the Stanford University Graduate School of Business in 1994.

same period in which ATM sales have failed to live up to expectations, LAN switches, unheard of as recently as 1992, have enjoyed almost unprecedented market acceptance, growing to a $3+ billion market in 1996, according to the Dell'Oro Group (Portola Valley, California).

Given the rapidly changing networking market and a culture emphasizing the "technology of the moment," how do network managers evaluate the latest such technology—gigabit Ethernet? Is gigabit Ethernet the next advancement in the line of Ethernet successes that will further cement Ethernet's position as the dominant Layer 2 networking technology? Or is it the next ATM, a promising technology that has thus far failed to live up to expectations in the LAN? On the surface, it seems logical to suggest that gigabit Ethernet is bound to succeed because, as Andy Bechtolsheim, a gigabit Ethernet pioneer, claims, "the most obvious thing to do to make networks work better [is] simply to scale the speed of existing technology." However, are there routing issues that might slow the acceptance of gigabit Ethernet? Is the speed and capacity of the Layer 2 technology (Ethernet, fast Ethernet, gigabit Ethernet, or ATM) vastly outstripping the capabilities of the Layer 3 forwarding devices, namely conventional routers?

IP Switching has emerged as an architecture to scale routing to switching speeds, alleviating the problems currently plaguing conventional routers and potentially accelerating the acceptance of very fast Layer 2 technologies such as gigabit Ethernet. IP Switching achieves this result by adding Layer 3 intelligence (fully functional IP routing) directly to high-speed switches. Through an intelligent flow classification process, IP Switches are able to use existing routing protocols such as RIP, OSPF, and IGRP to dynamically control packet forwarding while using a high-performance switch as a packet forwarding engine. The resulting IP Switch is therefore able to offer an order-of-magnitude increase in IP forwarding performance. Initial IP Switching implementations have used ATM switches as the underlying switching engine; however, the technology is equally applicable to other high-speed switching fabrics, such as frame relay or gigabit Ethernet. In this manner, IP Switching may actually become a catalyst for the acceptance of gigabit Ethernet as a technology.

This chapter examines the emergence of gigabit Ethernet technology and products and suggests that one of the barriers to the acceptance of gigabit Ethernet may be the lack of an acceptable solution for routing IP traffic at gigabit Ethernet speeds. It then presents the IP Switching architecture pioneered by Ipsilon Networks, before it was acquired by Nokia Telecommunications Inc., explaining how this technology currently operates on ATM switches and how it could be deployed on gigabit Ethernet switches. Finally, it discusses the potential advantages and disadvantages of these different switch fabrics when used in the IP Switching architecture.

11.2   Increasing Returns and the Acceptance of Networking Technologies

In a perfect world, it would be easy to understand and predict the acceptance of new technological advancements in networking: network managers would simply pick the superior technology (i.e., let the best technology win). However, as anyone familiar with high-technology markets and products knows, the "best" technology does not necessarily win in the marketplace. If it did, we might all spend our days working on Macintosh computers interconnected via ATM networks to NeXT workstations, while spending our nights watching videotapes on our Beta VCRs.

On the contrary, high technology seems to operate according to a relatively new economic theory known as *increasing returns*. As stated simply by the originator of the theory, Brian Arthur, increasing returns is the tendency for "that which is ahead to get further ahead, for that which loses advantage to lose further advantage." The idea is that because of very low marginal costs of production, product compatibility considerations, and customer training issues, once a high-technology product or standard gets ahead in the market, even if purely by chance (as is often the case), it is likely not only to stay ahead in the market but even to increase its lead over competitors. Ultimately, over some period of time representing the life cycle of the technology, the market changes due to an external event or a significant technological breakthrough, and the process is repeated. So, how does this apply to networking, and specifically to gigabit Ethernet? The widespread acceptance of new networking technologies such as gigabit Ethernet hinges on the ability of these technologies to either (1) work seamlessly with existing technologies and standards or (2) provide a compelling reason for users to abandon the old technologies and standards.

11.3   The Emergence of Ethernet and TCP/IP

Over the past decade two very powerful examples of increasing returns and positive feedback have emerged in networking: Ethernet and TCP/IP. Because of their significant momentum, these technologies will prove very difficult to dislodge, for the simple reason that customers do not want to see them dislodged. Customers simply have too much vested in their success:

- Many mission-critical applications depend on either or both of these technologies.
- Customers are very familiar with the operation of these technologies.
- There is a significant amount of industrywide investment in these technologies.

Because of these factors, new networking technologies must interact seamlessly with Ethernet and TCP/IP to have any realistic chance of success. For a new networking technology to fly in the face of TCP/IP, for example, requires that the perceived value of that new networking technology exceed the perceived value of applications (such as the World Wide Web) that depend on TCP/IP. Clearly, this latter course of direction is very difficult to foresee, because if anything is true in networking it is that the applications drive the network, not the reverse.

Ethernet and IP thus represent the dominant de facto networking standards in the LAN, and both of these technologies are in a very powerful positive feedback loop (i.e., the more they get ahead in the market, the more likely they are to stay ahead). To illustrate the power of this feedback loop, Fig. 11.1 shows a graph of the percentage of worldwide hub and switch port shipments that are Ethernet (as opposed to token ring, FDDI, ATM, etc.). As the graph indicates, over the last five to six years, Ethernet's share of this market has increased from roughly 50 percent to over 90 percent. With incredible economies of scale driving down costs of Ethernet network equipment, with very high customer familiarity, and with proven scalability from 10 to 100 Mbit/s, it would require a very powerful application indeed to dislodge Ethernet from its place as the dominant Layer 2 networking technology.

Similarly, TCP/IP appears to be where Ethernet was in 1991 or 1992— on the cusp of tremendous growth and market domination. As Fig. 11.2 shows, the percentage of desktops running TCP/IP is forecasted to rise from roughly 20 percent today to almost 70 percent by 1999. Granted, market statistics and forecasts should always be viewed with skepticism

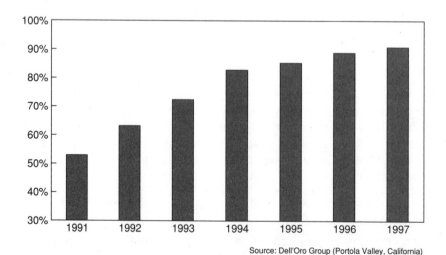

Source: Dell'Oro Group (Portola Valley, California)

Figure 11.1 Ethernet market domination (Ethernet ports as a percentage of worldwide hub and switch market).

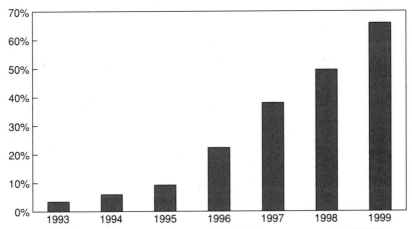

Figure 11.2 The emerging dominance of TCP/IP (desktop systems with TCP/IP as a percentage of total desktop systems).

(remember the ATM forecasts from a few years back). Nonetheless, these statistics have some very powerful drivers behind them:

- Microsoft began shipping TCP/IP networking software bundled with the Windows 95 operating system, giving every user of the dominant desktop operating system the ability to take advantage of the capabilities of TCP/IP networking (to surf the Web, for example).

- NT and UNIX-based systems are dominating in the server market, displacing NetWare servers, and both of these systems primarily rely on TCP/IP for networking.

- Access to the Internet (the world's largest TCP/IP-based network) is increasingly becoming a "must-have" tool for business professionals, and a very popular consumer service.

- Companies are aggressively deploying, or planning to deploy, internal TCP/IP-based Web server farms—Intranets—for a variety of functions including basic document distribution, process automation, and electronic commerce (70 percent of Fortune 1000 companies, according to one study by the Business Research Group, Newton, Massachusetts).

The effect of these trends is easy to predict. As network managers become more and more familiar with TCP/IP networking and users become more dependent on TCP/IP-based applications, TCP/IP will become further entrenched as the dominant Layer 3 networking technology. For network managers, this dominance means that efficiently routing TCP/IP traffic in a network will become an increasingly important issue.

11.4   Market Acceptance
of Advancements in Internetworking

Reflecting the status of Ethernet and TCP/IP as the de facto Layer 2 and Layer 3 networking standards, market analyst John McQuillan wrote the following: "The diversity that has been so endemic to nearly every corporate network is lessening somewhat—there is surprising homogeneity as the dominance of Ethernet and TCP/IP grows." Because of the increasing returns phenomenon, over time this homogeneity will become more powerful and less surprising. In light of this trend, it is instructive to analyze some of the more recent advances in networking not only from the standpoint of performance and technology but also from the standpoint of integration with Ethernet and TCP/IP. In many cases, these integration issues have had as much to do with the success or failure of these technologies as any real or perceived performance advantages.

The dramatic success of Ethernet switching (10 Mbit/s) is perhaps the easiest to understand against these factors. Not only do Ethernet switches offer an immediate and demonstrable performance advantage over shared hubs, they require no change in hosts or routers at Layer 2 or Layer 3. When Ethernet switches were first shipped, there was already a defined mechanism of routing IP traffic over Ethernet, there were plenty of routers capable of routing between several Ethernet interfaces, and Ethernet host network interface cards (NICs) were already widely deployed. Thus, Ethernet switches offered a compelling performance advantage over the incumbent alternative (shared hubs) in addition to seamless integration with Ethernet and TCP/IP. It is therefore no surprise that the rapidity and extent of the success of Ethernet switching has been almost unprecedented.

Fast Ethernet has also enjoyed tremendous success for similar reasons. Like Ethernet switches, fast Ethernet (switches, hubs, and NICs) also offered immediate and demonstrable performance advantages, especially for LAN-to-router connectivity and for server-to-backbone connectivity. However, in contrast with Ethernet switches—which required no change in hosts or routers—deploying fast Ethernet in any form required the use of new network interface cards in the host/server, new switch/hub interfaces, and new router interfaces. While on the surface these requirements may have seemed like an obstacle to the acceptance of fast Ethernet technology, the Ethernet NIC, hub/switch, and router vendors all were able to leverage their (and the industry's) existing Ethernet expertise to quickly release cost-effective fast Ethernet interfaces on these products. One key element was the development of autosensing 10/100 Ethernet NICs and switch interfaces, which made the process of integrating fast Ethernet into an existing network very straightforward. Finally, from the perspective of TCP/IP, there was effectively no change in migrating from Ethernet to fast Ethernet. Routing IP over fast Ethernet was the same as routing IP over Ethernet, although the former began to tax the performance of conventional routers because of its 10-fold increase in bandwidth. Thanks to its immediate performance advantages and integration ease, fast Ethernet is quickly becoming a default standard for high-speed (>10-Mbit/s) campus networking, displacing FDDI.

In contrast to Ethernet switches and fast Ethernet, ATM was presented with a series of challenges in its attempt to penetrate the LAN market. ATM was conceived as a "brave new world" networking technology, in that the underpinnings of ATM (connection-oriented "telco-style" signaling) had almost nothing to do with existing LAN internetworking (connectionless packet forwarding). As a result, when initial ATM products were released, users found little use for them, as there were no native ATM applications and no ATM interfaces on any existing networking equipment. Consequently, initial ATM test bed implementations were limited to statically configured workgroup islands, that is, groups of hosts with ATM NICs connected to an ATM switch via permanent virtual circuits. These prototype ATM networks saw limited success in terms of migrating into production LAN environments.

In an attempt to lower the barriers to ATM acceptance, the ATM Forum created the LAN Emulation (LANE) specification, which allows an ATM switch and ATM-connected hosts to emulate existing LAN technologies such as Ethernet. Although somewhat difficult to install and configure, LANE allowed for the demonstration of dynamic switched virtual circuits rather than statically configured permanent virtual circuits. Unfortunately, LANE had the rather perverse effect of hiding any of the perceived benefits of ATM from the user. After all, if LANE simply allows an ATM switch to emulate an Ethernet switch, then why not buy an Ethernet switch (or fast Ethernet switch)? The performance is comparable, the price is lower, and network managers already know how to install it.

Not only did ATM face Layer 2 integration issues and stout price/performance competition from fast Ethernet, but ATM also posed certain difficulties in integrating with IP. When ATM was first released, it included no mechanism for routing IP. LAN Emulation from the ATM Forum and Classical-IP-over-ATM from the IETF offered only single-subnet solutions, and it was only the introduction of the ATM router interface by existing router manufacturers that truly addressed the problem of routing IP over ATM. Once router vendors began shipping ATM interfaces on their existing routers, it became possible to route between standard IP subnets, emulated LANs, or logical IP subnets. However, the speed of ATM (predominantly 155 Mbit/s initially) and the relative immaturity of the ATM hardware severely tax the capabilities of existing routers.* In short, conventional routers simply cannot keep up with the raw link speed of ATM. Somewhat predictably, even with the availability of router interfaces, many users faced with expensive and relatively slow ATM router interfaces and the inherent difficulties of LAN Emulation and Classical-IP-over-ATM have simply opted for Ethernet/fast Ethernet solutions instead.

11.5   Gigabit Ethernet and the Need for a New Router Architecture

On the surface, gigabit Ethernet appears to share many of the characteristics of fast Ethernet and seems therefore likely to avoid many of the problems faced by ATM in penetrating the LAN market. With equipment vendors having already successfully scaled Ethernet from 10 to 100 Mbit/s and with a defined standard for routing IP over Ethernet frames (running at 10 Mbit/s, 100 Mbit/s, or 1 Gbit/s), gigabit Ethernet should be in a strong position.

In fact, gigabit Ethernet will exacerbate the performance problem already beginning to plague IP routers. While conventional routers have successfully scaled from lower-speed WAN and 10-Mbit/s Ethernet links to higher-speed fast Ethernet and FDDI links, they are straining to keep up with ever increasing traffic demands and faster data link rates. Today's routers are roughly four to five times as fast as routers five years ago, while transport rates and switching capacity have increased at much faster rates over that same time period (e.g., gigabit Ethernet is 100 times faster than conventional Ethernet). Quite simply, the processing power and bus capacity of routers have not improved fast enough to keep pace with network traffic demands. Figure 11.3 shows this disparity by comparing the relative increase in microprocessor performance (using the Intel processor architecture) versus the growth in traffic on the Internet. As Fig. 11.3 points out, the growth in traffic on the Internet has exceeded (by almost an order of magnitude) even the almost unprecedented increase in processor performance. As a result, a router architecture that relies on a central microprocessor to process each packet will simply not scale to meet the needs of the Internet or other large IP networks.[†]

---

* The very first ATM router interface shipped was capable of forwarding about 30 percent of the theoretical maximum number of packets per second on an OC-3 link, and there was a recommended design limit of only two of these interfaces per router.

[†] Interestingly, most conventional routers have not been based on the Intel processor architecture and have not even kept up with the Intel performance curve. Furthermore, most of the minicomputer-style router architectures have notoriously long lead times between major performance enhancements, a function of the proprietary nature of the architecture.

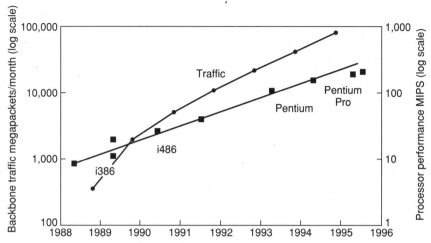

Figure 11.3   Conventional router architectures cannot keep up (comparison of processor performance increases versus Internet growth).

While routers cannot keep up with increasing traffic demands, the need for IP routing in large networks remains. The deployment of corporate Intranets is forcing a change in the network infrastructure. While networks were historically designed under the assumption that 80 percent of network traffic would remain within a given LAN (intrasubnet) and 20 percent of traffic would traverse the network backbone (intersubnet), centralized servers are reversing this trend. In fact, a highly utilized internal Web server farm can dramatically increase the amount of intersubnet traffic on a corporate network, straining the network backbone and increasing the need for high-speed routing. The success of high-speed LAN technologies requires a solution to the problem of routing IP at switching speeds, a problem that will no doubt worsen with the deployment of higher-speed Layer 2 technologies such as gigabit Ethernet or OC-12 ATM in campus backbones.

11.6   The IP Switching Approach

Ipsilon developed a unique approach to integrating fast switching technology with IP routing, with the goal of enabling IP routing to scale to meet the requirements of increased traffic demands and faster data link technologies such as ATM and gigabit Ethernet. Ipsilon's IP Switching approach centers on two key ideas. First, standard IP routing protocols (RIP, OSPF, BGP, IGRP, etc.) can be used in conjunction with standard high-speed switching hardware. Second, IP packets sharing certain characteristics (header fields) can be classified as flows, reducing the amount of forwarding work required of a central processor and dramatically increasing overall system packet throughput. Combining these two points, IP Switching marries IP routing functionality and high-speed switching performance to create a new class of networking device known as an IP Switch. Depending on the flow characteristics of the traffic, an IP Switch can dynamically shift between forwarding packets via standard hop-by-hop connectionless IP routing and forwarding packets via the high-throughput switching hardware.

The building blocks of an IP Switch consist of a high-speed switching fabric and a so-called switch controller. The controller, which is based around the Intel microprocessor architecture (allowing the controller to take advantage of the aggressive Intel price/performance curve), performs all route calculation functions and controls the attached switching fabric. However, the controller only forwards a relatively small percentage of the packets itself, leaving the majority to be forwarded by the switch. In the IP Switching architecture, the underlying switch fabric can be ATM, frame relay, or LAN switching fabrics such as gigabit Ethernet. Initial implementations of IP Switching have focused on ATM because of the compelling price/performance, multicast, and quality of service (QOS) characteristics of ATM switches. As gigabit Ethernet switching fabrics become available, the IP Switching architecture will be extended to support these switch fabrics as well—the basic concepts of IP Switching remain the same and are independent of the underlying switch fabric.

11.7   Flow Classification

The fundamental idea of IP Switching is to leverage the performance of the Intel-based system by requiring that it forward only a small fraction of the traffic, offloading the majority of the packet forwarding work to the switch, which is more suited to this task. To achieve this result, the IP Switching architecture uses the concept of IP flows. Simply put, an IP flow consists of multiple unidirectional IP packets that share certain TCP/IP header information and receive the same routing treatment. Ipsilon's IP Switching software tries to recognize longer-lived flows (flows consisting of many packets rather than a very small number of packets) and map these flows to specific 32-bit flow labels. These 32-bit flow labels uniquely identify IP flows on a point-to-point link. In the case of ATM, the flow label corresponds to a specific VPI/VCI. In the case of gigabit Ethernet, the Ethernet MAC address can be utilized to carry the flow label.

When a given flow has been classified and given a particular flow label, the IP Switching software generates a message using a peer-to-peer protocol known as Ipsilon Flow Management Protocol (IFMP), which is explained in Request for Comment (RFC) 1953. IFMP is a protocol for instructing an adjacent node to attach a Layer 2 label to a specified IP flow. These so-called IFMP redirect messages are receiver-initiated and sent over point-to-point links "on top" of IP. The content of an IFMP redirect message is fairly simple: it specifies the particular 32-bit flow label to be used for the particular flow, which is uniquely identified by the IP header fields defining the flow. Recipients of IFMP redirect messages thus will use the specified flow label for future packets matching the flow identifier.

When a flow has been identified and redirected on both the upstream and downstream point-to-point links of an IP Switch, the IP Switching software is able to set up a unique connection across the switch to handle the forwarding of subsequent packets belonging to the specific flow. Figure 11.4 displays this process using the very simple example of a traffic flow from an upstream node to an IP Switch and on to a downstream node. The upstream and downstream nodes could be any of a number of IFMP-compliant devices including edge routers, access multiplexers, LAN switches, host/servers, or other IP Switches.

Through this flow classification process, the processing power of the controller can be significantly leveraged, because for each packet that is processed by the controller, a much larger number of additional packets is forwarded exclusively through the switch. The resulting throughput

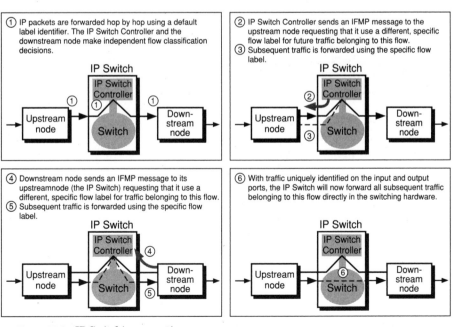

Figure 11.4   IP Switching operation.

of the IP Switch is therefore the product of the forwarding performance of the controller and a factor based on the percentage of the traffic that is classified as suitable for cut-through switching through the switch. Simulations by conducted by Ipsilon but based on actual traffic traces have indicated that approximately 80 to 85 percent of packets and 90 to 95 percent of bytes would be recognized as suitable for cut-through switching by the IP Switching flow classification software.

11.8   IP Switching Flow Types

The IP Switching specifications have defined several flow types that correspond to the specific TCP/IP header fields used to uniquely identify the flow. Flow type 1 specifies that two packets must have the same IP source and destination addresses, the same TCP/UDP source and destination port numbers, the same IP protocol type, and the same time to live (TTL) before they are considered by an IP Switch to belong to the same IP flow. This flow type provides a significant degree of granularity, because, for example, two IP hosts could have multiple flows between them representing different applications (the TCP/UDP port field gives an indication of the application being used). This granularity can be very useful in specifying QOS policies that apply to specific applications.

Flow type 2 is less granular, because it specifies that two packets must have the same IP source and destination addresses, the same IP protocol type, and the same time to live (TTL) before they are considered by an IP Switch to belong to the same IP flow. Continuing from the previous example, a type 2 flow would correspond to all traffic between two IP hosts, regardless of the TCP/UDP port (application) being used. This type of flow is useful as a catchall for general packet forwarding optimization in a network and does not take into account application-specific QOS policies. Note that the use of flow type 2 does not preclude the use of flow type 1 for a particular application. Note also that the trade-off between using flow type 1 or flow type 2 exclusively in a network is a trade-off between granularity and scalability. Flow type 1 provides application-specific granularity between hosts, but requires that more flow state be kept in the network (multiple flows between two IP hosts). Flow type 2 reduces the amount of state in the network (one flow between two IP hosts), but does not provide the application-specific granularity.

Together, flow types 1 and 2 provide enough granularity to allow even the largest corporations and many Internet service providers (ISPs) to develop certain QOS policies while still allowing for significant network scalability. However, the largest networks may require still more traffic aggregation. To support these customers, Ipsilon is defining additional flow types (flow types 3 and beyond) that provide for varying degrees of subnet aggregation and allow for a single flow to include all traffic between two networks (rather than between two hosts). Figure 11.5 provides an illustration of the different IP Switching flow types and the trade-off between increased traffic aggregation and increased flow granularity.

Importantly, by introducing the concept of IP flows and using IP flows in conjunction with IP routing, the IP Switching architecture accommodates both increased granularity and increased traffic aggregation depending on the unique needs of the particular network. Increased granularity allows application- or host-based quality of service policies to be implemented, and increased traffic aggregation allows enhanced network scalability. Note that the use of flows allows aggregation up to and even beyond a one-to-one correspondence between flows and routes. This architecture stands in contrast to exclusively route-based architectures, which are unable to offer application- or host-specific granularity or aggregation beyond the number of routes in the network. Figure 11.6 shows the increased granularity and aggregation options available by virtue of using flows as the trigger for soft-state caching decisions versus routes.

11.9 Switch Fabric Considerations

One of the core aspects of IP Switching is obviously the use of a switch fabric as a backplane, a decision that was made for the simple reason that a switch fabric offers greater aggregate throughput than a conventional bus backplane. However, rather than simply developing a

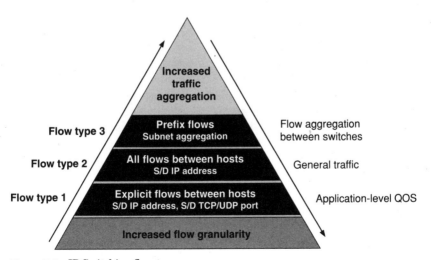

Figure 11.5 IP Switching flow types.

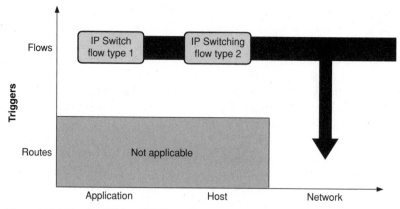

Figure 11.6 The flexibility of IP flows.

proprietary switch fabric to replace a proprietary bus as a backplane, Ipsilon designed its IP Switching architecture to accommodate different types of standard switches based on different technologies. In this manner, IP Switching can be deployed on ATM switches (as is currently the case), frame relay switches, and LAN switches (with gigabit Ethernet the obvious candidate). The only real difference in the IP Switching architecture from one type of switch to the next involves the mapping of 32-bit flow labels into the specific Layer 2 switching technology (for example, VPI/VCIs for ATM, DLCIs for frame relay, destination MAC address for Ethernet, etc.).

Within a given technology category (ATM, frame relay, or gigabit Ethernet), the IP Switching architecture has been standardized by developing protocol specifications, so that, for example, the specifications describing the IP Switching over ATM architecture can be utilized on different ATM switches from different manufacturers. This approach allows users to choose among different switches with different price and functionality characteristics while staying within a consistent routing architecture. For IP Switching vendors, the use of standard off-the-shelf switch fabrics can shorten the time between significant product enhancements by allowing the vendor to leverage industrywide efforts to develop these switch fabrics. While there is only one innovator working to enhance the architecture and performance of a proprietary conventional router, there is an entire industry seeking to develop different types of high-speed switching hardware.

Ipsilon developed three publicly available specifications to define the operation of the IP Switching architecture.* The first specification is

---

* <http://www.ipsilon.com/ipswitchtech/ips.html>

IFMP, which has already been discussed. This protocol is independent of the underlying Layer 2 technology and thus requires no change from one switch type to the next. The second specification covers the transmission of flow-labeled IP on ATM links, and is laid out in RFC 1954. It specifies how the 32-bit flow labels are transmitted over point-to-point ATM links using the VPI/VCI fields. The third specification, known as the General Switch Management Protocol (GSMP), enables the IP Switching software running on the controller to communicate with and control an ATM switch. It's spelled out in RFC 1987. Ipsilon also completed a draft specification that governs the operation of IP Switching over an Ethernet (or gigabit Ethernet) switch.

The ability for IP Switching as an architecture to be used not only on multiple switches within a technology category, but also across these technology categories, raises an interesting question, especially in the context of the emergence of gigabit Ethernet. What are the advantages and disadvantages of IP Switching on different switch fabrics? More specifically, what are the relative advantages of IP Switching over ATM versus IP Switching over gigabit Ethernet?

The differences between IP Switching over ATM and IP Switching over other switches basically boil down to the differences between the various types of switch fabrics. At the limit, there may be no differences whatsoever between these types of switches, other than that ATM switches always switch fixed-length cells while gigabit Ethernet switches typically switch Ethernet frames. However, even this distinction is artificial, because the actual switch fabric in some high-end switches is cell-based ATM even while the interfaces are Ethernet, Token Ring, FDDI, and so on. In this case the only distinction between an ATM switch and a mixed-media switch is that the former does not have a segmentation and reassembly (SAR) function because the ATM switch assumes it is receiving and transmitting 53-byte ATM cells.*

11.10   Quality of Service

Of course, as any switch consumer knows, there are significant differences between different switch fabrics. However, it is important to recognize that these differences stem primarily from a cost/functionality trade-off at the design phase of the switch rather than because of unique technology limitations. One clear example of the cost/functionality trade-off involves QOS capabilities. Because the very foundation of ATM dictated that ATM would be a multiservice technology sup-

---

* This latter point is notable because the SAR function can adversely impact performance for frame-to-cell-to-frame transmission, although this may improve with better SAR chips.

porting voice, video, and data,* ATM switches are typically designed to support different types of traffic through software constructs such as constant bit rate, variable bit rate, available bit rate, and unspecified bit rate. These software constructs are implemented via special queuing hardware, enabling the switch to support traffic shaping and other forms of QOS (priority queuing, per-VC queuing, etc.). There are, however, only a very small number of applications that are ATM aware, enabling them to actually specify a desired bit rate or QOS level. Even though the addition of QOS features to ATM switches no doubt adds to the cost of the switch, these QOS features are inherent, in one form or another, in virtually every ATM switch because they are viewed as fundamental to ATM technology.

Unlike ATM switches, Ethernet switches (specifically Ethernet and fast Ethernet switches) have generally been designed with cost as the overriding consideration. Although this is not exclusively the case, the price per port of Ethernet and fast Ethernet switches is one of the primary considerations for the customer. Because QOS is a concept orthogonal to Ethernet and Ethernet's carrier-sense multiple access with collision detection (CSMA/CD) technology, QOS features are not assumed to be an inherent feature of an Ethernet or fast Ethernet switch.[†] In practice, the addition of QOS features to these switches will only raise the cost of the switch—an often unacceptable alternative in a price-driven market.

Because IP Switching utilizes QOS features inherent in the underlying switching engine, the QOS features of IP Switching depend on the capabilities of the switch. IP Switching enables the user to take advantage of the QOS features of a switch by mapping specific IP flows (based on IP address, application, etc.) to specific QOS parameters. For example, a network manager could prioritize one application above all others using the priority queues inherent in the switching hardware. Or, an Internet service provider (ISP) could specify the exact bit rate a particular Web server will receive using the traffic-shaping features of the switching hardware. If gigabit Ethernet switches are designed according to the same assumptions under which fast Ethernet switches were designed (i.e., a focus on low cost), then the inherent QOS capabilities of gigabit Ethernet switches will be limited or nonexistent. In this case, IP Switching using an ATM switch would offer the customer significant

---

* Excerpt from *Introducing the ATM Forum:* "ATM is designed to carry multiple types of traffic simultaneously—voice, data and video. It addresses today's needs now, while enabling completely new applications in the future." <http://www.atmforum.com/atmforum/atm_introduction.html>

† In fact, the efficient support of traffic shaping would, in fact, be very difficult for such switches unless these switches utilized fixed-length cells.

QOS advantages over IP Switching using a gigabit Ethernet switch. However, it is important to note that several gigabit Ethernet switch vendors have recognized the potential value of some QOS capabilities, and these vendors will be adding some QOS functionality to their products.

11.11 Multicast Support

Another feature that may create differences between IP Switching over ATM and IP Switching over gigabit Ethernet involves multicast. One of the key advantages of the IP Switching architecture is the ability (depending on the capabilities of the switch) to support IP multicast with no degradation in forwarding performance. As with a conventional router supporting IP multicast, IP Switches utilize standard IP multicast routing protocols to determine the multicast tree. However, unlike traditional routers, IP Switches are able to replicate multicast packets in the switching hardware rather than consuming additional processor resources. The result is a more efficient implementation of IP multicast that scales much better than the processor-based solutions implemented in conventional routers. This result requires, however, that the underlying switch is inherently capable of point-to-multipoint switching.

Because one of the foundations of ATM was the support of real-time multimedia traffic, most, if not all, ATM switches are capable of supporting multicast functionality (through point-to-multipoint switching) with no degradation in forwarding performance. In contrast, Ethernet switches, which have been traditionally designed as inexpensive Layer 2 devices, have limited multicast capabilities. If gigabit Ethernet switches are designed as Layer 2 devices primarily with cost in mind, then they too may be less optimal than ATM switches when used as an IP Switching engine in an IP multicast-intensive environment.

11.12 Price and Cost Considerations

The statement that ATM switches may have certain feature advantages over most gigabit Ethernet switches assumes that most gigabit Ethernet vendors are making a trade-off of cost versus functionality. As a result, gigabit Ethernet switch fabrics may be significantly less expensive than those of comparable-bandwidth ATM switches. This cost difference may in fact be even more noticeable given the significant market penetration of Ethernet. Because of sheer volumes, Ethernet components tend to be very inexpensive, such that when a new Ethernet internetworking device becomes available, it normally follows a steep, downward sloping price curve. As shown in Figure 11.7, Ethernet switches (10 Mbit/s) in 1993 had an average price per port of ap-

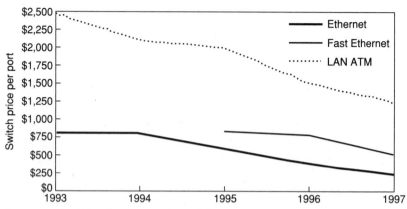

Figure 11.7   Switch prices per port by switch type.

proximately $800, and this price has since fallen below $300. Intro-duced a couple of years later, fast Ethernet switches are on a similar price curve.

However, because of the additional functionality required, and the relatively lower volumes (both of which result in higher component prices), ATM's price curve is both greater in an absolute sense, and less steep (on a percentage basis), than either of the Ethernet curves. LAN ATM per port prices were almost $2,500 in 1993, falling to approxi-mately $1,300 in 1997. Thus, a typical switched 155-Mbit/s ATM switch port is almost 2.4 times the price of a 100-Mbit/s fast Ethernet switch port, even though 155-Mbit/s ATM has only (theoretically) 1.5 times the bandwidth of 100-Mbit/s fast Ethernet. If this trend continues, gigabit Ethernet switch ports could actually be less expensive than 622-Mbit/s OC-12 switch ports, making IP Switching over gigabit Ethernet poten-tially less expensive than IP Switching over 622-Mbit/s ATM.

11.13   Conclusion

On the surface, gigabit Ethernet appears poised for tremendous growth:

- There is a very large installed base of Ethernet users.

- There is a proven track record for upgrading from Ethernet to fast Ethernet.

- There is no need to change existing IP routing protocols to accommo-date gigabit Ethernet.

However, this analysis misses a key point. Customers will want to use gigabit Ethernet as a backbone technology, yet there is no way conven-tional routers will be able to keep up with the speed of gigabit Ethernet links. Today's fastest routers are capable of less than 20 percent of the theoretical maximum packet forwarding rate for a single gigabit Ethernet interface. In order for gigabit Ethernet to become accepted and widely utilized, a new router architecture is required.

Pioneered by Ipsilon Networks, IP Switching defines a way to add complete IP routing functionality—not simply a limited subset of routing functionality—to high-speed switching hardware. Through an intelligent flow classification mechanism, IP Switching optimizes IP packet forwarding to create an order-of-magnitude increase in IP packet forwarding performance over conventional routers. One of the key aspects of the IP Switching architecture is the separation of the control plane from the data forwarding plane. This separation allows IP Switching to be implemented on different types of switching fabrics, from ATM to frame relay to gigabit Ethernet. In this manner, IP Switching can leverage industrywide advancements in switching technology and give users the flexibility to make cost/functionality trade-offs in choosing a switch fabric.

IP Switching initially has been implemented on ATM switches because of their price/performance characteristics, quality of service capabilities, and support for multicast. However, as gigabit Ethernet switch fabrics become commercially available, the IP Switching architecture can be extended to take advantage of the unique characteristics of these switches as well. At the limit, there may be very little difference between high-end ATM switches (i.e., >5 Gbit/s) and gigabit Ethernet switches, although typically Ethernet switches have been designed with cost as the primary consideration, while with ATM switches certain feature requirements, such as QOS and traffic shaping, sometimes take precedence over cost.

Rather than competing with gigabit Ethernet, IP Switching may be seen as the enabling technology for gigabit Ethernet, because the IP Switching technology provides a means of routing IP traffic at significantly faster link rates.

Cells in Frames and Gigabit Ethernet

Lawrence G. Roberts*

Chief technical officer of ATM Systems, an AMP company

12.1 Introduction

It has been suggested that providing lots of excess bandwidth will overcome all delay problems, making it unnecessary to add quality of service (QOS) and flow control in the protocol and the switch design. It has also been suggested that ATM is much too complex and expensive for the LAN.

Of all these claims, the only one with a grain of truth is that ATM has been more expensive than Ethernet. Complexity is clearly not a detriment in itself—only if it leads to expense or unreliability. Much of today's software is very complex but also cheap, functional, and reliable at the same time. The same is true of any of many high-volume electronic products.

As far as QOS goes, speed has never replaced and never will replace the basic requirement for multiple queues in switches in order to support quality voice and video. As for flow control, the first law of data networks is that flow control is required; otherwise the traffic will quickly overload the network, creating data loss and excessive delays. Of course, this appears to be the normal state of the Internet, since the flow control being used, TCP, is inadequate; but the goal should be to reduce the loss and delay, not allow it to get worse.

---

* Lawrence G. Roberts is currently chief technical officer of ATM Systems, an AMP company. Dr. Roberts, Ph.D. MIT, was responsible for the design and development of ARPANET, the world's first major packet network and the predecessor to the Internet. From 1973 to 1982, Dr. Roberts was CEO of Telenet (now Sprint). From 1983 to 1993, Dr. Roberts was CEO of NetExpress, Inc. Dr. Roberts' achievements include the AFIPS-Goode, IEEE-McDowell, and Ericsson awards.

In truth, there are serious problems with gigabit Ethernet as currently planned that could be eliminated through the use of ATM—or, if that is too expensive, with the ATM protocol used over Ethernet. Putting the ATM protocol over Ethernet takes advantage of the major advances in QOS and flow control that have been incorporated into ATM and eliminates the extra expense of cell segmentation and reassembly (SAR) and SONET line protocol. *Cells in Frames* (CIF), the protocol for putting ATM over Ethernet, has been specified by the CIF Alliance. CIF combines the cost and familiarity advantages of gigabit Ethernet with the advantages of ATM.

12.2 The Benefit of ATM (QOS and Flow Control)

ATM has been designed to provide the quality of service (QOS) and the flow control necessary for low-delay, interactive data transmission mixed with high-quality voice and video. ATM Forum members have put a large amount of work into specifying the ATM Forum 4.0 protocol suite, which superbly supports this capability. However, due to an artifact of the past, ATM still requires that data be cut up into little cells that are much too small to be processed efficiently by end stations and that provide a far less efficient transmission mechanism than larger packets. Thus, even though the ATM protocol would be far superior to TCP/IP, frame relay, or other current frame-based protocols, it has been all but rejected for desktop access due to its current high cost.

12.3 The Benefits of Ethernet

Because Ethernet is very simple and, in high-volume production, very low in cost, these days workstations often already have an Ethernet port. Thus, Ethernet to the workstation can be considerably less expensive than any new physical-level protocol. Simplicity is not a major issue with today's chip densities, but not having to buy and install a new NIC in each workstation is a major incentive to continue to use Ethernet interfaces. Thus, the question arises: why is a new physical interface required to support the ATM protocol at the desktop?

The reason a new line protocol is required is that ATM's physical layer subdivides data into small, fixed-length cells, which would require five times as many interrupts—totally overloading the typical operating system. There is no reason, however, to mandate that the ATM protocol must be used with small, fixed-length cells. Fixed-length cells were decided on in the 1980s when it was believed that high-speed switching would require fixed-length cells. In those days, the small size resulted from the need to minimize the delay for voice. Today, as a result of great progress in silicon, there is no need to have fixed-length cells in order to switch at the highest speeds. And, so long as the line speed is at least 10 Mbit/s, there is no serious delay impact from mixing short and long packets on the same line. Thus, it is time to reconsider the marriage of ATM to cells. We need to free ATM from the requirement that the line protocol must use the small cells that ruin line efficiency and require expensive segmentation and reassembly (SAR) hardware at every end point. When this is done, we can realize the true value of the ATM protocol suite without the high cost of new end-station hardware network interface cards (NICs) and the extra line overhead, dubbed the *cell tax.*

12.4   The Best of Both Worlds—ATM Protocol Using Frames (CIF)

Cells in Frames (CIF) uses the ATM protocol, retaining all its features and benefits, but sends the data in variable-size packets. The CIF Alliance (a group of about 30 interested organizations) has initially agreed on and published (http://cif.cornell.edu) the specifications for CIF over Ethernet and token ring facilities. This permits the ATM protocol to be made available to a workstation that already has an Ethernet or token ring NIC installed by simply downloading some software to process the protocol and provide support for the ATM QOS and flow control. One of the main reasons ATM has not been used for workstation access is the high cost of buying and installing an ATM NIC. There never was any good reason to require a new hardware NIC when protocol processing at 10 Mbit/s is such a small load on today's computers. The only reason ATM has required a new NIC is because the cells had to be segmented and reassembled. But at 100 Mbit/s and below, new, special hardware should not be necessary as long as the SAR function is not required at the workstation.

12.4.1   Workstation access

For workstation access, CIF uses packets constructed with an Ethernet or token ring packet wrapper filled with an ATM cell header and up to 31 cell bodies. On the network side, a CIF switch terminates the Ethernet or token ring line. If the trunk that the traffic is to take next is an ATM trunk, then the CIF switch does the segmentation and reassembly at the ATM trunk interface. In most external respects, the CIF switch might look like an Ethernet switch, but inside there is a critical difference; the CIF switch supports the ATM protocol, which means it supports QOS and explicit rate (ER) flow control. (See Fig. 12.1.)

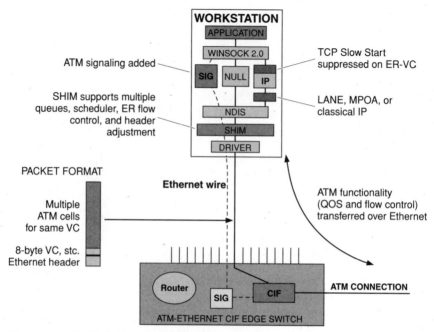

Figure 12.1   CIF Workstation and CIF switch, including protocol stacks.

12.5 CIF on Gigabit LAN Trunks

Mixing real-time traffic, like voice and video, with bursty data is not easy and requires many hardware and protocol features to achieve low delay variation for the real-time traffic, good separation, fairness, and good channel utilization. ATM has been planned to achieve this.

ATM has also been planned to support very high-speed channels by utilizing hardware packet switching. All previous packet switching, from the first Advanced Research Projects Agency Network (ARPANET) switch (Roberts 1970) to the routers operating the Internet today, used software packet switching. The ITU specified that all ATM packets would be fixed-length (53 bytes) in 1988. This was decided because, at that time, the use of fixed-length cells was thought to be necessary to achieve hardware switching at all.

Today, however, hardware switching is becoming easy for either fixed- or variable-length packets due to the introduction of large application-specific integrated circuits (ASICs). Ethernet switches have shown that switching variable-length packets is not only doable but quite inexpensive. Thus, ATM has no lock on hardware switching today. Soon, all routers and switches will use hardware switching.

The last eight years of hardware advances has taken away ATM's unique advantage for high speed through hardware switching. Now, hardware switching can and will be used for all switches, no matter what the protocol. Therefore, if the benefit of the extensive work done to perfect a protocol that would support mixed voice, data, and video is to be utilized, either the cost of ATM hardware must fall to that of other hardware or the ATM protocol must be used on the cheaper hardware.

Unfortunately, ATM hardware has so far proven to be considerably more expensive than Ethernet switch hardware. The best price today for a 25-Mbit/s ATM switch is roughly $300 per port, whereas 10-Mbit/s Ethernet switches are being sold for $150 per port or less. The speed difference may be part of the cost difference, but users do not currently believe they need more than 10 Mbit/s, and ATM has no standard speed that slow. Another difference is the complexity of the line interface; ATM has at least twice the complexity of the Ethernet MAC. The third clear cost difference is the volume of Ethernet versus ATM. Ethernet now has, and will continue to have, a major volume advantage. Thus, even for the switch port, ATM has a cost disadvantage today and will continue to have it for at least two years.

The most significant cost disadvantage for ATM is the workstation network interface card (NIC), which must be replaced for ATM but need not be replaced for older protocols like Ethernet. If the old NIC can be used, then ATM is at another disadvantage. Except for increasing the line speed, there is no basic reason why the old NIC cannot be used. And even when the user wishes to increase speed, the 100-Mbit/s Ethernet NIC is likely to stay priced far below any speed of ATM, based purely on volume, and will also be built into the motherboards of some PCs and workstations.

The ATM cell size was chosen so that at 64 Kbit/s a voice cell would not exceed 6 ms. In fact, the delay for voice is probably the most stringent delay requirement on a switch. In 1988, it was also important that no longer packets be used because the voice packet might have to wait for the longer packet to be transmitted. However, once voice has been packetized into 53-byte packets, there is no need to control the size of other packets to 53 bytes on fast channels like the 10-Mbit/s Ethernet line. Only the transmission time of the longest packet need be controlled to ensure that the voice is not excessively delayed. On a 10-Mbit/s Ethernet channel, the maximum packet takes 1.2 ms, which is not a problem for voice so long as only one or two of these long packets can get in front of the voice cell.

12.6   The Need for End-to-End Quality of Service

The ATM protocol supports many capabilities that are not available in current LANs and Ethernet switches. One of the most important additions is quality of service (QOS). QOS must be signaled with call setup and is almost impossible to achieve without specifying it for a virtual circuit—as is done in ATM. LANs today hide the virtual circuit at the TCP level and only send datagrams. QOS must be established on an end-to-end basis for the entire flow, and thus a serious change is required to today's LANs. For TCP/IP, this change is being planned with a protocol called Reservation Protocol (RSVP). RSVP will signal the QOS of a virtual circuit at the start of the call in much the same way as is done by the ATM Forum's Signaling 4.0 (SIG 4.0), the call-setup protocol. With either, it is possible to specify that a call needs a specific QOS to achieve its delay or bandwidth requirement. Voice and video are clear cases where QOS must be signaled if any reasonable quality is required. Less understood is the need for QOS on data calls to obtain a guaranteed response time. But in all these cases, the key issue is that QOS must be signaled completely from end to end. Also, all network elements—like routers, bridges, and switches—must implement multiple queues for packets with weighted fair queuing so that the packets that need to move on first can do so without waiting behind large collections of packets that do not need to move yet. To achieve QOS in a network, all network components must be upgraded or replaced to incorporate QOS signaling and weighted fair queuing with multiple queues. If one segment of the path is not upgraded, QOS can be totally lost.

12.7   The Need for End-to-End Explicit Rate Flow Control

The second major addition that is supported in ATM is explicit rate flow control. Flow control is used to control the traffic sources so that they do not send too much data into the network at any moment. If a trunk in the network is overloading, all the sources using that link must be told to slow down. This is absolutely necessary for a data network because the characteristics of data traffic are such that simply underloading the network will not work. The critical difference between flow control techniques is the time it takes to tell the source about the congestion and to get it under control. Today, TCP is typically used in LANs to control the data flow. TCP was created 15 years ago, when networks were much slower. It operates a binary rate flow-control algorithm between the two end stations, slowing down the data flow if the return path indicates that data was lost. If data is not being lost, TCP speeds up the flow. It always oscillates with a period of 1–2 s on the Internet, losing data each cycle and underutilizing the network on the other part of the cycle. Its characteristic time is about 1 s—the time it takes TCP to stop the data flow due to increased congestion. (See Fig. 12.2.)

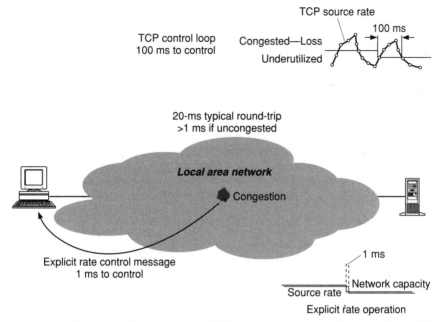

Figure 12.2   Typical oscillation cycle for TCP flow control versus explicit rate. ER is 100 times faster than TCP.

The ATM Forum has now completely specified a new generation of flow control—explicit rate flow control—which operates 100–1000 times faster than binary rate flow-control protocols like TCP. Explicit rate flow control operates with small control messages called resource management (RM) cells being sent around the network to convey the flow-control information back to the source. This way, a higher-priority path can be used for the RM cells so that they do not need to be delayed by the data. They typically consume 3 percent of the network capacity. When a switch sees an RM cell, it marks the cell with the highest data rate it can currently support for the associated source. This way, when congestion occurs, the switch can quickly mark RM cells with a new rate that will eliminate the congestion, and these cells can move at the speed of light back to the sources, telling them exactly how quickly to send data. The delay comparison for ER and TCP is:

Cause	Speedup for explicit rate over TCP
Priority for RM cells over data	Typically, in Internet today, a factor of 10
Explicit rate rather than oscillation	Typically, oscillation is another factor of 5
Switch to source distance versus round trip	Anywhere from a factor of 2 to 10

Thus, explicit rate can stop the source 100–500 times faster than TCP. This means that the buffer storage at the switch can be reduced by the same factor—100–500.

Even though there tends to be statistical smoothing of the data transients at a switch for both techniques, the reduction of delay is directly reflected in reduced buffer size. This is critical because high-speed hardware switches like ATM switches and 100-Mbit/s Ethernet switches cannot afford the same memory ratio that the older software switches and routers had. A 100-Mbit/s router often has 12 Mbytes of buffer storage or about 1 s of buffering at full input rate. A 10-Gbit/s ATM switch typically has about 20 ms of storage at full input rate and could not afford to increase this to 1 s since the cost would be impossibly high for the very fast synchronous random access memory (SRAM) needed for hardware switches. With 50 times less memory, a flow-control delay gain of at least 50 is needed to make high-speed hardware switches work. Explicit rate flow control has even more delay gain, which may help fix the current brownouts and overloads in the Internet if it can be incorporated.

As with QOS, explicit rate flow control is only useful if the RM cells flow end to end and all the congestion points arrange to mark the RM cells with the correct rate. If a current-day router or bridge is in the path, the RM cells will be delayed behind data and no marking will be done. This will not have a major effect if the device is always significantly underloaded, but in cases of overload TCP will need to be operated on top of the explicit rate flow control. Further, both end stations must participate in the explicit rate flow control in order for it to work. If one or both do not support explicit rate, TCP will be required, and all the benefit will be lost.

12.8 ATM Protocol—A Complete Protocol Suite

The ATM Forum finished a complete 4.0 protocol suite for ATM in April 1996. The previous 3.1 protocol did not have complete QOS signaling, had no flow control, and had no QOS-based routing. ATM switches without 4.0 have really not had any significant advantage over today's LANs except speed. However, ATM switches with the complete 4.0 protocol are able to support mixed voice, data, and video very effectively over both the LAN and the WAN. The elements of 4.0 are as follows:

- UNI signaling 4.0   Allows signaling of bandwidth and delay requirements for QOS
- TM 4.0   Specifies explicit rate flow control and QOS functions
- PNNI 1.0   Specifies QOS-based interswitch routing

No other protocol suite in history has existed that included signaled QOS, explicit rate flow control, or QOS routing. All are critical in operating a network with any quality of service other than one class. However, this protocol suite need not be locked to ATM switch hardware; it could be used on any switch platform and over any link protocol. All that is needed is to embed the ATM cells into whatever link protocol is used. This is what CIF is—the use of the full ATM protocol stack over legacy link protocols. If the use of these legacy protocols avoids the need to replace the end station NIC or reduces the switch port cost, or both (as seems be true for Ethernet and ATM), then it is extremely valuable to use the protocols but to embed the ATM protocol suite. Combined with an ATM backbone network, this achieves end-to-end QOS and explicit rate flow control, which is critical to enhancing the functionality available to the user. The capabilities that then become available at the desktop are:

- Toll-quality voice to the workstation over the Internet and the company's Intranet and from the PSTN

- Broadcast-quality audio to the workstation over the Internet

- Studio, broadcast, and conference-quality video to the workstation

- Guaranteed response time for Web access and server access; for example, 5 s per page

- No network collapse or brownout; no infinite delays due to network

- Fair service within price class; that is, the same data rate independent of network hop count

12.9   The IETF Alternatives: TCP/IP and RSVP

The Internet Engineering Task Force (IETF) supports the TCP/IP protocol and is currently drafting RSVP, a signaling protocol for QOS-based traffic. TCP/IP is used in the Internet and many LANs and consists of several parts. IP addressing is a good addressing plan and will continue to be used with or without ATM or CIF. IP addresses can easily be translated into ATM addresses through an Address Resolution Protocol (ARP) request that can be handled by a router incorporated in the CIF switch.

Today, TCP/IP just sends packets with IP addresses, and these are routed across the network with no knowledge of the QOS desired or the network path that might have the bandwidth or delay characteristics necessary to meet it. The IETF plan is to use a QOS reservation protocol called RSVP, which will make it possible to specify the same sorts of QOS requirements as the ATM Forum's UNI SIG 4.0. The plan is that

RSVP should be implemented by all workstations, routers, bridges, and switches within three to five years of its ratification.

For most equipment, like Ethernet switches, new hardware will be required to support QOS and RSVP. Today's switches and bridges must be retrofitted with multiple queues and weighted fair queuing to obtain QOS functionality. Some routers may be able to be upgraded with software, but performing weighted fair queuing would sorely tax their processing capability and degrade their throughput. Thus, most network equipment will have to be replaced to support QOS—no matter whether RSVP is used when it's ratified or ATM SIG 4.0 is used today.

The sad part is that the IETF and the ATM Forum/ITU are competing and not cooperating. Only one of the two protocols is required, and the user doesn't need the half-truths and rhetoric currently being pushed by companies that will profit from one outcome or the other. The truth is that either RSVP or SIG 4.0 could be used, and the network will be totally replaced in either case. RSVP could be used in routers and Ethernet switches or in ATM switching hardware, as is proposed by some for IP switching. ATM SIG 4.0 can be employed in ATM switches or in Ethernet or token ring switches through the use of CIF. As far as QOS is concerned, all four options are similar. The differences are in product cost and flow control. (See Table 12.1.)

12.9.1 QOS-based routing—avoiding the infinite delay syndrome

One major problem that plagues the Internet today is the infinite delay syndrome. This occurs when a trunk in the Internet is too small for the peak-hour traffic. So long as the trunk stays, and it is the shortest route, IP routing will keep sending all calls across it. As the load builds up, all packets are lost, all the sources back off and send more slowly, and the lost traffic is retransmitted. The retransmission increases the load further. The trunk loses more packets, and all the calls go more slowly yet. This continues until no one has any throughput and everyone is waiting forever for response (infinite delay). The trunk stays up and full. Users abort and try again, but the same route is assigned. This continues until the peak hour is over. The solution is to use QOS-

TABLE 12.1   Options for QOS Signaling

Switch type	Using RSVP	Using SIG 4.0
Current routers	Software upgrade	Software upgrade—CIF
Ethernet switches	New hardware—not available	New hardware and CIF
Token ring switches	New hardware—not available	New hardware and CIF
ATM switches	Software upgrade—IP Switching	Standard in 1998

based routing, such as ATM PNNI 1.0, which looks at the trunk usage before routing a call over it. The ATM protocol suite used by ATM and CIF has this capability. No similar capabilities have been defined or even started in the IETF, which means a pure TCP/IP solution will continue to have this problem for years. It was a very complex problem to define PNNI 1.0, requiring several years of work. Any approach undertaken by the IETF would most likely take just as much work and a similar time to complete. (See Fig. 12.3.)

12.9.2 TCP flow control

A second major problem in the Internet today is the slow speed of the TCP flow control mechanism. TCP ships about 1 s worth of data into the network before congestion can be stopped. This 1-s time is more or less fixed by the speed of light, the operating system delay, and the queuing delay multiple. So long as TCP operates only in the end stations, its operation cannot be substantially improved. When the router or switch memories in the data path all have sufficient memory, TCP works very well. However, the memory required is the delay-bandwidth product of TCP's cycle time and the full-input bandwidth of the router or switch. As the new hardware switching routers and switches are added to the network, their memory is not large enough for TCP's 1-s cycle. Most hardware switches in the 2- to 10-Gbit/s range—like Ethernet switches, IP switches, and ATM switches—have perhaps 1/50 of the memory required for TCP operation. Therefore, if TCP is not replaced, it will cause major overloads and outages on long-haul networks like the Internet. This is one of the unrecognized problems with IP switches, which use ATM switches and the IP protocol.

The IETF has not even considered revising TCP. In fact, no study has been done by the IETF on flow control because everyone seems to believe that what has worked in the past will continue to work. Nothing could be

Upper path unused since more hops

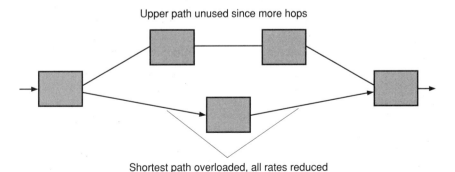

Shortest path overloaded, all rates reduced

Figure 12.3 IP routing today—shortest hop.

further from the truth in an environment like the Internet, where growth is over a factor of 2 per year. The Internet will collapse if TCP is not changed out for some flow control as good as explicit rate flow control.

12.9.3 IETF protocol conclusion

The IETF protocol stack is today grossly deficient for any QOS except bad. In fact, as the Internet grows, both TCP and the lack of QOS routing will cause even worse network congestion problems and collapses than are occurring today. The user community now wants QOS capabilities for voice and video, as well as guaranteed service. The IETF has responded by working on a draft of RSVP that is billed as fixing all the problems, but in fact will only address the signaling issue. Unless flow control and QOS-based routing are addressed, there will not only be no effective QOS, but network collapse also will be imminent. (See Table 12.2.)

There is no easy way for the IETF to come up with new, unique solutions to these protocol issues in less than four years. And by the time all the network equipment is replaced to incorporate the new protocols, eight years will have passed. By then the Internet will be in ashes. Thus, the logical solution is for the now-complete ATM protocol suite to be used, either on ATM switches or with CIF on the older legacy protocols. This can either be achieved by the IETF adopting the ATM protocols (unlikely) or by the users and Internet service providers (ISPs) making the decision to use CIF and ATM.

12.10 Cells in Frames Design Overview

The CIF specification (available at http://cif.cornell.edu) defines a CIF attachment device (CIF-AD), a switch that sits on the network side of a CIF connection. It defines a CIF end station (CIF-ES) as the device (typically a workstation) on the user side of the CIF connection. These definitions are sufficient for protocol definition but tend to simplify the picture too much if multiple stages of CIF switching are used. To identify the possible configurations more clearly, refer to Fig. 12.4.

TABLE 12.2   Support for Major Functional Requirements by Protocol

Functional requirement	ATM protocol	TCP/IP
QOS signaling (draft)	UNI SIG 4.0	RSVP
QOS routing	PNNI 1.0	None
Explicit rate flow control	TM 4.0	None

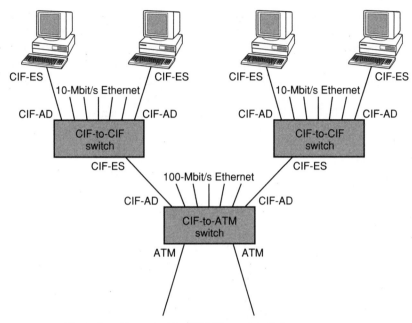

Figure 12.4   Typical configuration for CIF Ethernet switches.

In Fig. 12.4, the CIF-to-CIF switches must support both the CIF-ES and the CIF-AD protocol. The workstations support the CIF-ES protocol, and the CIF-to-ATM switch supports only the CIF-AD protocol. There does not need to be a CIF-to-ATM switch at all in a network so long as access to an ATM backbone is not required. Access to the Internet or other sites could be by CIF over Point-to-Point Protocol (PPP) if desired. Also, CIF can be used for token ring as well as Ethernet in the LAN.

12.11   The CIF-ES

The design of CIF is such that the end station need not have hardware additions but can be fully upgraded to ATM functionality with only software. The software required includes a shim to support the CIF interface, ATM signaling software, and slight modifications to the TCP module.

12.11.1   End station shim

The CIF end station has a software shim loaded between the NDIS and the driver. The shim adds the CIF header to packets before they are transmitted and removes the header when the packets are received. It organizes outgoing data into multiple queues for QOS management. For example, voice might go in one queue, video in another, and normal data into a third. Data with a guaranteed response might be in a fourth queue. Packets should be removed from the queues based on how urgently they need to be sent. A scheduler, much as is specified for NICs in the ATM Forum TM 4.0 document, would be quite effective for this task. To support CBR, VBR, and ABR traffic, the scheduler must be capable of sending the cells for each VC at any rate specified. The ABR rates are determined from the resource management cells being received, whereas the CBR and VBR rates are determined at setup. Figure 12.5 shows the protocol stacks in the CIF-ES in more detail.

12.11.2 ATM signaling in end stations

The end station must also have an ATM signaling package in order to set up calls. The signaling package is the same as would be required if an ATM NIC card were present. It is expected that Microsoft will include a signaling package in future releases of NT and Windows, but for now the package must be loaded just like the shim. If an application needed QOS, it would signal through Winsock 2 to the signaling package to specify the QOS desired. If old applications were run, they would not specify any QOS and would be sent as ABR traffic (with no minimum rate) or as UBR traffic.

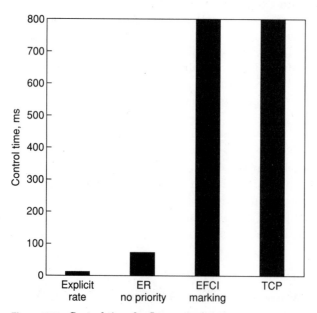

Figure 12.5   Control time for flow-control options.

When any new connection is to be established, the ATM signaling package should be called on to set up a VC, and the VC number should be given to the shim. Then, whenever data for that connection is passed to the shim, the shim inserts the CIF header in front of the data in the packet. This passes the VC number in the CIF header to the CIF-AD so that it may forward the packet without any necessity for further routing or bridging. If the original data is legacy data, where no ATM address was known, the shim would recognize the new address (IP, IPX, MAC, etc.) and, where necessary, the socket number, and use either classical IP or LAN emulation to establish a connection. From then on, the process is the same as it would be for an ATM-knowledgeable application.

12.11.3   TCP in the end station

When CIF software (or ATM software) is loaded into the end station, it may also be advantageous to modify TCP so that, when a fully flow-controlled connection is established from end to end, TCP will not try to guess at the sending rate and operate its slow start mechanism. Rather, TCP should have its parameters adjusted to send blocks of data to the shim only when the shim requests a new block. This will take maximum advantage of the ABR flow control and not force data loss to occur, as would normally happen with TCP and its slow-start technique.

12.11.4   QOS-based queuing

As data packets arrive from the CIF-ES, the CIF-AD must maintain multiple queues and provide a suitable QOS-sensitive mechanism for putting the data in the right queue and for selecting the most time-critical data to send on to the next node. The only known method for achieving delay and bandwidth guarantees for this process is to use weighted fair queuing, a technique where each VC is forwarded at or above the rate it was guaranteed. This can be done by sorting cells with time tags or by comparing countdown counters on each queue to determine the most time-critical data to send. FIFO queuing is sure to totally destroy any QOS that the end station might have maintained with its scheduler. Thus, a standard Ethernet switch (which uses FIFO queuing) will require major surgery to its hardware logic in order to support QOS. Even if the intent was to obtain QOS with RSVP (the draft IP QOS protocol), an Ethernet switch would require the same multiqueue logic upgrade to support QOS-based queuing. This change to the switching hardware is not expensive and thus should not add to the cost of the switch.

12.11.5   ABR flow-control marking

CIF attachment devices, which could possibly be a congestion point in the data path, must also differ from simple Ethernet switches in that they should mark the available bit rate (ABR) resource management (RM) cells when they are congested with either the rate that they could maximally support on the VC (explicit rate marking) or with their congestion state (CI marking). A third alternative, marking data cells or packets with the explicit flow-control indication (EFCI) state marking, is permissible for ATM switches but is no easier than CI marking and is about 10 times less functional. This technique is not recommended for new switches.

Table 12.3 outlines the time to control for the three different flow-control techniques (and TCP) for a connection with a 20-ms round-trip time, data queuing delays of 10 times the speed of light, and the RM cell priority and congestion point midway between the source and the destination.

Clearly, using EFCI or TCP provides dramatically worse results than explicit rate marking. The memory required for a switch is approximately the input bandwidth multiplied by the time to control (delay-bandwidth product). For TCP, it will be very expensive to incorporate sufficient memory into an Ethernet switch as one increases the bandwidth. However, using explicit rate flow control with RM cells and multiple queues to support RM cell priority, the buffer storage is about 100 times less, the packet loss rate is near zero, and the network utilization is considerably better.

The CIF end station may not be able to support the full rate of RM cells that would normally be sent by a typical ATM network—1 RM cell for every 31 data cells—because this would typically translate into one extra RM packet for every six packets (14 percent overhead). The CIF protocol permits the CIF-AD and the CIF-ES to negotiate a larger value of the number of cells per RM cell (Nrm) for the CIF network if they can both agree. The CIF-ES would normally request a value of Nrm = 128 in order to reduce the ratio to 1 RM packet every 24 packets—an overhead of 3.9 percent. If the CIF-AD can support a change of rate of RM cells, typically through the use of a virtual source/virtual desti-

TABLE 12.3   Time to Control for Various Flow-Control Techniques

Technique	Time to control, ms
Explicit rate marking	6
Explicit rate—no RM call priority	70
EFCI marking	800
TCP	800

nation (VS/VD), then it might accept the larger Nrm value. Otherwise it would be forced to reject the request because there is no assurance that any value but Nrm = 32 can be signaled on any ATM call. If it can act as a VS/VD, the result would be as shown in Fig. 12.6.

12.12   Delay Variance

The second area of performance difference could be in the delay variance incurred, particularly with half-duplex 10-Mbit/s Ethernet. When a voice packet arrives at the shim for transmission out to the network, the shim can schedule the voice packet right after the packet currently being transmitted if the voice has the highest time sensitivity. The maximum delay accrued by a single packet is 1500 bytes at 10 Mbit/s, or 1.2 ms. In addition, if the line is half-duplex, the other end may have a packet to send, and it could also take 1.2 ms. Thus, the maximum delay variance would be 2.4 ms for a high-priority packet. This has no significant impact on either voice or video. Voice can tolerate 24 ms of round-trip delay before echo cancelers are required and 100 ms one way for full-duplex voice before humans find it objectionable. Voice is the most time-critical application that is likely to be utilized on typical workstations, and toll-quality voice is quite feasible with CIF, even over standard Ethernet.

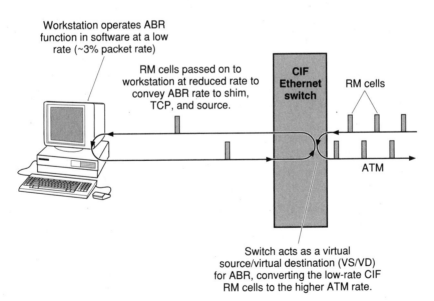

Figure 12.6   RM cell flow across the CIF link and the ATM link.

12.13   Workstation CPU Overhead

Since CIF requires that the flow-control function, CIF header processing, and multiqueue function all be in software, there will be some increase in the workstation CPU utilization. The most significant impact will be from the additional operating system interrupts due to flow-control packets being received. Most operating systems incur significant overhead processing an interrupt—far more than the extra processing to insert or remove the CIF headers would cause. However, if the CIF-AD can support the Nrm change to 128, then the rate at which RM packets would be received would be one for every 128 cells (6144 bytes) of data. If the average packet size being sent and received by the workstation is the same as the Internet average of 220 bytes, then the RM packets will cause a very minor 3.6 percent increase in the interrupt rate. If the CIF-AD cannot perform as a virtual source/virtual destination to reduce the RM cell rate, then with Nrm unchanged from the ATM standard setting of 32, the interrupt increase would be 14 percent a significant but not crippling increase.

The processing increase for supporting multiple queues is most likely necessary for any system where QOS is desired and is therefore not a disadvantage specific to CIF. The processing required to support adding and removing of CIF headers and to perform the flow control process for each packet is likely to be something less than 200 instructions, which at 100 MIPS for a 10-Mbit/s line with 220-byte packets would result in a 1 percent consumption of the CPU when the line was loaded. Thus, for standard Ethernet and token ring, the total CPU overhead impact of CIF is likely to be less than 5 percent. But, for 100-Mbit/s Ethernet, the impact might increase to 14 percent until faster CPUs are incorporated. Clearly, for gigabit Ethernet, the code must be optimized and the processor speed increased before a workstation could support even the interrupts at this line speed. CIF would be a minor increase over the interrupt overhead.

12.14   Conclusion

CIF provides a way to obtain the full functionality of ATM flow control and QOS over legacy frame-based media. The economic benefit of using CIF with current Ethernet or token ring installations is significant due to the fact that the NIC does not need to be replaced. This typically reduces the cost of upgrading to ATM functionality by at least a factor of 2. In addition, due to the higher volume of Ethernet components and the lower switch bandwidth required, a 10-Mbit/s Ethernet switch port will also be less expensive than a 25-Mbit/s ATM switch port. All of this means that full ATM functionality can be supported at nearly the same cost with CIF as with a non-QOS, no-flow-control Ethernet switch. (See Table 12.4).

TABLE 12.4 **Pros and Cons of Various Switches in the LAN**

Capability	IP routing	IP switching	ATM switching	CIF
QOS support	With RSVP	With RSVP	Yes, with SIG 4.0	Yes
QOS routing	No	No	Yes, with PNNI 1.0	Yes
ER flow control	No	No	Yes, with TM 4.0	Yes
Use existing NIC	Yes	Yes	No	Yes
Cost to desktop	Low	Low	High	Low

The importance of supporting ATM protocol across the final workstation link is that QOS and explicit rate flow control must be implemented end to end in order to be effective. If the final link is too expensive and users opt to use simple Ethernet switching with no QOS or flow control, then the benefit of ATM in the backbone is lost, and ATM could wind up relegated to the WAN. However, with CIF, the added cost over Ethernet switching is likely to be so small that the benefits of QOS and flow control far outweigh any cost difference. Since the performance of the typical workstation with standard applications running CIF over 10-Mbit/s Ethernet or 16-Mbit/s token ring is almost identical to that of 25-Mbit/s ATM, then CIF provides a very attractive path to supporting full ATM QOS and explicit rate flow control on an end-to-end basis.

References

Roberts, Lawrence G.: "Request for Coordination of Cells In Frames Specification," *ATM Forum 96-1104,* August 19, 1996. See http://www.ziplink.net/~lroberts.
———: "Communications and System Architecture," *Proceedings of the 1970 IEEE International Computer Group Conference,* June 16, 1970, Washington, DC.

NT-Switches: Layer 4 Gigabit Ethernet and Windows NT Services

Donal Byrne*, Cuneyt Ozveren†, Ravi Sethi,‡ and Robert Thomas§

Berkeley Networks

13.1 Introduction

The corporate computing landscape is undergoing dramatic change, which has far-reaching implications for the supporting network infrastructure. Today's enterprise environments are witnessing the emergence of a new diversity of applications based on IP networking protocols that depend on the network like never before. While network administrators continue to perform heroic acts in keeping enterprise networks stable and operational in the face of relentless change, this new wave of diversified, network-focused applications presents a new set of challenges: how to support application service needs in a controlled and secure fashion without adding complexity or compromising network availability.

---

* Donal Byrne is vice president of marketing and product management at Berkeley Networks. Previously, Mr. Byrne served as director of network architecture at Bay Networks, where he was responsible for defining and communicating Bay's Enterprise Network strategy. He also served as product line manager in charge of Bay's Layer 3 switching products. Mr. Byrne joined Bay Networks in 1994 from Digital Equipment Corporation, where he led the development of a number of switching and routing products, including a personal Ethernet routing switch for the DEChub 900 and the LAN-to-ATM line card for DEC's GigaSwitch platform. In 1987, Mr Byrne worked in the CNET research laboratories in France, where he focused on high-speed optical transmission technologies. He holds an M.Eng.Sc from University College, Dublin, Ireland.

† Ravi Sethi is the chairman, CEO, and president of Berkeley Networks. Mr. Sethi has more than 12 years of marketing and general management experience in networking, MPEG, CPU, microcontrollers, and memory ICs. Prior to joining Berkeley Networks, Mr. Sethi was the director of the networking business unit for Toshiba. He also served as product marketing manager at Fujitsu. Mr. Sethi started his career in the high-tech industry in

An NT-Switch is a new class of switched internetworking platform that provides the foundation for building high-speed, intelligent networks in response to these trends. (*NT-Switch* is a commonly used industry term referring to network switches that use the Windows NT® operating system.) An NT-Switch combines Layer 4 gigabit Ethernet switching hardware with a rich set of integrated networking services made available through the industry's fastest-growing extensible network OS, Windows NT. An NT-Switch delivers gigabit Ethernet wire-speed performance up to Layer 4 of the OSI stack, allowing it to take full advantage of NT's routing, management, directory, and security services while providing high-performance, scalable network bandwidth with full application awareness.

NT-Switches enable direct interaction between the computing and networking environments, resulting in the following tangible benefits for today's enterprises:

- Total control over networked application performance and responsiveness in the face of unpredictable traffic loads and patterns

- Support of delay-sensitive collaborative applications and services that improve workforce productivity without added networking complexity

- Reduced cost of operations through directory-service-based policy management, autoconfiguration, and centralized administration

This chapter describes the architecture, deployment, benefits, and future impact of the NT-Switch. It begins with a discussion of the motivating factors that prompted the evolution to NT-Switches.

1982 at Intel, where he worked in finance for two years and for a short time in marketing. Mr. Sethi holds both an MBA and a Ph.D. from University of California at Berkeley.

‡ Cuneyt Ozveren is the vice president of engineering for Berkeley Networks. Before joining Berkeley Networks, Mr. Ozveren worked for Fore Systems as an architect and project leader for their 80-Gbit/s switch development. Prior to his work at Fore Systems, Mr. Ozveren served at DEC, assuming various leadership roles for DEC's Gigaswitch and ATM switch development. Also, Mr. Ozveren has extensive experience in system-level design and architecture, including advanced optical switch development. Mr. Ozveren holds a Ph.D. and an MBA from MIT.

§ Bob Thomas is the CTO for Berkeley Networks. Before joining Berkeley Networks, Mr. Thomas was a consulting engineer with DEC, where he attained 13 years of networking design and architecture experience. At DEC, he worked on high-speed switch, adapter, and ASIC products. He was the system architect for the 6.25M pkts/s GIGAswitch/FDDI, which won the 1994 R&D 100 award and Data Communications' best product of the year. Prior to his successes at DEC, Mr. Thomas did three years of postdoctoral research at MIT. Mr. Thomas holds 8 patents, 12 patents in progress, and a Ph.D. in Computer Science from the University of California at Irvine.

13.2   The Need for Application Awareness

Today's enterprise LANs are undergoing tremendous change. Collaborative and distributed applications based on IP networking protocols are proliferating across the network. Centralized server farms consolidate network traffic at the backbone, creating critical network convergence zones. Fast Ethernet is migrating out to the desktop, while gigabit Ethernet is adding more horsepower to the enterprise backbone.

At the same time administrators are coping with these changes, they are struggling to keep pace with the demands of users, applications, and networking systems that already exist on their networks. As a result, today's networks are fast becoming a mélange of different and often overlapping networking models, creating new challenges for network administrators to reduce operational costs, predict and optimize traffic flow, and extend their control of the network without increasing administrative overhead and complexity (see Fig. 13.1). Simply adding more raw bandwidth fails to address fully the problems facing today's network administrators. What is needed is a straightforward and flexible way to scale the bandwidth of the network infrastructure to handle future needs while implementing and enforcing enterprise-wide policies and controls cost effectively.

One novel approach to achieving these goals blurs the boundaries between networking and computing in such a way that networks gain full knowledge of applications. Implementing such an infrastructure requires a solution that looks beyond Layer 2 and Layer 3. Application awareness requires support at Layer 4, as described in the following sections. In this context, the term *Layer* refers to the well-known stratifications of the open systems interconnection (OSI) model: Layer 1 is the physical layer; Layer 2 is the data link layer; Layer 3 is the network layer; and Layer 4 is the transport layer.

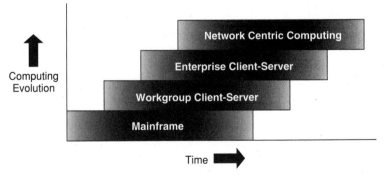

Figure 13.1 Evolution of the computing environment indicating constant change through the addition of new computing models while maintaining the older models.

13.2.1   Layer 1 requirement:
fast Ethernet to the desktop

Given widespread vendor adoption and highly attractive port prices, there seems little doubt that 10/100-Mbit/s Ethernet will be the dominant network connection for desktops. The price and performance of Ethernet, and in particular fast Ethernet, as a desktop connection is difficult to beat. And now that PC motherboards ship standard with a 10/100-Mbit/s Ethernet connection, any uncertainty about the desktop connection of choice has been removed.

Of late, there has been much discussion and energy devoted to thin clients, also known as NetPCs and network computers, motivated by the desire to reduce the total cost of ownership of traditional thick clients—PCs and workstations. Because thin clients store a lot less data locally, the success of the thin-client model depends on a network infrastructure capable of providing fast, responsive, and reliable connectivity between these fast Ethernet desktops and centrally located servers. Thick clients running Web-based applications also rely on network performance and availability for successful operation. Thus, either way, desktop computers are pumping more bits onto the network. The next-generation infrastructure must provide the raw bandwidth to accommodate this increase in client-server data traffic.

13.2.2   Layer 2/Layer 3 requirement:
server centralization without bottlenecks

The high cost of independently managing distributed workgroup servers has prompted an industry-wide move toward the consolidation of general-purpose servers into centrally located server farms. Further centralization of the IT infrastructure is promoting the use of large enterprise-class servers in place of multiple small servers residing on the backbone.

Traditional workgroup client-server computing created many localized pockets of heavy traffic in the network, the origin of the so-called 80:20 rule for LAN traffic. Fast, inexpensive Layer 2 Ethernet switches were a perfect match for this situation, because they increased the bandwidth significantly between clients and servers within the same subnet. Traffic that crossed subnet boundaries, such as Internet, WAN, and mainframe traffic, was easily accommodated by routers and shared FDDI backbones.

However, the move to centralized server farms, together with the wide adoption of Web-based collaborative applications, reverses the network traffic pattern of the workgroup client-server computing model, resulting in any-to-any connectivity. This sudden demand for increased bandwidth across the network has given rise to many elaborate and proprietary schemes for accelerated routing in the network. All of these schemes employ new protocols to acquire routing information, which then allows them to forward packets across subnet boundaries while leveraging the speed of a Layer 2 switching infrastructure. Administrators' concerns about adding new protocols to their networks, and the slow pace of some vendors in actually delivering this protocol support across their installed base, have impeded the proliferation of these new types of switches.

Gigabit Ethernet routing goes a long way to address the bandwidth scalability issues of today's internetworks, particularly with respect to the any-to-any traffic patterns. Wire-speed gigabit Ethernet routing can provide the bandwidth needed to solve many of today's networking bottlenecks. And because its basis is in Ethernet, the migration issues are minimal, compared with many of the proposed alternative schemes for high-speed routing.

Where gigabit Ethernet routing breaks down is in its ability to support a diversity of networked applications, simplify network administration, and provide control over network services. Solving these problems requires more than high-speed hardware.

13.2.3 Layer 4 requirement: support for new classes of applications

While networks today have no knowledge of the set of applications flowing through them, the most important and popular new enterprise LAN applications are slaved to network performance, responsiveness, and availability.

- Collaborative applications: audio and video conferencing, shared whiteboards, interactive document editing, distance learning (e.g., Microsoft NetMeeting, Vocaltec IP Phone, Intel Proshare)

- Broadcast services: real-time broadcast of financial data, noninteractive audio/video broadcast (e.g., IP TV/radio, Slingshot)

- World Wide Web publishing applications

- Distributed applications: Microsoft BackOffice suite, Oracle databases, SAP information management applications

For optimized performance, delay-sensitive and mission-critical applications require special treatment by the network. To deliver this treatment, the network must be able to identify the application, retrieve the appropriate policies, and then enforce these policies in real time.

To understand what applications are flowing through the network, switches must operate at Layer 4. By examining the TCP and UDP ports of IP packets, the switch can determine what applications are in use and enforce class-of-service policies as specified by the network administrator.

Achieving this level of application awareness should not introduce new performance penalties into the network. Thus, the natural integration point for Layer 4 switching is within the gigabit Ethernet backbone itself.

Table 13.1 summarizes the bandwidth requirements for application-aware networking.

13.2.4   Beyond L2/L3/L4 bandwidth

Today's networks are controlled and managed by software systems that are divorced from the operating systems controlling the desktops and servers of the computing infrastructure. But to fulfill the requirements of application-aware networking requires not only scalable bandwidth at Layers 1, 2, 3, and 4 but a much closer interaction among the desktop, the server, and the network. To realize fully the promise of application-aware networking requires a policy infrastructure for administering, controlling, and operating network resources for both applications and users. With such an infrastructure, the network administrator can supply application intelligence to the network, thereby allowing the network to participate actively in and influence the following tasks:

- Global policy enforcement for networked applications, providing mission-critical applications such as SAP R/3 priority over Web traffic

- User policy enforcement, allowing authorized users to reserve resources on the network using protocols like RSVP, further qualified with time-of-day and location-dependent policies

- Policy authentication, ensuring that users requesting to reserve network resources are authenticated before the requested policy is en-

TABLE 13.1   Bandwidth Requirements for Application-Aware Networking

OSI layer	Purpose	Requirement
Layer 1/Layer 2	Add raw bandwidth	• High-density 10/100 Ethernet to the desktop • High-density gigabit Ethernet for collapsed backbone
Layer 3	Scale the bandwidth	• Very high performance IP and IPX routing in collapsed backbone • Bridge remaining protocols in campus LAN
Layer 4	Control the bandwidth	• Wire-speed Layer 4 switching for TCP and UDP protocols over a gigabit Ethernet backbone

forced or that anyone specifying network policy has the proper privileges

Directory services are the logical source for such an intelligent infrastructure. Directory services provide an efficient mechanism to enable network administrators to manage relationships among devices, addresses, users, network services, and applications from a central location. Vendors such as Microsoft Corp. (Redmond, Washington), Novell Inc. (Orem, Utah), and Banyan Systems Inc. (Westborough, Massachusetts) have embraced the Lightweight Directory Access Protocol (LDAP) and offer directory services that comply with this standard interface. Berkeley Networks, Cisco Systems Inc. (San Jose, California), and Microsoft are currently defining schemas for controlling policy through directory services supported by Windows NT, the de facto standard operating system (OS) in today's computing environments.

Support for an industry de facto software standard such as Windows NT adds another benefit to application-aware networking: widely available, standardized tools and management interfaces for lower-cost operations. Most network administrators have long since learned that the cost of operations far outweighs the cost of equipment. By implementing a common OS across the desktop, server, and network, these administrators can increase the intelligence of the enterprise while simplifying their management burden through a common, easy-to-use toolset.

13.3   NT-Switches

NT-Switches are a new class of switched internetworking platform that combines the Windows NT operating system with intelligent, state-of-the-art switching hardware to transform today's networks from passive data conduits to application-aware communication and control infrastructures. NT-Switches are the first fully integrated high-performance switching platforms that support the full complement of networking services available via Windows NT while delivering wire-speed transmission at OSI Layers 1, 2, 3, and 4.

NT-Switches are unique in their ability to provide wire-speed switching up to Layer 4, even at gigabit Ethernet speeds. In this way, NT-Switches provide network administrators with complete control of network resources without added complexity or performance degradation.

The packet latency for a Layer 4 NT-Switch is the same as that commonly expected in a Layer 2 switch. NT-Switches employ dedicated hardware for packet switching, enabling extremely high aggregate performance. In contrast, traditional routers incur a significant performance and latency penalty for Layer 4 lookups. Layer 4 switching does not require any new protocols in the network and is fully compatible with existing bridging and routing techniques. In fact, NT-Switches can bridge and/or route packets while simultaneously performing Layer 4 switching operations. The bridging and routing functions determine the outbound port for the packet being processed, while Layer 4 gigabit Ethernet switching hardware determines the service class and its associated queue on the outbound port.

Finally, NT-Switches support the rich set of fully integrated and extensible routing, management, security, and directory services available via Windows NT.

13.3.1   Deploying NT-Switches

NT-Switches can easily and cost effectively eliminate bandwidth bottlenecks in enterprise campus LAN backbones while providing service-class support for mission-critical and/or delay-sensitive applications associated with today's networking environments. In addition, Layer 2/4 functionality in NT-Switches can be used in wiring closets to simplify management, increase bandwidth, and provide guaranteed end-to-end network performance for mission-critical applications while efficiently carrying best-effort traffic.

Figure 13.2 shows the topology for an enterprise campus network that has migrated from a workgroup client-server model to a centralized enterprise client-server model. The network has been upgraded to a collapsed gigabit Ethernet backbone with switched 10/100 Mbit/s to power user desktops. The NT-Switch, powered by gigabit Ethernet hardware, is the main point of aggregation for server, wiring closet, backbone, and WAN router connections. Existing hubs and Layer 2 switches in the wiring closets connect directly to the NT-Switch, while the legacy router connects WAN links to the NT-Switch. Since an NT-Switch directly supports multiple subnets per link (i.e., multinetting or secondary addressing) and multiple links per subnet (i.e., bridge port groups), there is no need to change IP addresses in the network except to give the existing central router a new address to connect to the NT-Switch.

Subnet 2 in the diagram uses a stackable NT-Switch to handle the bandwidth-intensive computing needs of power users. The NT-Switch operates at Layer 2 to minimize the latency of the switched 10/100-Mbit/s Ethernet desktop links while providing application awareness by processing Layer 4 information at gigabit Ethernet wire speeds. A gigabit Ethernet downlink connects to the NT-Switch backbone. Because both the stackable and the backbone NT-Switch platforms use gigabit Ethernet hardware and introduce no new protocols into the network, the migration issues are minimal.

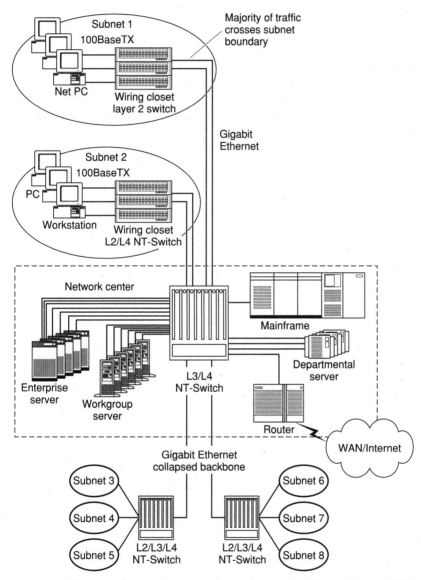

Figure 13.2   Diagram of enterprise computing environment with a mixture of thick and thin clients, centralized server farms, and a collapsed gigabit Ethernet backbone.

NT-Switches simplify network performance analysis because the throughput and latency is identical at Layers 1, 2, 3, and 4 of the OSI stack. Table 13.2 shows the calculated number of NT-Switches and the number of ports per switch that are needed for four different network designs. It assumes the packet-processing performance of the enterprise-class NT-Switches on the order of 25–50 million pps, a conservative estimate.

These examples assume in the short term that fast Ethernet will be used for server links. (It makes sense to migrate servers to gigabit Ethernet links as the servers become more powerful and network adapters become more efficient.) The bandwidths shown reflect data transmitted from servers to users, since this is the most likely case. However, the same amount of bandwidth can flow from users to servers at the same time, since all links are full duplex.

13.4   NT-Switch Architecture

The NT-Switch contains three major architectural components, as shown in Fig. 13.3:

1. Windows NT
2. Dynamic switch compiler (DSC)
3. Layer 4 gigabit Ethernet switching engine

Windows NT is used exclusively for management and control functions. From a management perspective, Windows NT provides a flexible configuration toolset with comprehensive statistics and monitoring

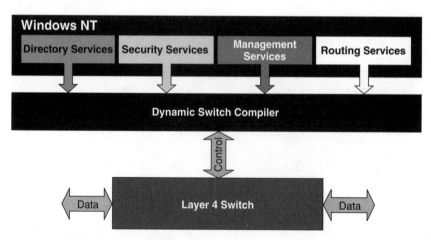

Figure 13.3 Block diagram of NT-Switch architecture.

TABLE 13.2 Example Bandwidth Design for 10K or 20K Users, with Either 1- or 10-Mbit/s Required Mean Concurrent Bandwidth per User, and 150 Users per Wiring Closet

Example number of users	Example mean busy hour Mbit/s to each use	Number of closets @ 150 users per closet	Total concurrent Gbit/s from servers	Total server farm size, four processors per server with four FE links each	Suggested number of enterprise NT-Switches	Suggested number of gigabit interswitch ports per switch	Suggested FE links to each closet	Suggested gigabit links to each closet	Required FE server ports per switch	Total required ports per switch Gbit	FE
10,000	1	67	10	25	1	0	3	0	100	0	300
20,000	1	133	20	50	2	2	3	0	100	2	300
10,000	10	67	100	250	10	4	0	3	100	24	100
20,000	10	133	200	500	20	6	0	3	100	26	100

capabilities. The simple and intuitive management interface has received high marks for ease of use and supports command-line scripting to automate repetitive tasks. Network administrators can obtain a systemwide view of the network from any networked location using either SNMP, HTTP, or traditional Telnet access. Support for remote procedure calls (RPCs) enables a fully secure, distributed implementation of the management functionality and allows the NT-Switch to take advantage of value-added management applications created by third-party developers. Windows NT provides additional tools that reduce administration costs, such as DHCP and dynamic DNS services.

The upcoming Windows NT 5.0 release will unleash the full power of the NT-Switch by allowing it to tie directly into the nerve center of the combined networking and computing environment, the directory service. The directory is the repository for all desktop, server, and network service and security policies. With direct access to the directory database, an NT-Switch can execute policies in concert with the desktop and server elements to set up secured, high-performance network zones.

Closely coupled to directory services are Windows NT's distributed security services, which hold the promise of allowing users, regardless of their network location, to be authenticated through a single network logon for appropriate and consistent access to system resources. The NT-Switch can enforce these policies networkwide at the highest level of performance.

The dynamic switch compiler is responsible for message-passing between the control software and switching hardware. Because Windows NT is not a real-time operating system (OS), the DSC adds a second level of processing to offload the OS and minimize latency.

The Layer 4 gigabit Ethernet switching engine handles wire-speed packet processing and forwarding, controlling applications as they flow through the switch. The switching engine applies policy filters downloaded from Windows NT, sets the policy, and assigns packet flows to different class-of-service (COS) queues dynamically. The switching hardware determines not only the output port to use for a given application flow, but also queue priority, based on the associated policy.

13.5   Windows NT and NT-Switches

An NT-Switch separates non-real-time control and management functions from packet forwarding and other data operations that must occur in real time. Windows NT is used exclusively for non-real-time functions, while dedicated switching hardware and low-level software handle all real-time control and management functions and packet forwarding.

The true value of Windows NT in the NT-Switch is its support of integrated routing, management, security, and directory services, as described in the following sections.

13.5.1 Routing services—Steelhead

Steelhead routing and remote access services, released in June 1997 as part of Windows NT Server 4.0, makes Windows NT a powerful and extensible internetworking platform. Steelhead supports various routing protocols, such as the Open Shortest Path First (OSPF) protocol by Bay Networks Inc. (Santa Clara, California) and the Routing Information Protocol (RIP1, RIP2 for IP, RIP for IPX). In an NT-Switch, Windows NT performs network topology acquisition using protocols like OSPF and bridge spanning tree. Figure 13.4 illustrates the elements of Steelhead's architecture, key members of which are described in the following list:

- Steelhead's Routing Table Manager (RTM) is the central repository of routing information for all protocols that operate with Steelhead. The RTM keeps independent route tables for each network layer protocol, provides routing information to all the interested components (such as management and monitoring agents), and determines the best route to each network destination.

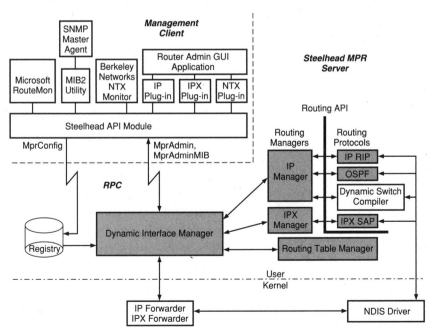

Figure 13.4 Block diagram of Windows NT Steelhead and integration with the dynamic switch compiler (DSC) in the NT-Switch.

- The Steelhead management APIs enable NT-Switches to be managed and configured like other legacy routers in a network. They also provide a standard interface for third parties to extend Steelhead's default administrative/management functions.

- Certain management and control packets in an NT-Switch are forwarded to the Windows NT host for processing. The IP and IPX forwarders, which in an NT-Switch contain a subset of the RTM database, are responsible for taking these packets from the NDIS driver and routing them to the host.

- The dynamic interface manager controls and manages the routing protocol components. It also implements an RPC server that provides a communication channel between the management client and the Steelhead routing protocols for remote management capabilities.

13.5.2 Management services

The NT-Switch leverages the management infrastructure provided by Windows NT Server 4.0 and provides further management support through custom NT-Switch plug-ins. Common management functions include support of standard SNMP MIBs, a command-line console, and a Web interface.

In addition, the NT-Switch management architecture supports a number of unique features, including status monitoring, remote management, and configuration wizards:

- NT-Switches can use Windows NT's built-in performance monitoring capability to set thresholds against any NT-Switch metric and to execute a program when the threshold is exceeded. Through the Windows NT performance monitor, network administrators can generate ad hoc reports for performance or fault summaries and export the results to Microsoft Excel if necessary. The Windows NT performance monitor is a distributed RPC-based application and can be run securely from any remote machine.

- The NT-Switch is unique in its support for secure remote management through OSF/DCE-compliant RPC-based APIs. A network administrator can operate the full suite of GUI-based configuration and monitoring tools from a remote location by simply running the management client on the remote machine. Traditional remote management schemes such as Telnet and SNMP are also supported.

- Finally, like most Windows applications that support installation and configuration wizards, the NT-Switch leverages the same set of libraries to develop switch-specific configuration wizards. Compared with conventional networking equipment, these wizards dramatically reduce the number of steps involved in configuring an NT-Switch.

13.5.3   Security services

The NT-Switch supports a number of important security features for safeguarded enterprise-wide administration, user management, and network authentication. Because NT-Switches process packets at Layers 1 through 4 at gigabit Ethernet wire speeds, implementing security policy does not impede the performance of the network in any way.

The Windows NT User Administration application handles security administration in an NT-Switch by using a distinct user group called *network administrators*. A network administrator assigned to this special group is authenticated through a Windows NT logon procedure before he or she can have access to the NT-Switch administrative applications. The NT-Switch supports full remote logon capabilities in addition to local logon.

Windows NT Server 5.0, available in 1998, promises to deliver a new distributed security architecture, which includes the following security enhancements:

- Kerberos V5 authentication protocol

- Public-key certificates for strong authentication

- Secure Sockets Layer (SSL) 3.0 for secure communications channels

- CryptoAPI V2.0 for encrypted communications channels

- Full integration with Microsoft's own directory services application, known as Active Directory

NT-Switches make full use of these protocols. Users requesting quality of service or service classes from the network are authenticated through Kerberos and public-key certificates. SSL and CryptoAPI secure communication between a remote administrator and the NT-Switch. Full integration of the distributed security services with Microsoft's Active Directory provides scalable, flexible, fine-grained management of network security policies for individual users.

13.5.4   Directory services

Vendors such as Novell, Netscape Communications Corp. (Mountain View, California), and Banyan offer directory service products that run on Windows NT. The NT-Switch is directory enabled through support of LDAP. When a directory contains an NT-Switch schema, the NT-Switch can use LDAP to access information such as default configuration profiles, network policies, and user bandwidth policies from the directory. The directory schema is typically vendor- and product-specific, but work is under way in standard bodies like the IETF to consider standard schemas for routers and switches.

In addition, the NT-Switch is unique in its ability to host a directory service natively. This provides reduced latency for directory queries and improves scalability.

13.5.5 NT-Switch extensibility

One of the most powerful aspects of using Windows NT in the NT-Switch is the ability for switch vendors or third-party ISVs to add value through service extensions and/or switch-specific applications. Examples of such extensions include the following:

- *Routing services.* Additional routing protocols, filtering for both bridging and routing, and open APIs for specifying filtering and policy tables.

- *Management services.* Product-specific plug-ins for RouterAdmin, installation and configuration wizards, additional SNMP MIBs, and Web-based GUI interfaces.

- *Security services.* Extensive filtering for policy enforcement and open APIs for integration of third-party security applications such as Internet firewalls and proxy servers.

- *Directory services.* Product-specific schemas and support for multiple directory service products.

13.6   The Dynamic Switch Compiler
of NT-Switches

The dynamic switch compiler is responsible for passing messages back and forth between Windows NT and the Layer 4 gigabit Ethernet switching engine of the NT-Switch. The DSC provides translation from high-level services in Windows NT to low-level hardware instructions in the switching engine. It also exposes intelligent hardware functionality in the switching engine to the integrated networking services in Windows NT. More specifically, the DSC performs the following functions:

- Feeds Windows NT with routing, bridging, policy and filter updates and network control traffic

- Performs information-gathering operations for the switching engine

- Compiles the results to respond to control and management requests

- Feeds the switching engine with compiled instructions for dynamic control of forwarding and filtering

The DSC is implementation specific. In Berkeley Networks' NT-Switch, the DSC masquerades as a routing protocol to the RTM as shown in Fig. 13.4, allowing it to obtain routes. The DSC compiles the information from the RTM and the network management modules to create a set of hardware instructions for managing the forwarding and policy tables in the Layer 4 gigabit Ethernet switching engine.

Figures 13.5 and 13.6 show two important examples of the DSC operation:

- OSPF link state updates (LSUs)
- Application-level policy setting

All OSPF LSUs are sent to Windows NT for processing. The Layer 4 gigabit Ethernet switching engine detects the OSPF LSUs and queues them to the DSC, as illustrated in Fig. 13.5. The DSC then hands off the OSPF LSUs to Windows NT. On reception of the updates, the Steelhead OSPF code processes the LSU and updates the tables in the RTM. The DSC then extracts the new routes from the RTM, computes new forwarding tables, and downloads new forwarding instructions to the switch hardware.

Similar to the update of OSPF LSUs, policy and service class requests are detected by the Layer 4 gigabit Ethernet switching engine and queued to the DSC as shown in Fig. 13.6. The DSC passes them to Windows NT for processing. Windows NT authenticates the request, and, on

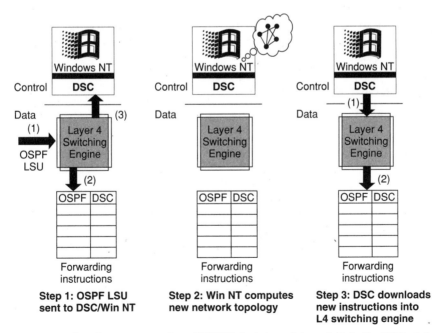

Figure 13.5   Step-by-step processing of OSPF link state update packets in the NT-Switch.

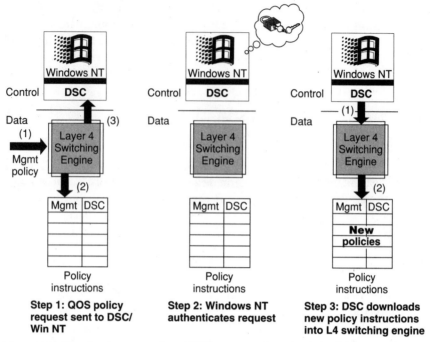

Figure 13.6   Step-by-step processing of QOS management policies in the NT-Switch.

successful authentication, the DSC downloads new policy instructions into the switching engine.

13.7   The NT-Switch's Layer-4 Gigabit Ethernet Switching Engine

The principal function of the switching engine in an NT-Switch is to handle packet processing and forwarding at OSI Layers 1, 2, 3, and 4, all at gigabit Ethernet wire speeds. Because Windows NT is a non-real-time operating system (OS), the Layer 4 gigabit Ethernet switching engine adds a second level of processing to offload the OS from time-critical control and management functions.

The switching engine controls application flow. When a packet enters the NT-Switch, the switching engine performs lookups at the MAC, network, and transport layers; consults forwarding tables stored locally in the switch; and checks dynamic traffic policy filters downloaded from the DSC. It then implements the policy in hardware.

The basic architecture of the Layer 4 gigabit Ethernet switching engine is illustrated in Fig. 13.7, which shows the steps involved in moving a packet through the switch. When a packet arrives at a port in an NT-Switch, a full lookup on the packet header is performed. The

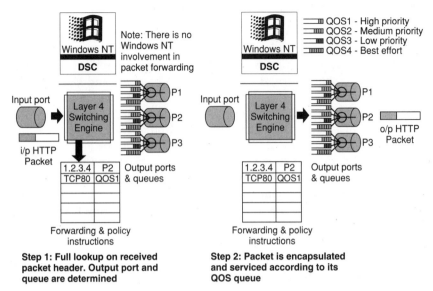

Figure 13.7 Step-by-step forwarding of IP packets in the NT-Switch.

lookup is based strictly on the forwarding and policy instructions that were downloaded by the DSC. As a result of the lookup, either the packet is dropped or a destination port is generated. The packet is then dispatched into a COS queue, which is also specified as a result of the lookup process. Finally, the packet is sent to its destination port and serviced according to its COS queue. Note that Windows NT never becomes involved directly in packet-forward processing. This distinction allows the forwarding rate of an NT-Switch to be scaled up to very high performance levels through hardware, independently of Windows NT packet-processing rates.

13.8   The NT-Switch Networking Model

NT-Switches enable direct interaction between the computing and networking environments for the first time, allowing the network to become aware of user and application needs. In this new networking model, the desktop, the servers, and the NT-Switches play an active role in configuring and administrating the network infrastructure, global policies, and individual user privileges. The NT-Switch networking model has three principal directives:

1. Simplify networking

2. Increase network intelligence and control

3. Address user-specific demands

13.8.1 Simplify networking

Migration to NT-Switches can significantly reduce the cost, complexity, and time spent on network administration and configuration. Microsoft already has made significant strides in reducing the total cost of ownership of desktops and servers with its Zero-Admin PC and EasyNet initiatives. NT-Switches naturally leverage these initiatives through their use of Windows NT as a network operating system. Using specialized wizards for installation, software upgrades, and application-specific configuration, NT-Switches can reduce complex configuration tasks to a small set of point-and-click operations. In many cases, the number of steps required for common procedures is lowered by a factor of 10 to 20 compared with conventional networking devices. Some tasks have completely automated options. NT-Switches can further simplify network administration by automating repetitive tasks, such as IP addressing, as described below.

NT-Switches use auto-IP addressing—implemented through direct integration of DHCP and D-DNS services—to tackle the issue of moves, adds, and changes. Unlike VLANs, this solution is scalable and does not destroy subnet summarization capability or create distributed IP subnets.

Much energy has been spent on investigating the use of VLANs as a technology for simplifying moves, adds, and changes in corporate LAN environments. The main idea behind VLANs was to keep a station or server in the same subnet with the same IP address after a move, add, or change. This had the desired effect of keeping client-server traffic within the same subnet. Therefore, the majority of traffic could be switched at Layer 2.

In practice, the distributed subnets created by VLANs are difficult to manage and diagnose. They also create an explosion in router table sizes because one loses subnet summarization capability when advertising subnet reachability.

NT-Switches perform routing functions at gigabit Ethernet wire speeds with no delay degradation. Therefore there is no performance penalty for crossing a subnet boundary if a station happens to end up in a different subnet after a move.

13.8.2 Increase network intelligence and control

NT-Switches provide network administrators with full control over applications flowing through the network, independently of desktop and server participation. The directory service enables centralized administrative control through support of a number of network-specific objects:

- User objects, which store attributes relating to users, such as privileges for managing network components, dial-up network access, and network resource reservation

- NT-Switch objects, which store configuration parameters and other attributes of an NT-Switch

- Network objects, which store network attributes, such as global policies on the priority of applications or network switch topology.

These objects allow network administrators to set global policies in the NT-Switch with simple point-and-click operations. Global policies can be placed in the directory service network object, automatically downloaded at system startup, and enforced by all NT-Switches in the network. User-specific network policies can, for the first time, be managed with the same set of tools used for specifying user-computing policies, all in a centralized fashion.

Figure 13.8 illustrates a network administrator creating three global policies to specify application priority in the network:

- *Goal.* Disable Doom on the network. It is not allowed.

- *Policy 1.* Block Doom traffic.

- *Goal.* Make the mission-critical SAP R/3 application higher-priority than Web traffic. Manufacturing always complains about

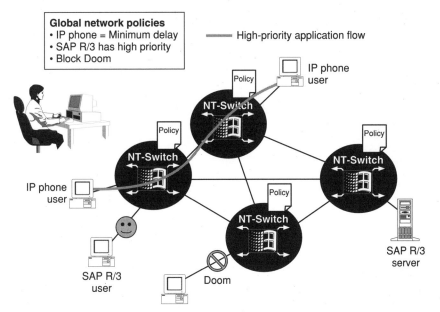

Figure 13.8 Application of global network policies for total-control networking.

slower response times at 8:00 A.M. when everyone launches their Web browser to download the latest news and stock reports.

- *Policy 2.*   Set SAP R/3 to high priority (HTTP is best-effort by default).

- *Goal.*   Add a new collaborative application that supports video and audio streaming, such as Microsoft's NetMeeting.

- *Policy 3.*   Set NetMeeting to minimum delay.

NT-Switches differentiate applications based on the TCP and UDP port numbers. Each of the three policies shown above uses transport-layer ports in slightly different ways. Doom and HTTP use well-known port numbers, while SAP R/3 port numbers are programmable by the network administrator. NetMeeting, based on H.323, dynamically negotiates UDP ports for audio and video streaming. The NT-Switch is capable of handling all three cases with equal facility.

As each application is identified, the NT-Switch checks its tables for associated policies. Once a policy is found, it is executed networkwide. Policy execution typically entails either blocking the packet or queuing it on the appropriate service class queue. Otherwise the packet is forwarded on the default best-effort queue.

13.8.3   Address user-specific demands

In NT-Switch networks, individual network users explicitly signal for prioritization or resource reservation for their networked applications. Before the COS request is granted, the user is authenticated and the associated privilege level is checked against the request.

Figure 13.9 shows how RSVP, Kerberos, and the directory service can be used together to address user-specific demands in a controlled manner. Two different users on the network are attempting to run a delay-sensitive conferencing application that supports RSVP. One user happens to be the CEO, while the other is a contractor. The network administrator inputs the user privileges to the directory service, thus describing each user's policy for reserving network resources; the CEO is allowed to run RSVP applications with unlimited bandwidth reservation privilege; the contractor is not allowed to reserve network resources.

When the CEO launches his or her application, the application gets a Kerberos ticket from the key distribution center (KDC). The application then sends an RSVP reserve message to the first NT-Switch with the embedded Kerberos ticket. Once the NT-Switch validates the ticket and authenticates the use, it looks up the directory service to retrieve the associated policies for the CEO. If the requested action is in agreement with one of the policies, the RSVP request is honored. If the con-

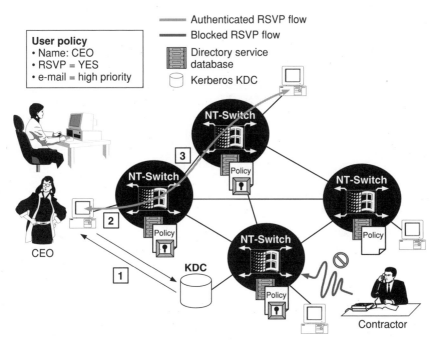

Figure 13.9   Application of user-specific policies in personal-privilege networking.

tractor launches a RSVP application, the reserve request is denied. The lookup of the contractor's policies in the directory service results in a denial of service.

13.9   Conclusion: The Future of NT-Switches

Current trends in networking demand simpler operation, scalable bandwidth, and full application and user awareness. NT-Switches uniquely address these demands through the fusion of Windows NT and Layer 4 switching. Windows NT provides a rich set of networking services that are fully integrated with each other and also with the computing environment. Layer 4 switching provides a new benchmark in high-performance switching intelligence with full application awareness.

NT-Switches represent a fundamental shift in the business model of networking vendors. The evolution of the computing industry provides a good analogy. In the early 1970s and 1980s, a single vendor with strong account control dominated the computing industry. Customers were forced to consider only single-vendor solutions. Complexity prevailed and innovation was stifled through closed proprietary lock-ins. Unix workstations and the Windows PC changed the game. Customers were given greater choices, better investment protection, and an all-around better value proposition. The computing industry of the 1990s delivers lower-cost, simpler extensible platforms. Nonoverlapping, best-in-class strategic partners dominate the industry. Value-add is created through speed, flexibility, and richness of open development.

The networking industry is similarly poised. To propel the industry forward, an analogous move to the business model that has been so successful in the PC industry must occur. The NT-Switch is the first of a new generation of internetworking products that makes this industry shift a reality. Berkeley Networks, Inc. has embraced, refined, and applied the computing business model to networking. Other companies will adopt this approach and add their own unique twist. The end result is an expanded industry that offers better choice, lower cost, and simplified networking solutions.

Gigabit Ethernet Versus Other High-Speed LANs

Purpose

To compare and contrast gigabit Ethernet technology with existing high-speed LAN standards—including ATM—and to highlight its advantages and disadvantages against each.

What Is Covered

Network managers now have to choose between an overabundance of high-speed LAN technologies. Gigabit Ethernet stands out among all of these not only for its low cost and its high performance but also its simplicity. But is it always the best option?

This section compares and contrasts gigabit Ethernet with the other available high-speed LAN technologies, pointing out both those areas where it is stronger and those areas where it falls down.

Chapter 14 compares gigabit Ethernet to six key competitors: ATM, FDDI, 100VG-AnyLAN, Fibre Channel, HIPPI, and fast Ethernet.

As befits ATM's status as the number one competitor to gigabit Ethernet, both Chapter 15 and Chapter 16 compare the new IEEE technology to ATM. However, the two chapters present different takes on the subject. Chapter 15 provides network managers with a methodology for choosing between gigabit Ethernet and ATM in network backbones, as well as design guidelines. Conversely, Chapter 16 describes how both technologies can be used to complement each other within the same network.

Contributors

Gigalabs Inc.; Madge Networks Inc.; 3Com Corp.

14

Comparing Gigabit Ethernet to Other High-Speed LANs

Simon Fok*

GigaLabs Inc.

14.1 Introduction

Gigabit Ethernet has received much attention since its inception. As Ethernet made up 68 percent of the total switch revenue in 1996 (Data Communications 1997 Market Forecast), it is no surprise that the industry is excited about another flavor of Ethernet being standardized. Further, with increasingly complex and multimedia-based applications on the horizon, and with escalating network traffic on local area network (LAN) backbones, gigabit throughput is in high demand. Gigabit Ethernet is still in development, with the standard scheduled to be finalized in the first quarter of 1998.

Many LAN technologies have been developed to date: Ethernet, fast Ethernet, token ring, and FDDI are all examples. Networking products have been designed that support one or several of these LANs. Some of these technologies directly compete with each other. However, many networks use several different types of LANs to complement each other and provide different levels of throughput or quality of throughput for different areas of the network.

---

* Dr. Fok became the CEO of GigaLabs in September 1995. Prior to that time, commencing in 1992, he had responsibility for the day-to-day operation of GigaLabs as vice president and chief operating officer, overseeing the engineering and the manufacturing of all products. Dr. Fok was one of the founders, in 1987, of Systolic Technology, which lead to the formation of GigaLabs Inc. (then Input Output Systems Corporation) in 1988. Dr. Fok received an A.B. in mathematics from the University of California, Berkeley in 1976, and a Ph.D. in applied mathematics from the University of California, Berkeley in 1980. His thesis covered computational analysis in fluid mechanics and was supervised by Dr. Alexander Chorin.

This chapter examines gigabit Ethernet in comparison to other high-speed LANs. Specifically, gigabit Ethernet is contrasted to other LAN technologies in terms of their inception, architecture, speed, technology, and benefits. The rationale behind this chapter is to allow the reader to weigh the benefits and disadvantages of using gigabit Ethernet as compared to implementing other LANs.

14.2   ATM

Born of the telecommunications industry as a wide-area communications network, ATM is being implemented as a high-speed replacement for traditional LANs such as Ethernet, fast Ethernet, token ring, and FDDI. ATM sends traffic across the network in fixed-length 53-byte-long cells. Because the cells are all the same size, they are transmitted at a constant, evenly spaced rate. This cell-based architecture lets ATM simultaneously support packet-based traffic such as AppleTalk and TCP/IP data as well as stream-based audiovisual traffic.

In contrast to ATM, other LANs packetize audiovisual (A/V) traffic and send it through the network inside IPX, NetBIOS, AppleTalk, and TCP/IP packets. However, packetized solutions are not well suited for A/V traffic because packets may be delivered unreliably and at varying intervals, which can make voice and video choppy. ATM's cell-based, in-sequence delivery guarantees smooth A/V traffic presentation to solve this problem.

Most of today's networks comprise a mixture of LAN technologies operating at different speeds. For example: 4- and 1.6-Mbit/s Token Ring, 10-Mbit/s Ethernet, 100-Mbit/s fast Ethernet, 100-Mbit/s FDDI, and 230-Kbit/s LocalTalk. In comparison, most ATM adapters today transmit at 155- and 622-Mbit/s speeds. The ATM Forum has also defined a 25-Mbit/s transmission rate for low-speed desktop links, and ATM can run at 1.544-Mbit/s speeds over wide-area T1-standard links. Therefore, with ATM, network managers will be able to build a homogeneous network using a single technology and yet support different line rates for different machines. In such a scalable network, desktop stations will all be running ATM—though at different speeds—making the network much easier to manage and troubleshoot.

In terms of architecture, ATM networks are switched, meaning that the full bandwidth of the network is actually multiplied or expanded when many nodes are connected. This is in contrast to shared-media LAN protocols such as Ethernet, fast Ethernet, 100VG-AnyLAN, Token Ring, and FDDI, which, in their original, nonswitched iterations, all divide the available bandwidth among all the nodes in the network. ATM switches typically have backplanes that can handle many gigabits per second of simultaneous traffic. ATM's switched architecture can dramatically increase overall network performance.

Furthermore, with ATM, users anywhere on the network can be software-assigned to different virtual networks (akin to zones) from the network manager's station. ATM lets managers assign users to different subnetworks through software. The physical topology of the network and the placement of switches and other devices has no relevance; the software (and, in turn, the administrator) is in complete control. By defining virtual networks through software, managers can frequently move users around without having to physically rewire. In large networks, this alone can relieve managers of substantial costs and hassles.

Thus, with pure ATM nets, the entire corporate network becomes one seamless whole. This homogeneity allows managers to assign subnetworks, addresses, and zone names without regard to geographical location. It also greatly eases troubleshooting and network administration.

LAN Emulation is the ATM Forum's standard for letting apps communicate seamlessly over ATM networks using native LAN protocols (TCP/IP, IPX, AppleTalk, etc.). The LAN Emulation client resides in the workstation adapter driver software. Adapters that support LAN Emulation will support existing applications without modification. Because LAN addresses are different from ATM addresses, a LAN Emulation server (LES) is required to map the two together. LAN Emulation configuration servers (LECSs) maintain initial setup information. Broadcast and unknown address servers (BUSes) are required by connection-oriented ATM networks to handle the LAN-like broadcast traffic frequently used in AppleTalk and Novell's IPX/SPX. One of each of the LAN Emulation server functions (LES, LECS, BUS, etc.) are required per emulated LAN. They can either reside in the ATM switch, in a file server (on MacOS, UNIX, NetWare, or Windows NT), or in any desktop station, and are critical to the network's operation.

ATM data travels over virtual circuits (VCs) between stations and through switches. There are two ways to create virtual circuits: permanent and switched. Permanent virtual circuits (PVCs) are a straightforward and easy, the least common denominator for interoperability since no signaling is required. They do, however, require manual setup and configuration when adding or removing stations from an ATM network. In small networks (3–10 stations), this may be the simplest option. Large networks require switched virtual circuits (SVCs). From an administrator's point of view, SVCs are much easier to work with, since no configuration is required.

ATM networks use signaling to set up and break down SVC connections between stations. The ATM Forum's User-to-Network Interface (UNI) 3.0 and 3.1 specifications define the standard for signaling between stations. If SVCs are used, signaling must reside in each station and in the switch. Savvy users have learned to check for interoperability and to make sure that only ATM Forum standard signaling is being used, if PVCs are not an option.

TABLE 14.1   Gigabit Ethernet Compared to ATM

	Gigabit Ethernet	ATM
Pros	Different levels of Ethernet, fast Ethernet, and gigabit Ethernet can be used to create an entire network for network homogeneity; reasonable cost; standardization process has gone relatively smoothly thus far.	Different levels of ATM can be used to create an entire network allowing for network homogeneity; network can be controlled via software.
Cons	New switching and routing hardware required; distance limitations over copper; workstations not able to take advantage of gigabit speed.	Not yet standardized and has been in standards process for longer than the estimated time; different levels of ATM expensive when compared to Ethernet counterparts.

For all its advantages, ATM is still a young and complex technology; and this has led to problems in getting products from different vendors to work together. In 1996, 15 ATM switches were evaluated at the interoperability test bed run by the University of New Hampshire (Durham). Only five could communicate using switched virtual circuits (SVCs), according to sources involved in the testing, who declined to be identified. Tests published since then in the trade press confirm that an interoperability problem still exists. Some ATM vendors are not yet implementing the ATM Forum's LAN Emulation standard for carrying legacy LAN traffic over ATM backbones, and the ability of rate-based flow-control schemes from different vendors to work together is an unknown.

One strong point of ATM is that it should integrate smoothly with future wide-area ATM services. However, it is difficult to predict when that may materialize. The major disadvantage of ATM is that it is an emerging system, and some specifications are not complete. But even when the ATM standards are finished, the benefits will have been eroded by advances in other network technologies. A good example is fast Ethernet, which was deployed later than 155-Mbit/s ATM but has enjoyed much better market penetration because it was perceived as an extension of Ethernet. At this point, history may repeat itself with gigabit Ethernet against 622-Mbit/s ATM. (See Table 14.1.)

14.3   FDDI

The Fiber Distributed Data Interface (FDDI) was designed in the mid-1980s to solve the network backbone issue. It was developed with management and fault tolerance built in. Unlike Ethernet, the FDDI specification uses a ring structure and timed token. The timed token mechanism allows each computer or workstation to send data to the

TABLE 14.2   Gigabit Ethernet Compared to FDDI

	Gigabit Ethernet	FDDI
Pros	Reasonable cost; easy migration path from Ethernet and fast Ethernet; same Management Information Base (MIB) as Ethernet and fast Ethernet.	High level of fault tolerance; proven and standardized technology.
Cons	New switching and routing hardware required; distance limitations over copper; workstations not able to take advantage of gigabit speed; standard not finalized until 1998.	Relatively expensive; slow compared to gigabit technologies.

next for a period of time negotiated when a station joins the FDDI ring. The packet size can be as large as 20,000 bytes; the most commonly used packet size is 4,500 bytes (as compared to Ethernet, which uses a maximum of 1,518 bytes).

FDDI can be implemented in two ways. The first implementation is as a dual-attached ring. The second implementation is as a concentrator-based ring. In the dual-attached ring environment, the workstations or computers are connected directly one to another. The dual counter-rotating ring provides fault tolerance in case a node goes down. If any of the nodes fail, the ring will isolate the failed node. There is one limitation to this design: if two nodes fail, the ring is separated in two places and creates two rings. Nodes on one ring are isolated from nodes on the other ring.

The second implementation is to use concentrators. Concentrators function like Ethernet hubs. Nodes are single-attached to the concentrator, which isolates failures that occur at these end stations. With this implementation, the nodes can be powered on or off without any interruption of the network services.

Because of the timed token scheme, multiple devices can talk at the same time. Collisions don't occur as they do in Ethernet, and, because of this, FDDI is capable of operating at 99 percent of its capacity.

Due to a lack of development in FDDI, some network managers may view the technology as not being sufficient for today's backbone bandwidth requirements. Even with the release of FDDI switches and FDDI-II, gigabit Ethernet will surpass FDDI as the backbone of choice among corporate networks. (See Table 14.2.)

14.4   100VG-AnyLAN

The 100VG-AnyLAN (100VG) standard was developed by the IEEE 802.12 committee. It can be deployed in both shared-media and switched implementations. 100VG does not impose distance limits more restrictive than those of conventional Ethernet. More important, 100VG-AnyLAN can support both Ethernet and token ring legacy applications, albeit not on the same network. A router is used to go between 100VG Ethernet and 100VG token ring. 100VG eliminates packet collisions and permits more efficient use of network bandwidth. It does this by using a demand-priority access scheme instead of the carrier-sense multiple access with collision detection (CSMA/CD) scheme used in 10BaseT Ethernet and fast Ethernet. Demand priority also permits rudimentary prioritization of time-sensitive traffic, such as real-time voice and video, making 100VG well suited for multimedia applications.

Although it supports a wide range of cabling options, 100VG's wiring requirements are not as flexible as those of token ring or conventional Ethernet. 100VG requires users to install new network adapter cards as well as new hubs or switches. Prices for this equipment are comparable to those for 100BaseT gear and are considerably lower than those for ATM workgroup equipment.

All 100VG networks make use of a scalable star topology. Its design rules allow 100VG to support all topologies used by both Ethernet 10BaseT and 802.5 token ring LANs. This means that any existing Ethernet or token ring networks can be upgraded to 100VG without changing topology or design. 100VG-AnyLAN hubs, like 10BaseT hubs, can be deployed in a cascading hierarchy, as in an Ethernet network. So long as all hubs on the network use the same frame format (Ethernet or token ring), a single 100VG network can contain several tiers without requiring a bridge.

Each hub in a 100VG network may be configured to support either 802.3 10BaseT Ethernet frames or 802.5 token ring frames. A single hub cannot support both frame formats at the same time, and all hubs on the network must be configured to use the same frame format. Because 100VG supports both Ethernet and token ring, a router is used to move traffic between a 100VG network using Ethernet frames and one transporting token ring frames. Going from either type of 100VG network and Asynchronous Transfer Mode (ATM), FDDI, or other topology requires either a router or a translating bridge.

Without question, the main article of contention for opponents of 100VG-AnyLAN is its unique media access control (MAC) layer. The demand-priority access scheme (which is defined in the IEEE 802.12 standard) replaces CSMA/CD. Instead of collision detection or circulating tokens, demand priority uses a round-robin polling scheme implemented in a hub or switch.

Unlike conventional Ethernet, 100VG is a collisionless technology. Using the demand-priority scheme, the 100VG hub or switch permits only one node to access the network segment at a time, precluding packet collisions. When a node needs to send something over the network, it places a request with the hub or switch. The hub or switch then services each node on the segment in sequence. If the node has something to send, the hub or switch permits access to the segment. If it does not, the hub moves on to the next node.

Each round-robin polling sequence provides every single-port node on the network with the opportunity to send one packet. Because all nodes requesting access are served during each polling round, all nodes are assured fair access to the network.

When a multiport hub or switch serves as a node in a larger 100VG network, it too must request network access from the next level up in the network's cascading hierarchy. This device is known as the *root hub*. When the root hub grants access to the larger network, the multiport node can transmit one packet from each port it supports.

In 100VG networks that use token ring frames, demand priority follows the same procedure—with the 100VG hub or switch essentially acting as the circulating token. Instead of waiting to catch the token before transmitting, a token ring node awaits permission from the hub or switch. As in conventional token ring, only one node is permitted to transmit across a network segment at any time. Multiport 100VG nodes adhere to the same procedures when using token ring packets as they do in Ethernet. When granted access to the root hub in a 100VG token ring, the multiport 100VG hub or switch may transmit one packet per port.

In addition to democratic polling of network nodes the demand-priority scheme permits rudimentary prioritization of LAN traffic. Time-critical network applications such as voice and video can be designated as high-priority. When the 100VG hub or switch polls the network segment, it first picks up high-priority traffic and then goes back to less time-sensitive traffic. In effect, 100VG allows applications to separate traffic into two classes: high-priority and normal.

Upper-layer applications at each node perform the actual prioritization of data. This information is passed down the 100VG stack to the MAC layer as part of the packet. If a packet is unlabeled, it is treated as normal traffic. Some 100VG hubs also can be configured to allow certain ports (such as the one to the server) to operate as high-priority all the time.

Each hub keeps track of high-priority requests. When such a request is made, the hub or switch completes the current transmission and then services the high-priority request. If the hub or switch receives more than one high-priority request, it services the requests in port order, or the order followed for normal traffic. Only when all high-priority requests have been satisfied does the hub or switch return to its normal round-robin polling sequence.

TABLE 14.3   Gigabit Ethernet Compared to 100VG-AnyLAN

	Gigabit Ethernet	100VG-AnyLAN
Pros	Ten times the speed of 100VG; multivendor support (both established corporations and start-up companies); both full- and half-duplex; can be deployed in a shared or switched network.	Permits the use of either the 802.3 Ethernet or 802.5 token ring packet format (though not simultaneously); incorporates a two-level priority scheme; supports a wide range of cables: four-pair category 3, 4, and 5 UTP, 150-Ω STP, multimode fiber.
Cons	Supports only 802.3z packet format; not yet standardized.	Limited vendor support; half-duplex; overly complex with two different packet frame formats and two priority modes; shared-media architecture; 100-Mbit/s bandwidth limitation.

Of course, when a tremendous volume of high-priority traffic is on the network, the hub or switch could take a long time to service normal requests, possibly causing time-outs and congestion. To avoid this, the 100VG hub or switch automatically monitors the period between the node's request for access and the provision of service. If too much time elapses, the hub automatically raises the priority level of normal requests.

The most important advantage of gigabit Ethernet over 100VG-AnyLAN is that it runs at gigabit speeds and not 100 Mbit/s. It can operate in both half- and full-duplex and in a shared or switched environment. Although it does not incorporate two-level priority schemes, the IETF is working on schemes such as RSVP to deal with priority issues in multimedia applications. The deterministic performance of 100VG-AnyLAN also loses much of its value when the configuration is used in a switched environment, which is the current industry direction. Last, gigabit Ethernet has support from a wide variety of vendors instead of only one. This last factor will prevent any widespread deployment of 100VG-AnyLAN equipment. (See Table 14.3.)

14.5   Fibre Channel

Fibre Channel is the general name of an integrated set of standards being developed by the American National Standards Institute (ANSI) with the intention of developing practical and inexpensive means of quickly transferring data between workstations, mainframes, supercomputers, desktop computers, storage devices, and other peripherals. It operates at speeds up to 1 Gbit/s.

There are two basic types of data communication between processors and between processors and peripherals: channels and networks. A channel provides a direct or switched point-to-point connection between communicating devices, while a network is an aggregation of distributed nodes that uses a protocol to support interaction among these nodes. Networks can handle a more extensive range of tasks than channels but have relatively high overhead, since they are software intensive and consequently slower than channels. Fibre Channel is an attempt to combine the best of these two methods of communications into a new I/O interface that meets the needs of channel users and also network users. However, the resulting technology cannot really be described as either a true channel or a real network topology.

One of the most salient feature of Fibre Channel is its flexible topology, which includes three connection methods: point-to-point, arbitrated loop, and switched fabric. The simplest of all connection methods is point-to-point, which involves a simple connection between two devices using a single full-duplex cable between them. This connection method provides the greatest possible bandwidth, since no other device could possibly cause delays.

Switched fabric provides the greatest connection capability and largest total aggregate throughput of the three topologies. Each device (computer, disk, etc.) is connected to a switch and receives a nonblocking data path to any other connection on the switch. This would be equivalent to a dedicated connection to every device. As the number of devices increases to occupy multiple switches, the switches are, in turn, connected together.

Arbitrated loop is similar to token ring, where each device arbitrates for loop access and, once access is granted, has a dedicated connection between sender and receiver. Arbitrated loop connects up to 126 devices in a ring. The available bandwidth of the loop is shared between all devices. The primary reason to use arbitrated loop is cost: since no switch is required to connect multiple devices, the cost per connection is significantly less. The most common application of arbitrated loop is storage interconnect.

Fibre Channel operates at a wide variety of speeds (133, 266, and 530 Mbit/s and 1 Gbit/s) and on a variety of both electrical and optical media. Transmission distances vary depending on the combination of speed and media. With single-mode fiber-optic media using longwave lasers, transmissions can reach 10 km at 1 Gbit/s. To accommodate a wide range of communications needs, Fibre Channel defines three different classes of service.

Class 1, a hard or circuit-switched connection, functions in much the same way as today's dedicated physical channels. When a host and device are linked using Class 1, that path is not available to other hosts. When the time needed to make a connection is short or data transmissions are long, Class 1 is an ideal link.

Class 2 is a connectionless, frame-switched link with an acknowledgment of receipt. There is none of the delay required to establish a connection as in Class 1 and none of the uncertainty over whether delivery was achieved. If delivery cannot be made due to congestion, a busy signal is returned and the sender tries again. The sender knows it has to retransmit immediately without waiting for a long time-out to expire. As with traditional packet-switched systems, the path between two devices is not dedicated, allowing better use of the link's bandwidth. Class 2 is ideal for data transfers to and from a shared mass storage system physically located at some distance from several individual workstations.

Class 3 is similar to Class 2 in that it is a connectionless service that allows data to be sent rapidly to multiple devices attached to the fabric. The difference is that no confirmation of receipt is given. By not providing confirmation, the one-to-many Class 3 service speeds the time of transmission. However, if a single user's link is busy, the system will not know to retransmit the data. Class 3 service is very useful for real-time broadcasts where timeliness is key and information not received has little value after the fact.

There are no limits on the size of transfers between applications. This is because with this technology it is the Fibre Channel's framing protocol layer's responsibility to break a sequence into a frame size that has been negotiated between the ports and the fabric. A 32-bit CRC (cyclic redundancy check) is used to detect transmission errors. Even if data arrives out of order in Class 2 or 3 transmissions because frames took different routes, the controls are in place to present data in the proper order to the receiving buffer of the application software.

When it comes to performance, Fibre Channel is a vast improvement over the simplex and half-duplex interfaces in use today. A 100-Mbyte/s Fibre Channel is capable of achieving data transfer rates of 100 Mbytes/s in both directions simultaneously. Thus, it is really a 200-Mbyte/s channel if usage is balanced in both directions.

Ironically enough, the shortcoming of Fibre Channel is due to its flexibility and complexity: it specifies four data rates, three kinds of media, four transmitter types, three distance categories, three classes of service, and three possible fabrics. Unfortunately, this has led to Fibre Channel products being slow to make it to market. And those that have emerged thus far generally run at 200 Mbit/s (266-Mbit/s signaling speed), a quarter of the standard's highest-rated speed. Fibre Channel has basically missed the window to be a gigabit network contender; however, it has begin to make progress as a gigabit channel solution. In this respect, it can complement instead of compete with gigabit Ethernet. (See Table 14.4.)

TABLE 14.4   Gigabit Ethernet Compared to Fibre Channel

	Gigabit Ethernet	Fibre Channel
Pros	Supported by many vendors; easy migration path from Ethernet and fast Ethernet.	Flexible; allows for interconnection between many different network components.
Cons	Not yet standardized; limited number of speed and connection options.	Complex; supported by a limited number of vendors.

14.6 HIPPI

HIPPI (High-Performance Parallel Interface) is a gigabit ANSI standard originally developed to allow mainframes and their supercomputing kin to communicate with one another, and with directly attached storage devices, at supersonic speeds.

HIPPI can be configured for either of two speeds: 800 Mbit/s or 1.6 Gbit/s, both either simplex or full-duplex. HIPPI connections can be set up with just three messages: request (which the source uses to ask for a connection), connect (which the destination uses to indicate that the connection has been established), and ready (which the destination uses to indicate that it's ready to accept a stream of packets). HIPPI channels can handle so-called raw HIPPI (data formatted with the technology's framing protocol without any upper-layer protocols), TCP-IP datagrams, and IPI-3 (Intelligent Peripheral Interface) framed data. (IPI-3 is the protocol used to connect peripherals like Redundant Array of Inexpensive Disks (RAID) devices to computers). Thus, HIPPI is equally adept at internetworking (with Ethernet and FDDI) and high-speed data storage and retrieval.

As for flow control, HIPPI also offers a credit-based system for reliable and efficient communications between devices operating at different speeds. In effect, the source keeps track of the ready signals and only sends data when the destination can handle them. The architecture of a HIPPI switch is based on connection-oriented circuit switching—nonblocking circuit switches allow multiple conversations to take place currently. Thus, the aggregate bandwidth of these switches can be very high—equal to 800 Mbit/s or 1.6 Gbit/s times the number of ports.

HIPPI switches are now an accepted technology for LAN interconnect. Furthermore, Request for Comment (RFC) 1347 from the Internet Engineering Task Force (IETF) specifies how these boxes are to be used on IP networks. All workstation vendors with HIPPI channels also have software drivers for TCP/IP.

HIPPI's streamlined signaling sequences allow connections to be set up and torn down in less than 1 μs. Thus, a single port on a HIPPI

TABLE 14.5   Gigabit Ethernet Compared to HIPPI

	Gigabit Ethernet	HIPPI
Pros	Reasonable cost; supported by many vendors.	Connection to many network components; mature, standardized technology.
Cons	Not yet standardized; workstations and other peripherals not able to take advantage of gigabit speed.	Expensive; supported by few vendors.

switch can deliver hundreds of thousands of IP datagrams every second, outperforming IP routers or hosts, which would typically be the source of this traffic. What's more, switch latency is on the order of 160 ns. HIPPI is so fast, in fact, that any bottleneck would be associated with protocol processing and data handling at the end devices.

HIPPI has long been employed as a high-speed channel to peripherals like disk and tape controllers. HIPPI switches can also be used to link diverse storage devices, thus allowing for shared access across a network. Further, IPI-3 is employed for host-peripheral communications. This protocol is specifically designed to deliver high I/O with minimal CPU overhead.

HIPPI has four major roles to play in such networked systems. First, it can link workstations and other hosts. Second, it can connect workstations with storage systems at higher speeds than any other currently available technology. Third, it can attach display peripherals for real-time visualization. Fourth, it can connect the entire system to another network.

Although HIPPI is a mature gigabit network technology, it has always been conceived as a supercomputer technology that belongs to the bygone days. Even though it can deliver gigabit speeds from supercomputer to the desktop, it is also extremely expensive. Network interface cards (NICs) typically cost between $4000 and $10,000. Even with the current decline in price per HIPPI link, it is unlikely that it will be lower than the gigabit Ethernet $2500 price range and, as such, will not be attractive enough as an alternative to gigabit Ethernet. (See Table 14.5.)

14.7   Fast Ethernet

Fast Ethernet, or 100BaseT, was developed by the 803.2 working group as an easy migration path for Ethernet users with increased bandwidth demands. Fast Ethernet is a 100-Mbit/s local area network that uses that same CSMA/CD media access control (MAC) as 10-Mbit/s

TABLE 14.6   Gigabit Ethernet Compared to Fast Ethernet

	Gigabit Ethernet	Fast Ethernet
Pros	Ten times the speed of fast Ethernet.	Reasonable cost; standard technology; widely supported; extensive interoperability between vendors.
Cons	Expensive; not yet standardized.	Relatively slow.

Ethernet. The only change is that MAC timing parameters are sped up by a factor of 10. Fast Ethernet uses the same types of topologies, wiring, and software protocols as 10-Mbit/s Ethernet. In other words, fast Ethernet provides a smooth upgrade to 100-Mbit/s performance from Ethernet. When configured as a shared-media network, it provides 10-fold increase in performance. In switch configurations, it provides another order of magnitude in performance.

The major advantage of fast Ethernet is its phenomenal support from vendors, which guarantees interoperability between their products. When configured as full-duplex links, fast Ethernet provides bandwidth of over 200 Mbit/s per link; in a switched configuration, it provides growth path to total bandwidths in gigabit range.

One major weakness of fast Ethernet is that it is a nondeterministic system. However, it should be emphasized that all switches, including ATM, cannot guarantee performance for individual packets in events such as buffer overflow. (See Table 14.6.)

14.8 Conclusion

For a comparison of major solutions, see Table 14.7.

Gigabit Ethernet is not yet a standard. However, many vendors have already announced products, and they have committed to interoperability. Vendors seem to understand that for a protocol to flourish in the marketplace, it must be able to work with other vendors' products. This all bodes well for a gigabit Ethernet standard in the near future.

Many network protocols are available to today's network implementers. There are benefits and disadvantages to be weighed when considering the various technologies as compared to each other. Increasingly, network managers are choosing to install several different LAN protocols in order to provide the right amount of speed where it is needed. Had ATM enjoyed a smooth standardization process, it may have been the clear winner today. One can conclude that the more vendors that support a network protocol, the better chance it has of succeeding in terms of market share.

TABLE 14.7 Cross-Comparing High-Speed LANs

	FDDI	ATM	100VG	Fibre Channel	HIPPI	Fast Ethernet	Gigabit Ethernet
Link speed	100 Mbit/s	25,51,155, and 622 Mbit/s	100 Mbit/s	1 Gbit/s	1600 Mbit/s	100 Mbit/s	1000 Mbit/s
Cost per (switch) port	$1500	$1500 (155 Mbit/s) $5000 (622 Mbit/s)	$300	$1500	$2400	$300	$3000
Media	Cat 5 UTP STP 62/125 fiber	Cat 3,4,5 UTP 150-Ω STP 62/125 fiber	Cat 3,4,5 UTP 150-Ω STP 62/125 fiber	STP 62/125 fiber	50-pair STP 50/62/125 fiber	Cat 3,4,5 UTP 150-Ω STP 62/125 fiber	62/125 fiber
MAC protocol	Token passing	Full-duplex	Demand priority	N/A	N/A (full-duplex)	CSMA/CD, full-duplex	CSMA/CD, full-duplex
Architecture/ topology	Ring/dual ring/switch	Switch	Shared media	Shared media/switch	Switch	Shared media/switch	Shared media/switch
Maximum diameter	200 km	Unlimited	12 km	10 m–10 km	25 m–10 km	412 m	200 m
MTU (bytes)	4800	53	4.96	2148	64K	1514	1514

Gigabit Ethernet may not be appropriate for all users and all network situations. However, it certainly has a place in today's networks as a backbone technology and will be in increasing demand as network traffic increases and applications grow more complicated and bandwidth intensive. Its link to Ethernet and fast Ethernet will appeal to many network implementers, as fast Ethernet did to those with Ethernet networks. One question remains, though: when the market demands gigabit Ethernet as a desktop connection, what will be the backbone connection?

15

Choosing Between ATM and Gigabit Ethernet in the Backbone

Martin Taylor*
Madge Networks

15.1 Introduction

On the face of it, gigabit Ethernet and ATM could not be more different. Gigabit Ethernet is a connectionless protocol with variable-length frames, whereas ATM is connection oriented and makes use of fixed-length cells. Gigabit Ethernet is intended to provide point-to-point communications, whereas ATM embodies sophisticated routing protocols enabling fault-tolerant, load-sharing mesh networks to be constructed. Despite these differences, the most popular application for both of these technologies is likely to be the same: high-performance LAN backbones.

This tutorial explains how network designers can make educated decisions between the two technologies based on factors including what type of network they currently have installed and their need for fault tolerance, quality-of-service capabilities, wide-area extensibility, and token ring compatibility. It starts by explaining the catalysts driving the need for higher bandwidth in corporate LANs, then explains and contrasts the different LAN backbone alternatives. It then uses cases studies to illustrate how a choice can be made between gigabit Ethernet and ATM.

---

* Martin Taylor is vice president of network architecture for Madge Networks. He is responsible for the company's strategic technology vision and for overall planning of product development across all product lines—including Ethernet, Token Ring, and ATM switching. He joined Madge in 1991 to head up the product marketing function; prior to that, he was a business development manager in LANs and in fiber-optic systems at GPT Ltd., the leading UK telecommunications equipment vendor. Martin graduated from the University of Cambridge in 1976 with an M.A. in Engineering.

15.2   Overview

The importance of the LAN backbone has increased dramatically in the last few years. Early LAN deployments were often driven by departmental or workgroup applications where all the users of the application were located on a single LAN segment. Connection via a local bridge to a backbone segment was almost an afterthought in many cases, and traffic over the backbone was typically very light. But the evolving nature of LAN usage has driven traffic loading on the backbone inexorably upward. The factors responsible for this include:

- The increasing control of central MIS organizations over LAN-based server resources, and the tendency to relocate these resources centrally

- The deployment of workflow and groupware applications that involve a number of different departments sharing common information resources

- The deployment of corporate-wide applications, such as e-mail, to which all departments need access

- The rapid uptake of Intranet technology, where departments mount local Web servers intended for information dissemination to other groups in the organization

- The growing size of data objects arising from enhancements to application software; for example, Powerpoint 7.0 files are typically twice the size of Powerpoint 4.0 files containing the same presentation material

- The increasing need to move large volumes of data around between servers for database replication or backup

- The increasing amounts of bandwidth being provided to end users through the deployment of workgroup switching technology

The growing pressure on backbones over the years has led, first, to the deployment of faster shared-media LAN technologies such as FDDI, and, more recently, to the use of switched high-speed LAN technologies including fast Ethernet and ATM. Gigabit Ethernet is a logical next step in this progression. In fact, for users with networks based on switched Ethernet and fast Ethernet, Gigabit Ethernet seems like so logical a step that it is tempting to ask why any other alternative should be considered.

15.3   LAN Backbone Technologies

Historically, FDDI has been the high-speed technology of choice for LAN backbones. FDDI has two attributes that are particularly valuable for the backbone. First, its dual-ring topology provides a high degree of fault tolerance. Second, FDDI provides deterministic performance, even when traffic loading approaches 100 percent, thanks to its token-passing access scheme.

Recently, FDDI has come under pressure from switched fast Ethernet in the backbone. While fast Ethernet's CSMA/CD access scheme may limit traffic loading to 50 percent of bandwidth when used with shared media access, this restriction does not apply when fast Ethernet is used as a dedicated link between Ethernet switches. Fast Ethernet has none of FDDI's built-in fault tolerance, but proprietary link backup schemes can make up some of this deficiency. Where fast Ethernet scores high is on cost: as an interface technology, it is inherently lower in cost than FDDI, while further gains are made by eliminating the need for frame format translation between Ethernet and FDDI.

Since gigabit Ethernet is essentially the same as fast Ethernet but runs 10 times as fast, these arguments in favor of fast Ethernet in the backbone are equally applicable to the deployment of gigabit Ethernet in situations where traffic loading demands something more than 100 Mbit/s. Just as Ethernet switches can be linked by fast Ethernet uplinks to a fast Ethernet switch, so fast Ethernet switches can be linked by gigabit Ethernet uplinks to a gigabit Ethernet switch.

This concept of a switched LAN with Ethernet frame formats throughout and a hierarchy of bandwidth from 10 Mbit/s at the desktop through 100 Mbit/s for workgroup uplinks and server connections to 1000 Mbit/s for trunk interswitch links is appealing for its apparent simplicity. However, a number of factors need to be considered that could point in a different direction for a backbone solution:

- *The presence of LAN technologies other than Ethernet at the desktop.* In LANs where token ring is deployed on all or a significant proportion of desktops, fast Ethernet and Gigabit Ethernet backbones may not be an appropriate choice because of the difficulty of translating between token ring and Ethernet frame formats.

- *The need for a very high degree of fault tolerance in the LAN backbone.* Fast and gigabit Ethernet do not have fault tolerance features built in, and it may be very difficult to create a standards-based backbone that does not suffer from single points of failure.

- *The need for future support of real-time communications applications.* New applications such as videoconferencing and voice telephony over the LAN require quality-of-service (QOS) capability in the LAN to deliver acceptable performance. Fast and gigabit Ethernet do not yet have the capabilities needed to guarantee quality of service for time-sensitive transmissions.

■ *The need to extend the LAN backbone over telecommunications facilities.* Fast and gigabit Ethernet do not map directly into telecommunications transmission technologies such as DS-3, SONET, and SDH.

In order to compare and contrast gigabit Ethernet and ATM as choices for LAN backbones, taking these factors into account, we need first to understand how ATM can act as a LAN backbone, bearing in mind that desktops are most likely to be served by Ethernet or token ring connections, and not direct ATM connections, in the future.

15.4   ATM as a LAN Backbone

It is not immediately obvious how a cell-based, connection-oriented technology like ATM can provide a backbone for frame-based, connectionless Ethernet or token ring communications. However, standards-based solutions do exist to do just this.

Standards for the use of ATM in the LAN are governed by the ATM Forum, a consortium composed of LAN and internetworking vendors, telecommunications carriers, and network users. Realizing that the acceptance of ATM as a LAN backbone would require a standard technique for interconnecting Ethernet and Token Ring with ATM, the ATM Forum formed a working group to devise a solution and issue a specification. The group delivered the first specification, LAN Emulation Over ATM Version 1.0 (LANE), in May 1995.

In essence, the LANE spec defines adaptation techniques that allow an ATM network to transport Ethernet and token ring packets from one point to another according to media access control (MAC) addressing information, just as if the packets had been transported over a conventional Ethernet or token ring network. In order to do this, the LANE spec deals with a number of issues, including:

■ How to set up ATM virtual channel connections (VCCs) between any two end points of the network that wish to exchange Ethernet or token ring packets

■ How to resolve Ethernet or token ring MAC addresses and token ring source routing information into ATM addresses in order to identify ATM end points so that VCCs can be set up

■ How to convert variable-length Ethernet and token ring packets into streams of ATM cells, and then convert such streams back into packets at the destination

■ How to ensure that broadcast Ethernet or token ring packets are forwarded to each and every end point that belongs to the emulated LAN

The LANE spec defines two distinct entities in the LAN Emulation process: LAN Emulation clients (LEC) and the LAN Emulation server and broadcast and unknown server (LES/BUS).

The LEC is the process used at every end point in the ATM network that wishes to send or receive Ethernet or token ring packets. An end station that is attached directly via an ATM adapter card to an ATM switch might run the LEC process as part of its driver software. The LEC receives Ethernet or token ring packets from the network operating system software and converts them to ATM cell streams for transmission over the ATM backbone. Incoming cell streams are reassembled into Ethernet or token ring packets and then passed to the network operating system software.

Similarly, an Ethernet switch with an ATM uplink may run the LEC process on its ATM interface. The LEC will receive Ethernet frames from the switch fabric and convert them to ATM cells to send across the uplink to the ATM backbone switch.

The LES/BUS is the process that provides central services, including address resolution and broadcast forwarding. The LES/BUS may run as software on the ATM switch processor or in a separate station attached to the ATM backbone.

Every LEC registers with the LES/BUS when it is initialized and opens a control channel to it. When a LEC has a packet to send, it checks to see whether it has a VCC already open to that Ethernet or token ring MAC address. If it has not, it needs to ask the ATM network to open a VCC, but before it can do this it must find the ATM address by which the target MAC address is reachable. It does this by passing a request to the LES via the control channel. The LES is responsible for providing the answer, either by using its own knowledge about which LECs are registered or by passing the request on to all the registered LECs. Once the LES has responded to the address query, the LEC can then use conventional ATM signaling techniques to request the ATM network to set up the VCC it needs.

This address resolution and VCC setup process sounds rather clumsy, but of course it is only used the first time a LEC has a packet to send to a particular destination MAC address. Once the process is complete, the LEC remembers the details of the VCC it needs to use for this particular destination address, and each subsequent packet can be converted immediately to ATM cells and sent on the appropriate connection.

The BUS part of the LES/BUS is concerned with the forwarding of broadcast packets to all registered members of the emulated LAN. The BUS is needed because ATM does not inherently provide a mechanism for a station to send a broadcast packet that will be received by all other members of the LAN, unlike Ethernet or token ring. When a station has a broadcast packet to send, it sends the packet directly to the

BUS via the control channel, and the BUS then sends it on to each and every registered member of the emulated LAN. Thus, broadcast protocol mechanisms such as Address Resolution Protocol, necessary for IP stations to resolve IP addresses to MAC addresses, are supported by LAN Emulation.

LAN Emulation is now well established as the most widely used technique for deploying ATM as a LAN backbone. It supports connectivity with Ethernet, fast Ethernet, FDDI, and Token Ring via edge switches, and direct connectivity of end stations to ATM. As of April 1997, the ATM Forum is in the process of approving a new and enhanced version of LANE, called version 2.0, which incorporates a distributed LES/BUS for greater fault tolerance and scalability, together with support for ATM quality of service (QOS), enabling real-time communications applications to take advantage of LAN Emulation. LAN Emulation 2.0 also provides the foundation for the ATM Forum standard solution for switching and routing multiprotocol traffic over an ATM backbone, known as Multi-Protocol Over ATM (MPOA). (See Fig. 15.1.)

15.5 LAN Backbones Compared

To illustrate the differing capabilities of gigabit Ethernet and ATM as LAN backbones, let us take two very different examples of network-

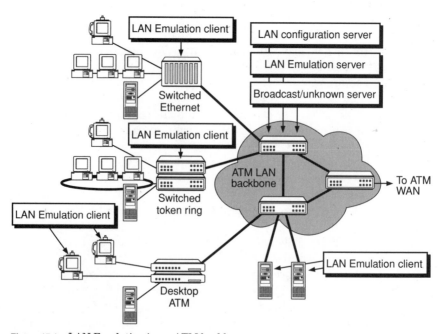

Figure 15.1   LAN Emulation in an ATM backbone.

ing environments to illustrate the extremes of suitability for each technology.

15.5.1   Case study 1: a large-scale CAD/CAM environment

Powerful UNIX workstations are deployed for engineering design and 3-D modeling and visualization. Data files are very large, typically tens or hundreds of megabytes. To achieve acceptable performance when opening and saving files, 100-Mbit/s connections are needed to most desktops.

CAD/CAM is the most demanding application that will be used. There may also be some requirement for LAN-based videoconferencing. However, high quality is not a requirement. The entire CAD/CAM facility is located in a single building, and no provision is made for disaster recovery. A reliable LAN is needed, but the cost of downtime is not so high that heavy investment in fault-tolerant networks can be justified.

The principal requirement in this case study is that large files can be moved between servers and workstations quickly, since making engineers wait minutes while files load could represent substantial lost productivity. However, the proportion of time that each workstation spends loading or saving files is only a few percent overall. Hence, the most appropriate LAN solution is shared fast Ethernet, where perhaps 20–30 workstations are connected to a fast Ethernet segment.

These shared fast Ethernet segments would be connected together with each other and with servers by means of fast Ethernet switches. For best results, the servers would be connected via dedicated fast Ethernet connections to the switches.

Assuming the site has about 500 CAD/CAM workstations and 25 servers, it will need about 40 fast Ethernet ports to support the shared hubs for workstation connections and another 25 ports for servers. Room should be provided for expansion, allowing perhaps 50 percent of additional ports. Therefore, the site needs, in round numbers, 100 fast Ethernet ports.

Currently, it is not possible to obtain fast Ethernet switches with this number of ports. Even if it were, it might not make sense to design this network around a single switch because of cabling densities, length of cable runs, and so on. The alternative might be to use four switches with 25 fast Ethernet ports each and a single gigabit Ethernet uplink. In this configuration a single, four-port gigabit Ethernet switch would be required at the center of the network.

It is probable that most of the traffic in this type of network will be Internet Protocol (IP) traffic. This being the case, it would be desirable to avoid the use of IP addressing schemes that require multiple IP subnets within the LAN. This can be accomplished if the organization is fortunate enough to own one or more Class B addresses, or, failing that, if a private IP addressing scheme is adopted. The reason for this is that routers would be required in the LAN to connect multiple IP subnets together, and, in most cases, this is likely to increase cost dramatically while seriously compromising performance.

The design proposed here is simple and cost effective, and it can offer exceptionally high performance. Speed and cost effectiveness come from avoiding frame format translations and carrying out all switching at Layer 2. The service delivered to users is a good match for the need to download and upload large files on a periodic basis, but other types of applications such as real-time videoconferencing would probably work quite well most of the time because there should be plenty of available bandwidth. However, at peak times, when several users at once are requesting files from different servers, there may be queuing and media access delays arising in the fast Ethernet switches that could result in intermittent breakup of video connections. Also, the fast Ethernet and gigabit Ethernet switches represent single points of failure that could affect large numbers of users for substantial periods of time if something goes wrong.

For this particular type of LAN backbone application, ATM offers little by way of incremental benefits. The fault tolerance and quality-of-service capabilities offered by ATM are not "must-have" requirements in this situation, and the use of Ethernet frame formats throughout means that ATM's ability to also handle token ring traffic is of no advantage. The location of all the CAD/CAM facilities in a single building, with no requirement to extend the LAN environment over the wide area, means that ATM's ability to extend across telecom facilities is not needed, either. Gigabit Ethernet is the obvious choice. (See Fig. 15.2.)

15.5.2 Case study 2: financial services environment

In this case study, the LAN supports both dealing room and back office applications. Token ring is used throughout as the desktop connection. Data objects are not unusually large, but constant high performance is a necessity. The likely future introduction of kiosk-based videoconferencing applications for customer interaction requires that the LAN backbone infrastructure be capable of delivering guaranteed quality of service.

The cost of network downtime is very high and justifies fully redundant configurations that limit fault domains to small groups of users and eliminate single points of failure in the backbone. Also, the disas-

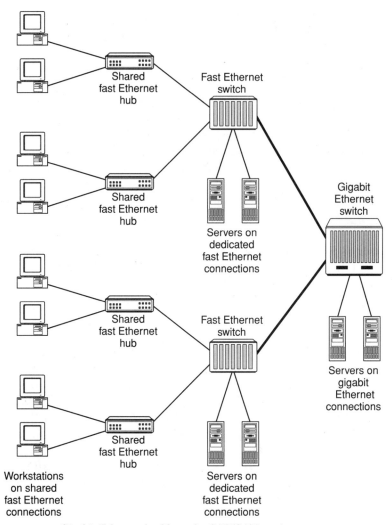

Figure 15.2   Gigabit Ethernet backbone for CAD/CAM environment.

ter recovery plan requires that LAN-based applications be capable of running at full performance across a wide-area link where telecom transmission facilities are available at DS-3 speeds.

Since the current desktop technology is Token Ring throughout, the use of gigabit Ethernet technology as a backbone is not a good fit. Numerous problems exist in attempting to translate token ring packets into Ethernet packet format. These include the need for reversing the bit order of addressing information in many types of packets, the mismatch in maximum packet size (token ring packets can be up to 4

Kbytes, while Ethernet is limited to 1.5 Kbytes), and the lack of support for source routing in the Ethernet environment. There are three types of solutions that could be used:

- *Packet format translation.* Each packet must be individually processed to carry out address bit order reversal where required and either to strip out or insert source routing information as needed. It may also be necessary to reconfigure token ring end stations to limit maximum packet size to 1.5 Kbytes. This technique involves a great deal of per-packet processing, which will add to the cost of the solution and is likely to degrade performance. In addition, it may also introduce compatibility problems with some types of applications that make particular use of the properties of source routing.

- *Encapsulation.* Token ring packets are encapsulated as the payload in gigabit Ethernet packets. No standards exist for this technique; hence, it is proprietary. It may require special gigabit Ethernet switches to accommodate larger packet sizes so that 4-Kbyte token ring packets can be carried. It also requires proprietary drivers on fast Ethernet or gigabit Ethernet server adapter cards to deal with the encapsulation.

- *Routing.* Token ring segments are linked to gigabit Ethernet via routers. The router terminates the token ring protocol and forwards packets onto gigabit Ethernet. Large IP packets coming from token ring can be broken down via standard IP fragmentation techniques to meet the packet size limitations of gigabit Ethernet. This technique is likely to be costly while providing relatively poor performance due to the large amount of per-packet processing. In addition, it does not necessarily provide a clean solution to the handling of nonroutable protocols commonly found in token ring networks such as NetBIOS or SNA. These may have to be carried via IP encapsulation.

On the other hand, ATM provides direct support for token ring packet formats via LAN emulation. The full range of token ring packet sizes may be accommodated, and both source routing and transparent modes of token ring operation are handled. Therefore, there is no packet format translation required (beyond the normal ATM segmentation and reassembly of frames to cells), there is no risk of application incompatibility being introduced, and the token ring characteristics of the network are not compromised in any way. (See Fig. 15.3.)

The network design is likely to consist of shared token ring hubs supporting perhaps 20–40 users connected to token ring switches with ATM uplinks. Each shared ring is connected to two token ring switches, and the uplinks of each pair of token ring switches are connected to a different ATM switch in the backbone. The backbone ATM switches are interconnected via ATM to provide multiple redundant paths through the network, eliminating any single point of failure in the backbone. In the event a token ring switch fails, stations will automatically reestablish sessions that are routed through the failed switch by means of standard source routing procedures. In the event of a backbone ATM switch failure, the LAN Emulation client process in the token ring switch uplinks will automatically reestablish switched paths via alternate routes through the ATM network. When all switches are operating normally, the full capacity of redundant paths may be utilized by taking advantage of the load-sharing capabilities of ATM's routing protocols.

High-capacity servers and power users can be connected directly to ATM using ATM adapter cards that support token ring LAN Emulation. These servers and workstations can communicate transparently with any station connected to token ring.

ATM meets the requirements for high capacity, high performance, and scalability in the backbone with interswitch or server connection speeds of 155 and 622 Mbit/s. Where quality of service is needed to guarantee the performance of real-time video or voice applications, the upcoming LAN Emulation 2.0 specification will support this. Real-time traffic will take advantage of token ring's existing eight-level priority scheme, and the token ring switches with ATM uplinks will map high-priority packets to ATM virtual channel connections with appropriate quality of service to ensure very low end-to-end delay, even in the face of peak load data activity on the backbone.

The ATM backbone is easily extended across wide-area telecom links between different sites or to disaster recovery facilities to support distributed operation of LAN-based applications with local performance. Such ATM links may be supporting voice connections between PBXs concurrently with the LAN data traffic.

While this network is described as exclusively token ring, the use of an ATM backbone does open up the possibility of integrating Ethernet connections. LAN Emulation supports both Ethernet and token ring connectivity, and while it does not in itself provide for translation between Ethernet and token ring frame formats, it does allow both token ring and Ethernet users to access common server resources that are connected directly to ATM—provided the ATM adapter card in the server is capable of supporting both Ethernet and token ring LAN Emulation simultaneously. This attribute provides an excellent migration path for token ring shops that would like the option of deploying Ethernet connections at the desktop without the cost and complexity of routing for Ethernet–to–token ring connectivity.

The scenario described assumes the use of single, flat network infrastructure that does not contain multiple IP subnets. This approach is commonly used in large token ring environments, and the use of an

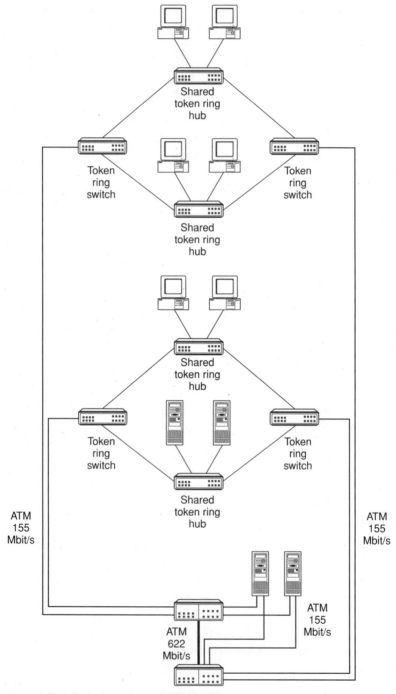

Figure 15.3 Switched token ring and ATM backbone.

ATM backbone with token ring LAN Emulation forms an integral part of this flat network domain. In the event necessity dictates the use of multiple IP subnets, the ATM backbone can accommodate this with the aid of enhanced software installed on the token ring–to–ATM uplinks and on ATM adapter cards in servers or workstations. This software will support routing between IP subnets over direct switched ATM connections by implementing the ATM Forum standard for Multi-Protocol Over ATM (MPOA).

15.6 Conclusion

The two case studies in Sec. 15.5 describe installations at opposite ends of the spectrum. They clearly illustrate the extreme cases where gigabit Ethernet and ATM are the obvious choice of LAN backbone technology. Real-world networks are likely to lie somewhere between these two extremes. Network planners who must make the decision between these two technologies need to establish where their networks lie in the continuum between the extremes and determine the suitability of one technology over the other based on needs for fault tolerance, quality-of-service capabilities, wide-area extensibility, and token ring compatibility.

But one thing is absolutely sure—both technologies have a valid place in the catalog of LAN backbone solutions, and each is worthy of close attention when LAN backbone investments are being planned.

Gigabit Ethernet and ATM: Complementary Technologies for Scaling Performance

Paul Sherer* and Bruce Tolley†

3Com

16.1 Introduction

Despite the claims by some industry pundits and proponents of gigabit Ethernet that ATM is dead, customer research by the Gigabit Ethernet Alliance and 3Com Corporation at Interop Las Vegas in May 1997 shows that network managers are evaluating both technologies as backbone solutions.

In fact, gigabit Ethernet and ATM are complementary rather than competing technologies. Gigabit Ethernet has emerged as a key technology to aid in preserving network simplicity while migrating to gigabit-per-second speeds. Gigabit Ethernet offers the bandwidth needed to aggregate fast Ethernet as well as provide high-speed server connections, switched building backbones, interswitch links, and support for high-speed workgroups. ATM technology, on the other hand, is ideal for robust network requirements such as building scalable campus back-

---

* Paul Sherer is vice president of technology development for 3Com Corporation. In this position he is responsible for bringing new networking technologies from conceptualization into product development.

Prior to this position, Sherer was director of technology development and director of systems architecture for 3Com's Network Adapter Division. He joined 3Com in 1984 and has been involved with the architecture and definition of several 3Com technologies including the Network Driver Interface Specification (NDIS); Demand Protocol Architecture (DPA); Heterogeneous LAN Management (IEEE 802. lb); Softhub/linkWatch; Priority Access Control Enabled (PACE) technology for multimedia over Ethernet; the Parallel Tasking Architecture (used in the EtherLink III product family); and fast Ethernet (IEEE 802.3u). He helped found the Fast Ethernet Alliance and served as chair-

bones, interbuilding backbone links, providing load sharing, supporting multiservice aggregation, and connecting to wide area network (WAN) services.

This chapter addresses how increasing demands from new applications have driven the need for higher-capacity networks. It examines the strengths of each technology and how these strengths are meeting the most demanding network requirements. It considers how both technologies can be leveraged within mixed environments. And it concludes with basic guidelines for their deployment.

16.1.1 New bandwidth drivers

Four classes of applications are driving the need for greater bandwidth (see Table 16.1). The first includes scientific modeling, publication, and medical imaging. The sizes of scientific modeling and medical imaging files, for example, are quickly growing from megabytes to gigabytes and terabytes. The next class comprises Internet and Intranet applications, which are delivering multimedia content to corporate and home desktops. Data warehousing and backup is another class of application where files are ballooning in size. The fourth class features bandwidth-intensive groupware functionalities like desktop videoconferencing (DVC), interactive whiteboarding, and real-time video streaming, which are becoming increasingly popular productivity and training tools.

In addition to requiring more raw bandwidth, some of these applications pose additional issues. Multimedia applications such as DVC and interactive whiteboarding, for example, demand low latency and, sometimes, predictable latency to be used effectively. Internet/Intranet applications create unpredictable traffic patterns that can place extraordinary bandwidth pressure anywhere on networks. The rapid proliferation of World Wide Web and HTML applications have broken the old 80:20 rule under which 80 percent of local area network (LAN) traffic was expected to remain local and 20 percent would be nonlocal. Increasingly, there is no way to predict where traffic will go.

---

man. An inventor, he holds several U.S. patents related to network technology and has several pending U.S. and international patents.

Sherer holds a bachelor of science degree in electrical engineering from the University of Alabama.

[†] Bruce Tolley is a business development manager at 3Com with responsibility for high-speed networking and other emerging markets. He also represents 3Com in the Steering Committee of the Gigabit Ethernet Alliance, of which 3Com is a founding member. Prior to his current position, Tolley held various marketing management positions at 3Com. Before joining 3Com in 1991, Tolley was a senior industry analyst and research director at Frost & Sullivan in Mountain View, California. He has also held teaching positions at Stanford University and San Francisco State University. A graduate of University of California at Santa Cruz, Tolley holds post-graduate degrees from Stanford University, Palo Alto, California, and an M.B.A. from the Haas School of Business, University of California at Berkeley, Berkeley, California.

TABLE 16.1   New Bandwidth Drivers

	Data types/size	Traffic implication	Network requirement
Scientific modeling, publications, medical	• Data files • Mbytes to Gbytes	• Increased bandwidth needed for large files	• Higher bandwidth for desktops, servers, backbone
Internet/Intranet	• Data files now • Audio now	• Increased bandwidth needed for large files	• Higher bandwidth for desktops, servers, backbone
	• Video future • High transaction rate • Large files	• Low latency • COS/QOS • Many data streams	• Low latency
Data warehousing, backup	• Data files • Gbytes to Tbytes	• Increased bandwidth needed for large files • Search and access require low latency • Backup performed during fixed period	• Higher bandwidth for servers and backbone • Low latency
DVC, interactive whiteboard	• Constant data stream	• COS/QOS	• Higher bandwidth for servers and backbone
	• Up to 3+Mbit/s to the desktop	• Many data streams	• Low latency • Predictable latency

16.1.2   Two high-speed solutions

Two technologies have emerged to meet the bandwidth challenge posed by the aforementioned applications: Asynchronous Transfer Mode (ATM) and gigabit Ethernet. Originally developed by the telephony industry, ATM is being deployed today for campus and wide area network (WAN) applications. ATM offers speeds of 1.5, 25, 100, 155, and 622 Mbit/s, and, soon, 2.4 Gbit/s. Gigabit Ethernet, which is expected to be ratified as a standard in June of 1998, features 1 Gbit/s of available bandwidth. It complements the 10/100-Mbit/s Ethernet products already widely available. When comparing the two technologies, it is vital to keep their different design goals, roots, and evolution in mind.

16.1.3   Complementary design goals

Gigabit Ethernet was designed to provide the next logical step in the expansion of Ethernet bandwidth. Ethernet itself was designed as a packet-based network to transport data traffic. As an evolution of Ethernet, gigabit Ethernet preserves Ethernet's simplicity and leverages Ethernet cost drivers. Its goal is to offer simplicity. Gigabit Ethernet leverages the installed base of existing Ethernet devices and IP-based traffic, as well as the existing industry expertise on how to make all these devices and technologies work together. It also leverages network managers' knowledge and familiarity with Ethernet.

In comparison, ATM was designed from the ground up as a scalable, resilient network backbone. Moreover, a key objective of ATM designers was to integrate telephony, data, voice, and video traffic into one consolidated ATM network for LANs and WANs.

16.2   Defining Gigabit Ethernet

Gigabit Ethernet is Ethernet, with all of Ethernet's strengths and weaknesses. On the plus side, Ethernet is today's dominant desktop connectivity technology. By the end of 1996, according to International Data Corp. (Framingham, Massachusetts), a market research firm, 83 percent of all networked desktops worldwide were Ethernet. Like Ethernet, gigabit Ethernet is a media access control (MAC) and physical-layer technology. Gigabit Ethernet specifies the data link (Layer 2) of the OSI protocol model, while TCP and IP, in turn, specify the transport (Layer 4) and network (Layer 3) portions to allow reliable communications services between applications (see Table 16.2). Like the 10- and 100-Mbit/s versions of Ethernet before it, gigabit Ethernet provides optimal bandwidth and connectivity for data transport. Yet by itself it is not optimized for voice or video traffic. More specifically, gigabit Ethernet does not offer the functionality required to support real-time voice and video traffic. In other words, it is not able to define and deliver class of service (COS) by prioritizing traffic or quality of service (QOS) by prioritizing traffic and delivering specific bit rates and jitter limits.

To obtain these higher-level services, gigabit Ethernet, like 10- and 100-Mbit/s Ethernet, can leverage other technologies and standards that are being developed to make Ethernet and IP-based networks more robust. This need for higher-level services is not unique to gigabit Ethernet, but is in fact applicable to most frame-based technologies on the LAN or WAN. The Ethernet and IP community is working on key open standards and protocols, such as 802.1p, 802.1Q, and RSVP. Some of these technologies are still in development by various standards committees within the IEEE, the IETF, and other forums, and they will be available soon. Since gigabit Ethernet is Ethernet, whatever higher-level service works for Ethernet will also work for gigabit Ethernet.

TABLE 16.2   Ethernet and Other Services

OSI layer	OSI name	Example
4	Transport	TCP
3	Network	IP, RSVP
2	Data link	Ethernet MAC, 802.1p 802.1Q
1	Physical	10BaseT, 1000BaseSX

16.2.1 802.1p, 802.1Q, and RSVP

802.1p and 802.1Q are proposed Ethernet specifications for switched LANs that will enable Ethernet networks to offer COS functionality. 802.1Q defines a tagging scheme and 802.1p uses part of the tag to define priority. These specifications will work at all Ethernet speeds, including gigabits per second, and are expected to become part of the Ethernet standard by the end of 1997.

Resource Reservation Protocol (RSVP) is an important, emerging standard being developed by the IETF. RSVP enables networks to reserve dedicated bandwidth. The RSVP protocol allows end systems to request a particular QOS from the network for a particular traffic flow. The request goes from end to end through the network and is seen by all RSVP entities in between. Each entity has the ability to confirm or deny the requested QOS. The initiating end system will then be provided with this information and has the option of proceeding with the session. The QOS level requested can be granted or denied. If it is denied, an end station may accept a lower level of service or terminate the session until the necessary resources become available.

Gigabit Ethernet can also leverage other technologies such as TCP/IP, multicast protocols, and routing protocols, or, more specifically, Layer 3 switching. At first it might appear that gigabit Ethernet provides so much bandwidth that IT managers will no longer have to be careful how they use and manage bandwidth. In fact, this is not the case; as networks migrate to the higher speeds of gigabit Ethernet, the issues of bandwidth management and control actually become more pressing.

Various implementations of gigabit Ethernet may include one or more of these standards or services to provide a more robust or functional networking connection. Yet, the overall success of gigabit Ethernet is not tied to any of them. In fact, gigabit Ethernet might actually spur the widespread deployment of other technologies, such as Ethernet COS and gigabit Ethernet Layer 3 switching.

16.2.2 Ethernet bandwidth migration

As the latest innovation in Ethernet speeds, gigabit Ethernet makes three basic promises: higher speed, continued simplicity, and lower costs. Gigabit Ethernet is 10 times faster than fast Ethernet and 100 times the speed of regular Ethernet. It provides the bandwidth required to meet the demands of current and future applications. Since the basic Ethernet packet format remains intact, gigabit Ethernet retains the very familiar Ethernet structure at the MAC and physical layers. Its implementation requires no retraining of MIS personnel in new technologies, which reduces administrative and maintenance costs. By relying on a very well-established technology, gigabit Ethernet is expected to offer competitively priced solutions. Based on the historical trends of 10BaseT and fast Ethernet pricing, the cost of migrating to gigabit Ethernet should decline over time, making the transition from 100 Mbit/s to 1000 Mbit/s as cost-effective as the transition from 10 Mbit/s to 100 Mbit/s.

Another major benefit of gigabit Ethernet is that it will enable the widespread deployment of switched 10BaseT and even fast Ethernet desktop connections. Presently, this market is small, despite popular misconceptions to the contrary, but it is growing rapidly. Leading networking vendors are providing gigabit Ethernet solutions for the aggregation of switched desktop connections. As aggressive pricing erodes the economic barriers, gigabit Ethernet is poised to radically accelerate this market for switched fast Ethernet.

Most desktops today, and the majority of large networks, use Ethernet. This is not expected to change. Because Ethernet users are already familiar with Ethernet technology and its maintenance and troubleshooting tools, the support costs associated with gigabit Ethernet will be lower than those of ATM. Gigabit Ethernet will require only incremental training of personnel and incremental purchases of gigabit Ethernet–specific products and tools. The personnel and training costs associated with upgrading to gigabit Ethernet are also expected to be relatively low. Network administrators who are comfortable with 10BaseT Ethernet and fast Ethernet will be able to install, troubleshoot, and support gigabit Ethernet with confidence.

16.3   Understanding ATM

ATM is designed to handle any type of traffic. It accommodates a variety of bit rates; supports the bursty communications behavior exhibited by voice, data, and video traffic; and possesses the ability to control jitter and limit latency across the network. The last feature is required to support real-time traffic. End stations can request a level of QOS from the network and ATM can deliver that service across a campus ATM network or a WAN. This ability to deliver quality of service and support real-time, interactive traffic is one of ATM's greatest strengths.

To summarize the history of ATM, when companies began interconnecting LANs across wide area networks in the early 1980s, two different approaches to WANs emerged. One was private WANs based on shared resources and optimized for data transmission. The other was the use of public WAN telephony networks, which are based on dedicated bandwidth and optimized for voice traffic. In 1986, the CCIT, now known as the ITU, formed a study group to explore the concept of a high-speed integrated network that could uniformly handle voice, data, and a variety of other services. The result was broadband integrated services digital network (B-ISDN). B-ISDN services require high-speed channels for transmitting digitized voice, data, video, and multimedia traffic. ATM is the switching and multiplexing technology for supporting B-ISDN services.

16.3.1 Scalability and cell switching

ATM is based on a highly scalable architecture. The size of an ATM network is increased by simply adding more switches. ATM is also inherently scalable, because one protocol works for all speeds, from 1.5 Mbit/s to 2.4 Gbit/s, both within the campus network and across the WAN.

It's not only scalability that sets ATM apart from other networking technologies. To date, packet switching has been the technology of choice for bursty data traffic because it consumes bandwidth only when traffic is present. However, ATM's cell switching architecture overcomes some of the limits of packet switching by offering fixed-length packets known as *cells*. Each ATM cell consists of a 48-byte payload and a 5-byte header. Fixed-length cells also allow cell relay switches to process cells in parallel for speeds that, until quite recently, have exceeded the limitations of bus-based, packet switching architectures.

Networking and switching queuing delays are more predictable with fixed-length data cells. Fixed-length cells are less complex and more reliable to process. The high predictability that results from using cells allows ATM hardware to function more efficiently because control structures, buffers, and buffer management schemes have been designed to known size criteria. Switch vendors can install mechanisms to ensure the appropriate level of service from all types of traffic, especially for such delay-sensitive services as voice and video.

16.3.2 LAN emulation

The strengths of ATM are its scalability, flexibility, and ability to handle any kind of traffic—voice, data, or video—as well as its promise to deliver QOS. Yet, when faced with interoperability with the installed base of packet-based, connectionless LANs, ATM faces some challenges. ATM is designed for point-to-point connections. Most networked desktops use Ethernet, which is a broadcast-based technology. The ATM community has developed technologies and protocols such as LAN Emulation (LANE), LAN Emulation server (LES), and broadcast and unknown server (BUS) to allow the ATM network to emulate a LAN and thereby enable connectivity to Ethernet desktops.

LANE version 1 is already in use by vendors and version 2 is expected in 1997. When these services are added onto ATM, they enable the ATM network to appear as a broadcast LAN to the Ethernet devices and end stations and deliver the same broadcast services. But these additional services add complexity.

16.3.3 ATM quality of service

Besides providing efficient bandwidth, ATM addresses another critical networking issue: the need to manage and guarantee certain performance levels of the network itself. Since various traffic types require different levels of jitter control, different delay behavior, different delay variation, and different loss characteristics, ATM provides various qualities of service to accommodate these variations.

To access the network, an end station requests a virtual circuit (VC) between the transmitting end station and the receiving end station. During the connection setup, the end station can request the QOS it needs to meet the transmission requirements and ATM switches will grant the request if sufficient network resources are available. The guaranteed QOS for cell-based switched access is extremely effective for transporting such real-time interactive communications as voice or data.

16.3.4 ATM solutions: a summary

ATM offers solutions for today's networking needs: switch to switch (including in the backbone) and switch to server. ATM-to-Ethernet edge devices support Ethernet desktops. ATM allows the seamless, cost-effective integration of all LAN technologies, including high-performance Layer 3 functionality, into data-centric networks. For multiple-services networks, ATM enables the delivery of full COS throughout an enterprise with support for QOS control while allowing for the complete integration with legacy WANs. Today's ATM products have features that can support a robust structure for LAN backbones. ATM offers multiple connections to the backbone core for wiring-closet switches (see Fig. 16.1). It also provides dynamic links that are load sharing and provide alternate paths, as well as switched virtual circuits (SVCs) that can reroute around failures. ATM provides standards-based virtual networks and end-to-end congestion control. Finally, the technology enables the deployment of multiple switches in the network core and many gigabits per second of total capacity.

Further, ATM products are rapidly becoming today's dominant WAN technology. Most frame relay services at the core of today's carrier networks rely on ATM switching fabrics. Native ATM private services are increasingly available, and soon publicly tariffed ATM services will be offered. ATM delivers scalable and resilient services in WAN backbones and supports such services as T1/E1, T3/E3, and native services OC-3 at 155 Mbit/s (see Fig. 16.2). As a result, ATM's dominance of the WAN market should actually increase in the future.

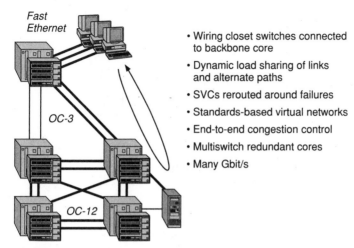

- Wiring closet switches connected to backbone core
- Dynamic load sharing of links and alternate paths
- SVCs rerouted around failures
- Standards-based virtual networks
- End-to-end congestion control
- Multiswitch redundant cores
- Many Gbit/s

Figure 16.1 ATM delivers scalable and resilient LAN backbones.

ATM's standards are well established (see Fig. 16.3), ensuring full interoperability for multivendor product environments. In fact, ATM's robustness is expected to increase due to additional technical work being performed in the ATM Forum.

16.4 ATM and Gigabit Ethernet

Any comparison of ATM and gigabit Ethernet must consider not only their different design goals, but also the respective roles the two tech-

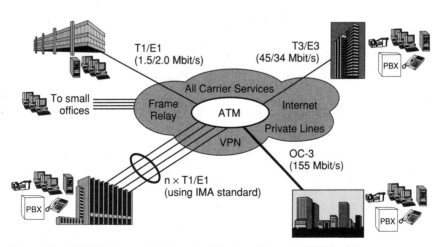

Figure 16.2 ATM delivers scalable and resilient multiservice WAN backbones.

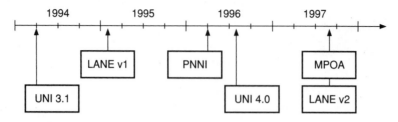

- LANE v1—highly scalable (as proven by 3Com)
- LANE v2—a software upgrade and backward compatible

Figure 16.3   Essential ATM standards: stable and complete.

nologies play in enterprise networks. As mentioned, most of today's desktops are 10BaseT Ethernet, and in corporate environments many of these are expected to migrate to 100BaseT. On WANs, most of the public and private transport services delivered today are supported by switches with ATM switching matrices. Private native ATM services are available today, and soon public native ATM services will be widely available and competitively priced.

Consequently, in terms of the overall enterprise network, it is not a question of choosing between ATM and Ethernet, but of where both technologies will be implemented. Both ATM and gigabit Ethernet will coexist and complement each other. The most likely scenario is that the enterprise desktop will remain Ethernet and the enterprise WAN will become increasingly ATM (see Fig. 16.4). The desktop and the WAN

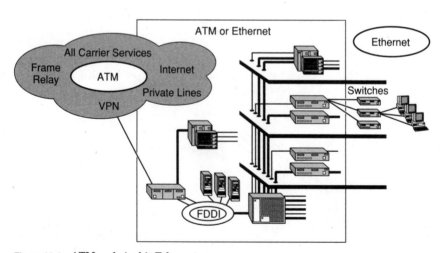

Figure 16.4   ATM and gigabit Ethernet.

meet at the backbone—both the campus backbone and the building backbone. This area is where network managers must decide to deploy the appropriate high-speed technology or technologies, keeping in mind the applications and services they want to support and the very different design objectives of the respective technologies.

From the ATM perspective, the issue of ATM versus gigabit Ethernet becomes a question of how far into the backbone net managers will take the ATM cloud. From the Ethernet perspective, the question becomes how far toward the WAN net managers need to take 10/100/1000-Mbit/s Ethernet. Where ATM and Ethernet meet depends on the applications and services required on the network.

16.4.1 ATM QOS and Ethernet COS

In summary, ATM was designed explicitly to offer QOS from the transmitting end station to the receiving end station over dedicated VCs—with the ability to deliver defined latencies, jitter limits, and committed bit rates. In comparison, gigabit Ethernet—and Ethernet at 10 and 100 Mbit/s—promises to offer class of service (COS). Ethernet, when combined with the emerging 802.1Q and 802.1p specifications, will be able to provide for the prioritization of traffic through the network. It will use RSVP to offer bandwidth reservation. Using these approaches, it will be able to provide high- (and low-) priority queues to support voice and video over IP in both non-real-time and real-time applications.

16.4.2 Delivering QOS in mixed
ATM/Ethernet environments

Today, high-end modular ATM switches can provide campuswide support for COS and QOS in mixed environment networks. These switches can be deployed at the core of hybrid gigabit Ethernet and ATM data center networks. They can provide full connectivity with gigabit Ethernet stackable and modular switches, allowing for simultaneous support of COS and QOS. Networking vendors will use the emerging 802.1p and 802.1Q standards to ensure a standards-based mechanism to allow multimedia applications to run over existing infrastructures. A combination of tagging and queuing techniques will ensure a standards-based, end-to-end solution for time-sensitive applications and the ability to support COS within frame-based technologies. ATM–to–gigabit Ethernet switches will provide ATM QOS support and translate Ethernet classes of service into the corresponding ATM quality of service, such as constant bit rate (CBR) and variable bit rate (VBR). They will also support best-effort services, namely available bit rate (ABR) and unspecified bit rate (UBR).

The ability to simultaneously support voice, video, sound, and data is important for delay-sensitive applications, such as videoconferencing, as well as for support of PBXs in the campus LAN. Voice traffic dialed into the PBX would be transported over the Ethernet or ATM network, not over dedicated voice wiring within the building or campus. This allows one wiring infrastructure to support both voice and data traffic within the campus.

ATM–to–gigabit Ethernet switches from leading vendors will enable the mapping of Ethernet COS to ATM QOS. Functionality within the ATM-to-gigabit switch will enable traffic requiring one of diverse classes of services in scalable Ethernet environments to be mapped to the corresponding ATM QOS in ATM environments and vice versa. Mapping will be in accordance with the ATM Forum LANE 2.0 standard and the IEEE 802.1p and 802.1Q specifications. This capability will allow customers to ensure higher priority for mission-critical applications, such as SAP applications, or appropriate qualities of service for videoconferencing and other delay-sensitive applications in a mixed ATM and Ethernet environment.

16.4.3 Optimizing network control

This chapter began by characterizing the four classes of applications and their demands for bandwidth. In addition to bandwidth, the new classes of applications also require network administrators to manage the bandwidth and optimize network control. Gigabit Ethernet gives IT managers the ability to scale Ethernet networks from 10 to 100 to 1000 Mbit/s. However, increasing network performance is more than an issue of merely increasing speed or increasing bandwidth. The bandwidth must be managed or controlled. Gigabit Ethernet by itself does not offer this functionality. It is provided by higher-level services such as IP switching and Layer 3 switching, both of which offer ways to optimize the IT manager's control over the network.

The various options for scaling Layer 3 control functions with gigabit Ethernet may be understood in a grid of cost versus levels of control (see Fig. 16.5). In the early days of LAN-to-LAN interconnections, many flat, bridged (or switched) networks were built. Today, for some implementations, the building of a flat, switched network or "switch everywhere," may be perfectly appropriate given the trade-offs of cost versus control and applications that need to be supported on the network.

Choices offering more control begin with the second option; the simplest, lowest-cost approach that can be summarized by the common advice of early switching advocates: "switch where you can,

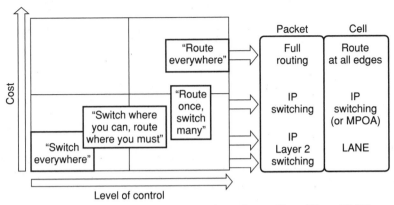

Figure 16.5 Network control optimizations for packet and/or cell-based LANs.

route where you must." This technique utilizes Layer 2 switches either in front of routers or without any routing at all. While suitable for many networks, this approach offers the lowest level of bandwidth control.

The next step up, typified by the so-called IP switching approaches, applies a "route once, switch thereafter" philosophy. IP switching, sometimes termed *cut-through routing,* enables Layer 3 scaling to gigabit Ethernet speeds by utilizing the existing routing infrastructure and Layer 2 switches.

Cut-through routing can work quite well in gigabit Ethernet networks. Certain implementations will scale the maximum control available with full routing to over 30 million packets per second. IP routing services will reside in existing routers and be supported by gigabit Ethernet switches. This will enable IT managers to deploy IP switching without forklift upgrades of existing routers.

For IT managers who require the greatest control with the full range of router security and functions, a few vendors are shipping Layer 3 switches today. Many established networking vendors, as well as start-ups, have announced plans to support Layer 3 switching at gigabit-per-second speeds. A brief definition of Layer 3 switching will be useful.

Layer 3 switches determine the eventual destination of a packet by looking beyond the MAC destination address in its header to the IP address, which is buried deep in the packet. This technique allows packets to be forwarded accurately to intermediate nodes for the most effective transport though the network while the IP subnet can be ascertained, allowing broadcasts to be contained within the appropriate subnets.

The classic Layer 3 device is the router, and while router vendors are finding ways to make routers faster, switching vendors have been building the ability to make Layer 3 decisions into the switching silicon itself. Such Layer 3 switch products accomplish many of the same tasks as traditional routers, but at superior price/performance ratios. In most cases, narrowing the protocol support to IP allows devices to optimize tasks and accomplish more with dedicated hardware. Network managers who are evaluating these technologies should consider three key decision criteria: the vendor's ability to provide a scaled 10/100/1000 solution, compatibility of the products with the installed base of switches and routers, and, when necessary, the vendor's ability to support other high-speed options such as FDDI and ATM.

A key strength of gigabit Ethernet will be its ability to leverage all the installed technologies developed for Ethernet and switched fast Ethernet. Because gigabit Ethernet retains the very familiar Ethernet structure at the MAC and physical layers, these technologies include routing protocols and multicast capabilities.

16.5 Conclusion and Application Guidelines

The comparisons between ATM and gigabit Ethernet are summarized in Table 16.3. Ethernet was optimized for simplicity, while ATM was optimized for scalability. Since gigabit Ethernet is Ethernet, it will offer bandwidth at a relatively lower cost than ATM. ATM was designed to offer guaranteed QOS from the transmitting end station to the receiving end station over dedicated VCs with guaranteed latency, jitter limits, and delivered bit rates. Gigabit Ethernet, on the other hand, offers COS. By relying on emerging standards and protocols, gigabit Ethernet will provide for the prioritization of traffic throughout the network and some bandwidth reservation. Gigabit Ethernet is optimized for high-speed data, but with Ethernet COS, the technology will provide high-priority queues to support voice and video over IP and latency-dependent multimedia traffic. ATM was designed to support voice, data, and video, and it will offer seamless integration to native ATM WAN services, whether private or public. Gigabit Ethernet LANs will require an Ethernet-to-ATM WAN edge device with Layer 3 functionality to access native ATM WAN services.

TABLE 16.3   Gigabit Ethernet and ATM Summary

Key features	Gigabit Ethernet	ATM
Design goal	Simplicity	Scalability
Bandwidth	Low cost	Moderate cost
Service	COS with 802.1Q/p, RSVP	Guaranteed QOS
Data types	High-speed data	Data, Some MM support Voice, video
LAN/WAN switching	Via edge devices	Native connections
Availability	2H1998	Shipping now

In conclusion, simple guidelines are available for the implementation of ATM and gigabit Ethernet technologies.

Gigabit Ethernet should be deployed where:

- Simplicity and cost are foremost concerns.
- Network traffic is primarily data and QOS guarantees are not required.
- Physical LAN distances are limited to a radius of 5 km, which is the distance goal for single-mode fiber.

Use ATM where:

- Network scalability and resiliency are foremost concerns.
- One infrastructure for all data, voice, and video needs is desired.
- LAN-to-WAN switching is required.

In addition, there are two basic hybrid models available for particular networking environments. These are:

- 10/100/1000 Ethernet connections on each floor of a building with ATM deployed for the building backbone
- 10/100/1000 Ethernet intrabuilding connections with ATM deployed for the campus backbone

Much has been written within the industry that extols the virtues of one technology over the other. The underlying realities, however, are that ATM and gigabit Ethernet are designed for very different needs and are two complementary technologies. They should be evaluated based on specific network requirements and they can be deployed together to optimize network performance. For many implementations, ATM and gigabit Ethernet will coexist within the same enterprise network. ATM and gigabit Ethernet each have a strong future in the marketplace.

Migrating to Gigabit Ethernet

Purpose

To provide guidelines on how to improve performance on a network by adding gigabit Ethernet while keeping disruption to a minimum.

What Is Covered

For many, choosing to migrate to gigabit Ethernet will be a simple decision. But actually rolling gigabit technology out in existing networks will be anything but simple. The three chapters in this section all deal with the knotty issue of how best to minimize disruption and maximize performance when deploying gigabit Ethernet in a legacy network.

Chapter 17 explains how gigabit Ethernet presents net managers with a dilemma. On the one hand, much about the technology is familiar. On the other, there are new rules that must be obeyed. Chapter 18 provides more advice on this subject. In Chapter 19, the authors present the concept of a bandwidth hierarchy—essentially, a model that can be used to work out where and how to install gigabit Ethernet.

Contributors

XLNT Inc.; Cabletron Systems Inc. (Digital Network Products Group); 3Com Corp.

17

Charting a Gigabit Ethernet Migration

Robert M. Grow*
XLNT, Inc.

17.1 Introduction

On the face of it, gigabit Ethernet looks like a natural evolution in network performance. But while the proponents of all new technologies promise the products will be easy to adopt, in the hard light of reality these promises can become ephemeral. Consequently, anyone who has participated in the adoption of any new networking technology will approach gigabit Ethernet warily.

The relentless increase in bandwidth demands created by improving desktop systems, new applications, the Internet and Intranets, relocation of servers, and myriad other forces make the adoption of higher-performance links inevitable. Because such change is unavoidable, the challenge to network administrators is how to get this higher bandwidth into the existing network with minimal disruption. Gigabit Ethernet provides a rich set of product options that in turn yield a variety of options for migrating this new technology into networks. Gigabit Ethernet also provides increased bandwidth in a familiar form, which eases its adoption.

---

* Robert Grow is cofounder and vice president of industry relations at XLNT, a high-performance networking company located in San Diego, California. Grow's career spans 22 years in communications and computer systems architecture and development for several technology companies. His principal expertise is in high-speed local area networks.

Grow has participated in the development of many ANSI, IEEE, and IETF networking standards. He is an editor and subtaskgroup chair for the IEEE 802.3z gigabit Ethernet standard currently under development. His work in the development of LAN product hardware and software has produced several patents, including the timed-token media access control protocol used in FDDI and IEEE 802.4.

Gigabit Ethernet has the same operational properties and can serve the same network applications as its 10-Mbit/s Ethernet and 100-Mbit/s fast Ethernet predecessors, but it also has some differences that must be factored into any migration plan. The following sections explain the major similarities and differences affecting a network migration, and explore migration paths for a disparate set of gigabit Ethernet uses.

17.2   Ease of Migration

Since its conception more than two decades ago, Ethernet has evolved significantly. This evolution includes increases in speed, adaptation to different cable types, and addition of full-duplex operation with flow control. Gigabit Ethernet shares its basic operational properties with 10-Mbit/s Ethernet and 100-Mbit/s fast Ethernet. It utilizes both the identical frame format of the lower speeds and the core CSMA/CD protocol that controls shared media access.

As was the case in adapting 10-Mbit/s Ethernet for operation at 100 Mbit/s, some changes are required to support transmission at 1000 Mbit/s. It is the changes that have the potential to create difficulties when adopting the new technology. The most significant changes are in the methods used to encode and transmit data on the media. Two enhancements to the CSMA/CD protocol that are unique to gigabit operation enable half-duplex operation on 100-m link lengths. This allows half-duplex operation over the horizontal cabling specified in building cabling standards. While these CSMA/CD changes might seem major to Ethernet purists, they in fact have little impact in the adoption of the technology. To understand better how easy it will be to migrate to gigabit Ethernet, it is useful to look at what stays the same and what changes in the 802.3 standard in order to add operation at 1000 Mbit/s.

17.2.1   It's still Ethernet

All speeds of Ethernet are specified in a common standard (even if the standard itself can't be bought as a single volume). With the addition of each new flavor of physical media, or increase in speed, the base standard is supplemented and, where necessary, changed.

Where possible, changes always take into account the need for backward compatibility with the installed base of equipment. And, despite all the improvements to the original technology, two things remain consistent for all varieties of 802.3 standard-based Ethernet—its frame and management.

17.2.2   Frame format

The basic frame format used by Ethernet's media access control (MAC) protocol is the same for all speeds and configurations of Ethernet operation. This includes address fields, the type/length field, and CRC generation and error checking. The frame sizes are also the same, with a 64-octet minimum frame and 1518-octet maximum frame applying to each speed and to both full- and half-duplex modes of operation. (This will remain true with the new 802.3ac project, which proposes to add a tag field to the 802.3 frame format. The tag field will be consistent for all speeds and modes of operation.)

Migration of gigabit Ethernet into existing networks is significantly simplified by this consistent frame format. It simplifies switching and forwarding of frames between Ethernets of different speeds, since no fields in the frame need be changed. Less processing on the packet allows less costly logic and lower latency in switch architectures. Switching between Ethernet and FDDI or token ring, conversely, requires conversion of address order and, on large frames, fragmentation of the frame to Ethernet sizes. It also requires insertion and deletion of Snap headers. Note, however, that all this is still significantly less complex than the protocols required for connection of LANs to ATM links.

17.2.3 Management

Consistent management is especially important to network administrators. Gigabit Ethernet possesses the same management attributes as its predecessors. Management attribute definitions related to the frame format are not affected by the increase in data rate. For example, a "runt" frame is the same for all Ethernets. Most attributes related to CSMA/CD operation also have common definitions independent of data rate. For example, a deferral of frame transmission is the same for all Ethernets.

Though most attributes do not change in meaning, the semantics of a few attributes will be affected by the CSMA/CD enhancements for gigabit operation. For example, a late collision has different effects in a frame burst (a MAC feature unique to gigabit operation).

The change in data rate does cause some noticeable changes in management. For instance, a higher data rate produces a corresponding increase in the increment rate of many management counters. The 100:1 ratio in data rates makes some of the counter attributes originally sized for 10-Mbit/s operation rollover at 1000 Mbit/s faster than is desirable for management sampling intervals. This applies to most of the octet and frame counters. Computed attributes are similarly affected. For example, utilization computations must recognize the 1000-Mbit/s rate of operation to produce proper results.

The counters are perhaps the most significant issue for installed management software. This is because the underlying counter type definition must be changed, for example from Counter32 to Counter64. This will be achieved by changing the underlying standard protocols like SNMP, modifying them as necessary to accommodate the higher increment rates.

Other management attributes must also be updated for gigabit Ethernet. For example, attributes enumerating PHY types (e.g., 100BaseTX, 100BaseFX) must enumerate additional values for the gigabit Ethernet PHY types (e.g., 1000BaseSX).

This lengthy set of individually simple changes could lead one to wonder why management is listed under the heading "It's Still Ethernet." But, because these changes do not change the meaning of the attributes, their effect is quite minor on the end user of Ethernet equipment. This consistency of management attribute semantics applies to the IEEE 802.3 MIB as well as IETF standardized MIBs for SNMP and RMON, and it is this consistency that will enable standard management platforms to include gigabit Ethernet with ease. For most users, the change will boil down to loading a new release of their existing management software package. Most important, addition of gigabit Ethernet to a network should not require changes to the management architecture, and it does not require large amounts of new learning from the people managing the network.

17.3   It's Still Ethernet, But . . .

While the fundamental characteristics of Ethernet are preserved in gigabit Ethernet, it's the things that are different that will shape the migration plan of the technology into a network. Gigabit Ethernet presents most of the same migration issues that existed when net managers were migrating from Ethernet to fast Ethernet. Most of these issues result directly from the increase in speed.

17.3.1   New link coding

The coding methods used at 10 and 100 Mbit/s are not suitable for operation at 1000 Mbit/s. As a result, 8b/10b encoding is used for the 1000BaseX PHY types. Though not needed or used by most network administrators, diagnostic equipment that generates or decodes the serial bit stream on the transmission media must be designed for gigabit Ethernet operation.

Because gigabit Ethernet's 8b/10b coding was adapted from Fibre Channel, some diagnostic equipment developed for Fibre Channel will be adapted to gigabit Ethernet. Two important elements of this adaptation are the increases in speed from Fibre Channel's 1062.5 Mbaud to gigabit Ethernet's 1250 Mbaud, and the fact that gigabit Ethernet and Fibre Channel use different sets of control symbols. A network administrator using low-level diagnostic equipment will have to determine whether existing equipment can be upgraded or whether it will need to be replaced.

17.3.2 Coping with data volume

While management protocols are affected in a minor way by gigabit Ethernet, the 1000-Mbit/s data rate will have a significant effect on the suitability of existing platforms to serve as gigabit Ethernet devices. The dramatic increase in network data rates has outpaced the processing capabilities of many management devices. For example, processor-intensive RMON implementations will not be able to stand up to the "fire hose" of data coming from a gigabit Ethernet segment.

This problem is a familiar one to network administrators, since at one time 100 Mbit/s was the fire hose. The introduction of switching presents a similar problem—replacement of a shared hub with a switch multiplies the aggregate data rate—and also compounds the problem by multiplying the number of segments and hence the number of locations in the network that must be probed for data. In response to these changes brought on by switching, many switches now feature hardware that accumulates basic management data, with software that interfaces to standard management systems.

Some management tools will adapt to gigabit link data rates and multigigabit switch data rates by sampling the information flow. For example, the basic Stats Group attributes in the RMON MIB are suitable for hardware implementation on a per-port basis in a switch, but more complex attribute groups defined for implementation in software algorithms must be adapted to reduce the flow of data. Port mirroring helps with the number of points on a switch that must be probed, but it does not solve the fire hose effect of a gigabit data link. This is appropriately implemented using sampling techniques, which will increasingly become a feature in management systems and the switches that feed the data into those systems.

17.3.3 Cable plant

One of the important features of gigabit Ethernet is its ability to operate on media specified in international cabling standards. A significant migration consideration in most networks is the suitability of the cable plant for the new technology. This is an issue for gigabit Ethernet just as it is for any technology that increases the speed of operation.

The increase of gigabit Ethernet's 1000-Mbit/s (1250-Mbaud) data rate results in limitations to the fiber-optic link distances supported. At 100 Mbit/s, 62.5-μm multimode fiber links can be up to 2 km in length. In contrast, a 1000BaseSX 62.5-μm multimode fiber link is limited to 220 m. Single-mode fiber can be used for links up to 5 km in length.

The specification of both 1000BaseSX (short-wavelength optics) and 1000BaseLX (long-wavelength optics) provides for a trade-off between device cost and link distance. As was the case in migration from 10 Mbit/s to 100 Mbit/s, the network planner must determine the suitability of the installed cable plant for gigabit links and select the appropriate gigabit Ethernet PHY variant. In some cases, this will affect the location of the gigabit Ethernet equipment.

17.4   Migration Case Studies

Most initial applications of gigabit Ethernet will be in building and campus backbone networks. These backbone networks are stressed by increasing traffic demands, which gigabit Ethernet is well suited to ameliorate. Gigabit Ethernet backbones will mirror current backbone architectures, but physical media limitations may affect the physical implementation.

Gigabit Ethernet will be applied to both collapsed and distributed switched backbones as well as to routed backbones. Gigabit Ethernet will also be applied to high-performance servers and ultimately to high-performance desktop workstations, just as has happened with 100-Mbit/s LANs.

The following examples illustrate these different types of migration, with examples of switch-to-switch, router-to-switch, and server-to-switch connections. As switches assume more router functionality through Layer 3 switching, the distinctions between some topologies begin to blur, but they provide a fairly rich mix of gigabit Ethernet applications. In these examples, the network operating system does not change, nor does the application software or driver software of existing NICs—that's part of the simplicity of migration to gigabit Ethernet.

17.4.1   Server farm

High-performance servers can now overload dedicated 100-Mbit/s links. A fast Ethernet–connected server cluster is illustrated on the left of Fig. 17.1. The 100-Mbit/s bottleneck can only be fixed with additional

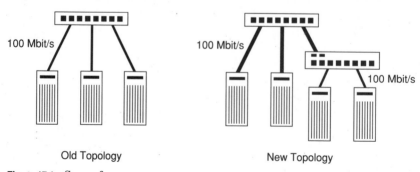

Old Topology   New Topology

Figure 17.1   Server farm.

bandwidth. A natural approach to this is to upgrade the interfaces on the high-performance servers with gigabit Ethernet NICs. These fast Ethernet switches are then replaced with either a gigabit Ethernet switch or repeater. Full-duplex repeaters (also called *buffered distributors*) and half-duplex repeaters both provide shared bandwidth for gigabit Ethernet and are well suited for this type of application.

In some cases it many be desirable to continue the use of existing servers/interfaces. This requires either a switch with both 100- and 1000-Mbit/s interfaces or a 100-Mbit/s switch with a gigabit uplink. The right side of Fig. 17.1 illustrates the latter case.

17.4.2 Switched backbone

Migration of a backbone to gigabit Ethernet is logically an uncomplicated process: Just replace fast Ethernet switch ports with gigabit Ethernet switch ports. Figure 17.2 illustrates this type of topology, with two fast Ethernet backbone switches connecting fast Ethernet workstations and servers and switches that aggregate from 10-Mbit/s links to a 100-Mbit/s uplink.

This network can be upgraded using a gigabit Ethernet switch or repeater connecting multiple fast Ethernet switches with gigabit Ethernet uplinks, as shown in Fig. 17.3. Once the backbone is upgraded, high-performance server farms can be connected directly to the backbone as shown previously, increasing throughput to the servers for high-bandwidth users. The network can now support a greater number of segments, more effectively use bandwidth on segments and, correspondingly, support a greater number of nodes or improve the throughput of existing nodes.

Use of the gigabit port on the collapsed backbone switch or repeater also requires an upgrade of the interface on the other end of the link.

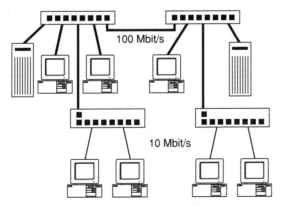

Figure 17.2 Existing fast Ethernet backbone.

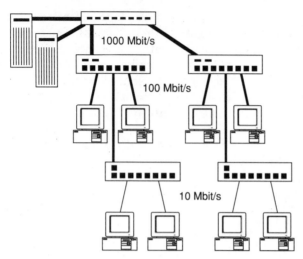

Figure 17.3   Stand-alone switched backbone.

The fast Ethernet switches in Fig. 17.2 must now have gigabit Ethernet interfaces. Depending on the flexibility of the existing equipment, it may be possible to upgrade it with a gigabit interface, or the switch may have to be replaced. As attached servers and workstations are upgraded, migration continues and they are in turn attached to higher-performance switches. This migration approach using stand-alone switches is illustrated in Fig. 17.3.

An alternative migration path for the same logical topology is to use modular chassis-based or stackable switches. This will, in many cases, provide a better result. The best architecture will often be affected by the link distance capabilities of the installed media. Figure 17.4 illustrates migration with a chassis-based switch that supports both 100- and 1000-Mbit/s interfaces. With this approach, the existing fast Ethernet switches can be directly connected to the collapsed backbone without upgrade. Obviously, this will not remove any bottleneck presented by the 100-Mbit/s uplink. Where this is a problem, critical devices can be moved to connect directly to the chassis switch. In addition to better extending the life of the installed equipment, this approach allows a flatter topology with lower latencies.

17.4.3 Desktop

Just as high-performance server technology stresses 100-Mbit/s networks, similar improvements to desktop systems will cause fast Ethernet or FDDI-connected desktops to run out of bandwidth. Gigabit Ethernet NICs are used to provide high-performance desktop computers

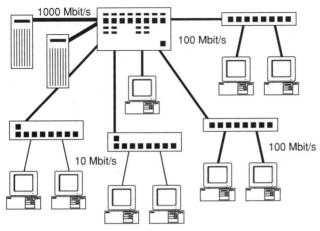

Figure 17.4 Chassis-based switched backbone.

with gigabit Ethernet connectivity. The high-performance desktop computers are then connected to gigabit Ethernet switches or full- or half-duplex repeaters. Figure 17.5 illustrates a shared gigabit-connected desktop system cluster connected into a gigabit backbone switch.

17.4.4 FDDI

The redundancy features of FDDI are an important consideration in many existing networks. Loss of connectivity guarantees a lot of user attention, none of it good. In mission-critical environments it can translate into irrecoverable losses. In recognition of this market need, some gigabit Ethernet equipment supports similar redundancy features.

In most cases, the investment in FDDI fiber-optic cabling can be retained, though long-length multimode links may need to be upgraded to single-mode fiber as explained earlier.

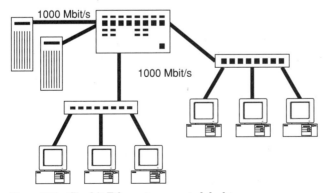

Figure 17.5 Gigabit Ethernet–connected desktops.

An FDDI campus or building backbone can be upgraded by replacing an FDDI concentrator with a gigabit Ethernet switch or repeater and upgrading the connected devices. Migration to an FDDI switch will often be a practical step in the migration to gigabit Ethernet, especially when the upgrade to the connected devices is expensive. Figure 17.6 illustrates a routed FDDI backbone, with the routers dual-homed into a shared FDDI ring.

A chassis switch supporting FDDI-to-FDDI and FDDI–to–gigabit Ethernet switching provides significant flexibility in migration. Figure 17.7 illustrates migration of the router-based FDDI backbone. To preserve the investment in the FDDI router interfaces, they are moved from the shared FDDI ring to dedicated FDDI switch ports. Again, some equipment will support redundancy through dual homing to independent switch ports. This migration also allows high-performance servers to be migrated to gigabit Ethernet. The redundant connections can be used only as needed. As routers are upgraded, they can be moved to gigabit Ethernet just as some servers have moved to get the higher performance of gigabit Ethernet.

17.5 Summary

The diversity of gigabit Ethernet products provides network administrators with a broad range of possibilities. Wherever the local network

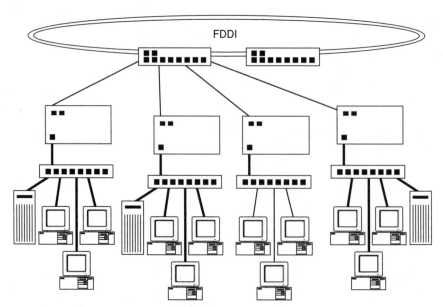

Figure 17.6   Routed shared FDDI backbone.

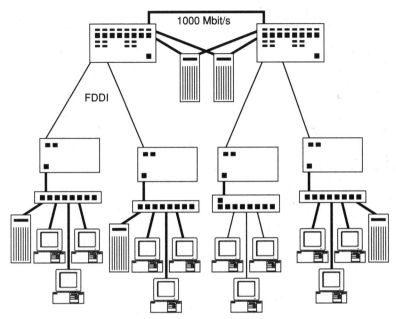

Figure 17.7 FDDI backbone migration.

requires a boost in bandwidth, gigabit Ethernet can be effectively applied. Most network migrations to gigabit Ethernet will include a combination of the cases described in the preceding sections. As is the case with any migration to a new technology, the most important factor in developing a plan will likely be the economic trade-offs; but, unlike some other networking technologies, gigabit Ethernet products allow implementation plans to vary from forklift replacement of equipment to incremental migration—preserving installed equipment and interfaces, maximizing benefits while minimizing the difficulties of migration.

18

Integrating Gigabit Ethernet into Legacy Networks

Anil Rijhsinghani*

Consulting Engineer, Digital Products Group, Cabletron Systems, Inc.

18.1 Overview

Does gigabit Ethernet's speed qualify it as the perfect solution for increasingly bandwidth-starved network environments? Not automatically.

Gigabit Ethernet isn't being installed in a vacuum. Rather, it is going into densely populated, complex computing environments. The vast majority of corporate PCs are already connected to local area networks. These legacy LANs frequently contain vast installed bases of cabling, routers, switches, and bridges.

Fortunately, the similarities between gigabit Ethernet and existing 10- and 100-Mbit/s Ethernet LANs promise to provide a relatively painless solution to the challenge of integrating new bandwidth-hungry applications into existing networks (see Table 18.1). Also, developers have addressed (or are beginning to address) a wide variety of other concerns: gigabit Ethernet should be able to accommodate facilities such as quality of service (QOS) and class of service (COS) and allow efficient management of traffic domains. Gigabit Ethernet's developers are also grappling with issues such as integration with virtual LANs (VLANs), multicasting, and flow control.

---

* Anil Rijhsinghani is a consulting engineer in the Digital Network Products Group of Cabletron Systems Inc. He has worked on the architecture and development of several networking products including switches, hubs, bridges, routers, and servers, and technologies such as SNMP, Ethernet, FDDI, ATM, and Virtual LANs. Anil also represents Digital in standards bodies including the IEEE and IETF. Anil earned his MS in electrical and computer engineering from the State University of New York at Buffalo and his BE in electrical engineering from the University of Bombay.

TABLE 18.1   Part of the Family

Gigabit Ethernet shares many technical characteristics with other forms of Ethernet.

Characteristic	Ethernet	Fast Ethernet	Gigabit Ethernet
Nominal speed	10 Mbit/s	100 Mbit/s	1,000 Mbit/s
Uses CSMA/CD access method	✓	✓	✓
Supports star topology	✓	✓	✓
Uses 802.3z Ethernet frame format	✓	✓	✓
Full- and half-duplex operation	✓	✓	✓
802.3x flow control	✓	✓	✓
Copper cable distance*	100m	100m	25m–100m
Multimode fiber-optic cable distance	2km	2km	220m–500m
Single-mode fiber-optic cable distance	60km	60km	5km

* Projected
SOURCE: Gigabit Ethernet Alliance

The sections that follow examine where gigabit Ethernet fits in legacy LAN environments and describe how network managers can most easily integrate it into their existing operations.

18.2 Legacy LANs

Today, the LAN is ubiquitous. Just a few years ago, only a quarter of an organization's staff was typically connected to a local area network. By 1998, more than 90 percent of staff will be linked, according to surveys from Sentry Research and Analyst Services (Westborough, Massachusetts).

Over the years, such networks have been deployed using a number of different LAN technologies. These technologies coexist today, and each features different speeds, frame formats, forwarding methods, interconnection solutions, and management capabilities.

The most common LAN connection today is shared 10-Mbit/s Ethernet, but that technology is being pushed to the breaking point. Users are demanding more bandwidth—for both old and new applications (see Fig. 18.1). Because there are more users, there is more use of e-mail, file sharing, and groupware applications taking place on the LAN than ever before. And the average size of files being shared has increased by an order of magnitude. Emerging applications such as graphics, audio, and video regularly require the movement of files of 10–100 Mbytes in size.

Because of the bandwidth crunch, desktop devices are increasingly being linked to LANs using switched Ethernet. This shift provides

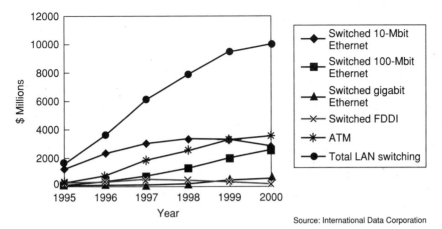

Source: International Data Corporation

Figure 18.1   Soaring bandwidth demand.

higher throughput without the need to change the adapter in the PC. Furthermore, many new desktop computers are being shipped with fast Ethernet cards, which will allow for data transmission at up to 100 Mbit/s. (Most desktops don't transmit data at that rate, however. Fast Ethernet cards are dual-speed cards, and the average LAN continues to operate at the more conventional 10-Mbit/s speed.)

At the backbone, fast and robust Fiber Distributed Data Interface (FDDI) technology has become popular for linking Ethernets. About 50 percent of large corporate networks use FDDI today, according to Sage Research (Natick, Massachusetts). High-end core FDDI switches are now being used to provide the interconnection between edge switches, which in turn provide Ethernet connections to the desktop (often via hubs). Servers accessed by a large number of users are connected directly into the backbone.

While FDDI is a token ring technology at the physical layer, it is very different in nature from the 802.5 access method that runs at 4 and 16 Mbit/s. It is usually used as a "transparent" network: Layer 2 switches and bridges are not visible to the end station, which means that to an end user, a conglomeration of FDDI and Ethernet LANs appears as a single "extended LAN."

FDDI switches use the spanning tree algorithm to connect with one another and to provide relay services. In contrast, 802.5 bridges typically use a method of forwarding traffic referred to as *source routing* that requires the end station to route its packets through a sequence of hops of which it has knowledge.

Finally, some LANs use asynchronous transfer mode (ATM) as a backbone technology, typically in LAN Emulation mode.

18.2.1   Access method

The access methods for each of these LANs are different: Ethernet employs a carrier-sense multiple access with collision detection (CSMA/CD) method, which uses collisions in a half-duplex network to transmit packets; token ring uses a token-passing method to share access to the medium; and FDDI uses a timed-token variation of this method. ATM has a virtual-circuit-based switched topology. A full-duplex mode of Ethernet has been developed for use at both 100 Mbit/s and 1000 Mbit/s; it eliminates collisions and enables communication to take place in both directions over the same link.

18.2.2   Speeds and feeds

The speeds of the various LANs available also differ. Obviously, when a high-speed LAN is connected to a lower-speed LAN, problems can ensue. Switches connecting these different LANs need buffering capability to contain traffic bursts, because the cut-through techniques used to control traffic when LANs of similar speeds are connected can't be used when the speeds don't match. Because even the buffers themselves can be overrun during a sustained peak interval, flow-control mechanisms can be implemented at various layers to slow down traffic where and when necessary.

Legacy LANs have different data transmission characteristics as well. There is variation in the following three areas: packet format, packet size (maximum and minimum), and bit order of the data.

Since the frame formats of different LANs differ from one another, switches that interconnect LANs must use "translation" to move packets from one type of LAN to another. There are two packet formats for Ethernet—Ethernet V2 and 802.3—and each of these requires a different translation mechanism when being converted to other LAN types.

Packet size varies widely among LANs. For instance, Ethernet has both a minimum size [64 bytes including cyclic redundancy check (CRC)] and a maximum size (1518 bytes). FDDI has no minimum size at all, and a significantly larger maximum size (4500 bytes including CRC). Because of these differences, packets sent from FDDI to Ethernet using the IP protocol must be fragmented. Fragmentation, which at one time was done only in routers, is now commonly implemented in Layer 2 switches as well. The Ipv4 protocol has a fragmentation procedure defined in RFC 791. Conversely, there are protocols—such as Internetwork Packet Exchange (IPX) and Ipv6—which do not allow for fragmentation.

The bit order of data also varies according to the different networking technologies. Ethernet uses an approach called least significant bit (LSB) first; FDDI uses most significant bit (MSB) first. Accordingly, switches that interconnect these LANs must rearrange the bits so they are in the appropriate order. Switches that connect 802.5 and Ethernet/FDDI networks may need to have the bit order of addresses embedded within the data field [such as for Address Resolution Protocol (ARP) packets] reversed for the addresses to be interpreted properly at an end station.

18.2.3 Switched versus shared

Network managers today have to make a choice: dedicate a LAN segment to a user or allocate a number of users to share a LAN. In other words, in most cases, they must choose between a switched and shared Ethernet environment. In the traditional 10-Mbit/s Ethernet hub environment, users share a port and effectively compete for available bandwidth using the CSMA/CD method. This is a low-cost solution, but one in which users compete for network access; and the more concurrent users on a LAN, the lower the effective bandwidth.

Users are unlikely to put up with this state of affairs. Indeed, most corporations already have moved to add switches to the network mix. Switches multiply bandwidth by the number of ports provided. The advantage is that this results in a collision-free environment. Because bandwidth is dedicated, four switched Ethernet ports provide four times the throughput of a shared network. More than 60 percent of all installed network ports will be switched ports, according to Business Research Group (BRG, Newton, Massachusetts), a market research firm.

While switched Ethernet has been effective in increasing bandwidth at the desktop, it doesn't help to mitigate the bottleneck for servers, which typically have connections to several users and need to send traffic at a much higher rate than individual stations. One solution is a fast Ethernet backbone. However, fast Ethernet is already available on some LANs, and its 100-Mbit/s capability is already being pushed to the limit.

An FDDI backbone, with servers connected via FDDI, is another alternative. FDDI is a mature and reliable solution, not constrained by Ethernet's distance limitations. However, FDDI's performance capabilities are also being pushed to the limit by fast Ethernet at the desktop. Hence the need for gigabit Ethernet as part of a corporate network.

18.2.4 Servers

The need for speed in server-to-server connections is intense. High-bandwidth applications at the desktop have forced a paradigm shift: instead of most traffic residing within a workgroup, most network transmissions today are between servers and clients. Dataquest, a market research firm in San Jose, California, says that the traditional 80:20 rule has been inverted. In the past, 80 percent of network traffic stayed in the workgroup and only 20 percent was server traffic. Today, only 20 percent is local.

The fastest connections available today come from gigabit Ethernet adapters. Under optimal conditions, gigabit Ethernet will give servers 10 times the bandwidth of fast Ethernet or FDDI server switches. Upgrading a switched backbone to gigabit Ethernet primarily involves swapping new switches for old. As a network engineering challenge, it is not a trivial problem—but it is a straightforward one. A fast Ethernet backbone switch that aggregates multiple 10/100 switches can be upgraded to a gigabit Ethernet switch supporting 100/1000 switches. This approach means that a network can be changed over to gigabit Ethernet in stages, and that the network can operate at two speeds at once.

Vendors are also shipping gigabit Ethernet adapters, which allow network managers to upgrade the end points in their networks. The latest adapters are designed not only for high-speed access but also for the ability to offload the host by performing zero-copy moves and pre-processing on packet headers. These *intelligent adapters* form the basis for a high-performance network infrastructure.

18.2.5   Layer 2 versus Layer 3

The long-standing debate of bridges versus routers in the networking world will be unaffected by the implementation of gigabit Ethernet: both technologies can be tweaked to handle the higher speeds involved. Those who favor bridges and switches can implement technologies such as virtual LANs and rate limiting to protect lower-speed networks from being swamped by floods of data from gigabit Ethernet.

From the router standpoint, it's simple to port a Layer 3 implementation to gigabit Ethernet. One of gigabit Ethernet's strengths is that it is Ethernet—using the same frame format as the more established versions of the technology. The only difference lies in the cost of the link, which has to be configured for both Layer 2 and 3. Since the cost of porting a Layer 3 implementation to gigabit Ethernet will be relatively low, it is an appealing choice.

18.2.6   Management

As noted, one reason gigabit Ethernet is gaining acceptance so rapidly is that it is still Ethernet, albeit a slightly modified version of the old industry standard. Gigabit Ethernet basically uses the same frame format, frame size, protocols, and management data as Ethernet and fast Ethernet. This is significant for network managers, who will find that they already have the knowledge and tools in place to cope with the new technology.

Simple Network Management Protocol (SNMP) has traditionally been the protocol of choice in managing LANs, though different types of LANs have different troubleshooting and management mechanisms.

The Remote Network Management (RMON) standard has become widely adopted in monitoring and debugging network problems. RMON probes and protocol analyzers can be embedded in the network to provide snapshots of network performance: they can track network traffic, bandwidth utilization, and device interaction. Specialized LAN analyzers can be used to investigate the problems in further detail. However, these devices are expensive and there is a steep learning curve involved in implementing them. Furthermore, it is not clear that the basic devices are fast enough to be upgraded to work in gigabit Ethernet environments. However, test and measurement equipment vendors are rapidly incorporating higher-speed processors—such as digital signal processors—into their devices to cope with the increased use of higher-speed networks.

18.3   New Technology and Applications, Multimedia

Today's users are pushing existing network capabilities to the limit. Increased user access, faster desktop machines, swollen data files, and multimedia applications have increased bandwidth requirements to the point where they threaten to overwhelm infrastructures that have a high-end speed of 100 Mbit/s. Furthermore, the demands for backup in an environment where more and more systems must be in operation around the clock, seven days a week, means that hundreds of gigabytes of data have to be transferred in a vanishing window of opportunity.

Yet, even as customers are pushing the bandwidth envelope, they're demanding more reliability and cost-effectiveness from those infrastructures. Essentially, the network has become the backbone of most businesses. Roughly 75 percent of network managers today consider their networks critical to their organizations, according to the market research firm International Data Corp. (Framingham, Massachusetts).

Emerging technologies will create even greater demands. The drive toward network computers (desktop devices with a minimum of computing power) will put even more strain on the networks: these thin clients need to be in constant contact with the server.

IP multicasting is being increasingly used as a method to broadcast data to a collection of clients in the Layer 3 network; video and voice are potential high-bandwidth users; and there's a huge pent-up demand for videoconferencing capabilities over the network (see Fig. 18.2). In addi-

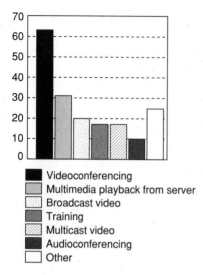

Video conferencing	63
Multimedia playback from server	31
Broadcast video	20
Training	17
Multicast video	17
Audioconferencing	10
Other	25

Source: Infonetics Research

Figure 18.2   Bandwidth drivers.

tion, these applications put limits on latency bounds and need a quality of service surpassing that of traditional data.

Today's networks simply weren't designed for such heavy loads. Higher-speed servers are needed for the network to run efficiently. Some vendors have used hunt groups or data links trunked together in order to achieve these speeds; others have used proprietary high-speed connections between switches. An effort is now under way within the 802.3 group to define a standard method of data link trunking. Gigabit Ethernet satisfies many of these needs for additional bandwidth in an industry standard manner.

18.3.1   IP multicast

New applications that require one-to-many communications use multicasting at the network layer, which in turn translates to multicasts at the datalink layer. A large amount of multicasting, such as a multimedia application might demand, can swamp a LAN not designed to handle it. New protocols designed to handle multicasting include Protocol Independent Multicast (PIM) and Distance Vector Multicast Routing Protocol (DVMRP). Corresponding protocols at level 2 are being designed within the 802.1p group of the IEEE in order to restrict multicasts to segments that have clients that need to see them.

Multicast pruning causes multicast traffic to be limited to the parts of the network where clients have requested such traffic. The packets will be treated like any other traffic by systems without multicast capability—that is, they will be forwarded. To accrue the benefits of multicast pruning, network designers will have to properly place switches that conform to the appropriate protocols.

18.3.2 VLANs

Virtual LANs let network managers configure forwarding domains regardless of the physical topology of the network. The VLAN allows segments across a switched network to be connected by causing switches to attach information to packets denoting the VLAN to which the packets belong. This information, also referred to as a *VLAN tag,* is carried on backbone LANs along with packets. In this context such backbone LANs are referred to as *trunks.* Using the VLAN tag, switches are able to dynamically change the reachability and configuration of LANs and to limit broadcast domains.

Gigabit Ethernet is ideally suited to being a trunk for any edge LANs. Its speed allows it to act as a highway for many attached LANs. The VLAN tag, when added to maximum-sized packets, causes the maximum transmission unit (MTU) to increase. The ability to deal with these larger-than-normal packets is currently being addressed in the development of gigabit Ethernet by specifying the MTU differently when the protocol type denotes a VLAN—as described in draft standard 802.3ac.

In addition, the use of gigabit Ethernet as a VLAN trunk allows the relay of priority information with the use of class of service information in packets. Gigabit Ethernet can also remove the need for translation between 802.5 and Ethernet by encapsulating token ring packets completely within a VLAN header. This allows gigabit Ethernet to be used as an efficient interconnect between 802.5 LANs even when they are running in source-routing (as opposed to transparent) mode.

18.3.3 COS/QOS

Class of service (COS)/quality of service issues have dogged Ethernet for years. Ethernet to date has not had a priority field in its packet header. This means that messages could not be differentiated by priority, even though such a priority field does exist for both 802.5 and FDDI.

In short, when the network gets congested, there is no way of letting an Ethernet network know what is priority mail and what is the electronic equivalent of the U.S. Postal Service's fourth-class bulk mail.

But Ethernet's historical inability to prioritize messages is finally being addressed. With the advent of 802.1p, class of service has been added as a component of the VLAN packet header. There is a 3-bit field in the packet header that denotes the traffic class of the packet. Based on this field, an end station may request a priority and a switch will treat packets in accordance with priority. Classes of traffic that may be differentiated include video, voice, and network-critical applications. This will allow latency to be bounded for packets that need it. The Resource Reservation Protocol (RSVP) will be used at a high level by applications to request bandwidth and latency requirements; this will eventually translate into a class of service contained within the header.

Gigabit Ethernet switches and end stations will be able to take advantage of this priority-based method. But new mechanisms such as RSVP and 802.1p will need to operate across legacy environments, even if the older equipment can't use them. Switches that can't differentiate between different classes of traffic must still be able to forward that traffic to newer network components that can do so. This creates a special difficulty for mechanisms based on 802.1p and 802.1Q: the additional fields needed for priority and VLAN information may increase the packet size beyond the maximum size the legacy systems can handle. If this occurs, end stations may need to have the MTU size reduced so that the aggregate package is comfortably below the maximum packet size.

18.3.4 Full-duplex

Although Ethernet has traditionally been used as a half-duplex access mechanism, the initial use of gigabit Ethernet will be as a full-duplex mechanism. Full-duplex connections establish a point-to-point link between two devices, which allows data to flow in either direction for an aggregate bandwidth of 2 Gbit/s. This eliminates contention on the wire in the form of collisions. With full-duplex, the distance between Ethernet devices varies from 200 m to 3000 m, depending on the type of optical fiber cable being used.

18.3.5 Flow control

When high-speed and low-speed LANs are mixed, excessive traffic from a particular LAN may cause performance degradation in the network. 802.3x has defined a means to control traffic on a particular full-duplex LAN using an XON/XOFF method that allows one side of the link to stop the other side from sending data for a specified period of time.

18.4   Deployment

Gigabit Ethernet will almost certainly not be deployed all at once. Companies will want to avoid the cost and stress of new equipment, training, and support. Even though such issues are typically less serious than might be expected (because gigabit Ethernet builds on existing Ethernet expertise), a cautious, phased approach is still recommended.

One reason for the caution is price. Gigabit Ethernet switches are still relatively expensive, although prices are declining rapidly. By the year 2000, according to projections by the Dell'Oro Group, gigabit Ethernet will be providing a theoretical 100-fold increase in performance over standard Ethernet for just over three times the price per port. In mid-1997 there was a fourfold premium (see Fig. 18.3).

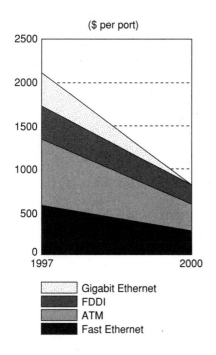

Technology	1997	2000
Fast Ethernet	551	265
FDDI	1727	810
ATM	1373	602
Gigabit Ethernet	2100	800

Source: IDC

Figure 18.3   Gigabit Ethernet to narrow price gap.

Interestingly, the price multiple also reflects real-world performance. Depending on the type of server, its internal bus structure, and the adapter cards, gigabit Ethernet performance ranges between three and five times the performance of fast Ethernet in testing done in mid-1997, according to an article in the October 1997 issue of *Byte* magazine ("Preparing for Gigabit Ethernet," by Mike Hurwicz, page 63). UNIX-based servers showed higher levels of performance than PC-based servers, the article noted.

However, box prices do not completely reflect the total cost of ownership. Relative to other high-speed technologies, gigabit Ethernet offers the potential for significant productivity benefits to IS managers and network administrators. Furthermore, the introduction of servers with the 64-bit PCI bus will boost the internal CPU throughput to 4 Gbit/s and take advantage of gigabit Ethernet's capabilities.

For now, however, gigabit Ethernet is a premium service. For this reason, as well as because of overall network performance considerations, the upgrades may proceed along the following path.

- Backbone (switch-to-switch or switch-to-server connection)
- High-performance server
- High-performance desktop

There's a natural inclination to first look at the backbone as a logical gigabit Ethernet opportunity. However, IS managers may want to consider first installing gigabit Ethernet as a link between high-traffic, high-volume servers. If there's a problem with the first gigabit Ethernet implementation, only a portion of the users and the organization will be affected if a server-to-server link is disrupted. Ultimately, though, gigabit Ethernet will be more popular as a backbone technology.

18.4.1 Backbone upgrade

An existing backbone built with several fast Ethernet switches or FDDI-to-Ethernet switches can be upgraded to use gigabit Ethernet links between the switches. This is logical initial step toward increasing network performance and can be accomplished without disrupting users. The gigabit links allow existing switches to communicate between themselves at speeds an order of magnitude higher than before. This link eliminates a potential bottleneck where backbone bandwidth could not accommodate the traffic between servers. This also allows a larger number of fast Ethernet or FDDI LANs to be added at the next level down.

The advantage of upgrading an FDDI backbone to gigabit Ethernet is that it's already operating on fiber-optic cable. That means that upgrading a centralized FDDI backbone is just a matter of replacing a concentrator with a gigabit Ethernet switch (routers may have to be replaced with switches as well). Upgrading a distributed FDDI backbone means replacing each concentrator and router separately with a switch, but this can be done in a staged fashion (see Fig. 18.4).

Upgrading a fast Ethernet backbone is even simpler: it's simply a matter of replacing one switch with another (see Fig. 18.5). With a fast Ethernet infrastructure, a backbone consisting of multiple switches with a high density of fast Ethernet ports is limited by the speed of interconnecting ports. With full-duplex fast Ethernet connections between switches, the maximum speed is still 200 Mbit/s, which can fall short given the larger number of other ports that can generate traffic that is nonlocal. In addition, servers are limited by the speed of the link into which they are connected. The headroom afforded by gigabit connections effectively expands the already available bandwidth in a fast Ethernet–based network by allowing it to flow more freely and without congestion at bottlenecks such as switch-to-switch connection and server connection. Hence the potential available bandwidth become more usable due to a speedup by a factor of 10 at these critical junctures. This will open up new applications that are currently limited by these bottlenecks while allowing many users to remain connected in the same way they are currently (e.g., with either direct 100-Mbit/s connections to the desktop or switched Ethernet connections).

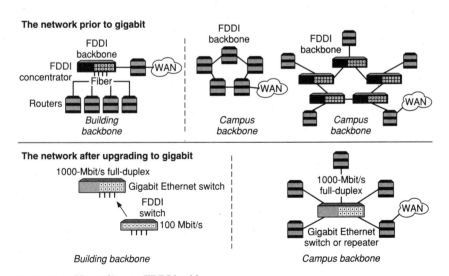

Figure 18.4   Upgrading an FDDI backbone.

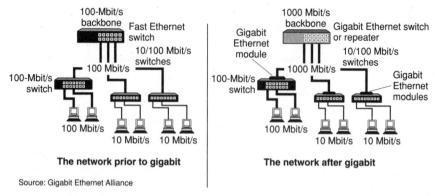

The network prior to gigabit | **The network after gigabit**

Source: Gigabit Ethernet Alliance

Figure 18.5   Upgrading a switched fast Ethernet backbone.

Switches used for gigabit Ethernet interconnection must not only have high performance and be nonblocking (that is, capable of running in worst-case traffic situations), but also must be robust and fault-tolerant. They will be expected to maintain protocol connectivity in the face of excessive traffic and to be capable of limiting floods and storms when they occur. Critical elements of the switch must not be single points of failures; they will be expected to be backed up by redundant elements. Modules must be hot-swappable to retain connectivity as physical configuration changes are made. The performance must scale well as the number of connections increases. Finally, gigabit Ethernet switches must be field upgradable to circumvent unforeseen problems.

18.4.2   High-performance servers

Gigabit Ethernet backbones alone don't solve the bandwidth problem; they just push it down the network. Once the gigabit backbone is in place, servers eventually start to feel the strain of the higher performance of the overall network. Traffic congestion will occur at server links that are running at 100 Mbit/s, much as it occurred in the connection between switches originally. While servers with multiple connections can be an initial solution, the simplest and best performance solution is to connect the server directly into a gigabit Ethernet switch port. This will increase capacity greatly, and also ease network administration in the process.

As mentioned earlier, intelligent adapters will have to be deployed in servers to make use of the increased bandwidth capacity. A design that does not offload the host CPU will in all probability swamp the host system, thus reducing its ability to take advantage of the speed. Offloading can occur at many layers: for example, processing and validation of the packet at the network and transport layer, as well as to analyze and appropriately handle VLAN and CoS characteristics. Elimination of data copying and duplication in transferring the data to the host will also increase efficiency.

18.4.3 High-performance desktop

As gigabit Ethernet becomes more widely deployed, it is conceivable that desktops with very high bandwidth requirements will be limited by 100-Mbit/s connections. These can be migrated to gigabit Ethernet switch ports or repeater ports as appropriate in the later stages of migration. However, this may well begin the whole cycle over again—with desktop bandwidth demand driving backbone performance requirements even higher.

18.4.4 Physical wiring

Currently, gigabit Ethernet over fiber has been defined for use over multimode fiber (MMF) as well as single-mode fiber. The maximum distance for 62.5-μm multimode fiber with 850-nm wavelength (the most commonly installed fiber in the United States for FDDI, according to *Byte*) is 220 m for gigabit Ethernet, while 50-μm MMF can handle distances of up to 550 m. With single-mode fiber, it is possible to have distances of 5 km between MACs. Note that some FDDI implementations have runs as long as 2 km, so check the installation plans before installing gigabit Ethernet equipment on existing FDDI cables.

Gigabit Ethernet over copper is still being defined, and various schemes are being evaluated to support link distances of 100 m for half-duplex mode using the CSMA/CD mechanism.

18.4.5 Timeline

The initial installation of gigabit Ethernet ports must take into account not only the benefits of gigabit Ethernet, but its limitations as well. Gigabit Ethernet is still an emerging technology, with standards that have not yet been granted final approval by standards bodies. The technology has distance limitations that will affect where it can be deployed. And not all the products are at the same level of maturity.

Typically, net managers will want to launch a pilot project first, perhaps involving several ports in a small gigabit Ethernet backbone, before upgrading wholesale to the new technology. Network managers should expect to spend about two months phasing in and testing this initial project, paying particular attention to interoperability among products from different vendors. Once the initial installation has been given a green light, additional implementations should proceed at a faster pace.

19

Scaling Ethernet Networks

Bob Gohn* and Paul Sherer†

3Com Corporation

19.1 Introduction

The pressure introduced by today's bandwidth-hungry applications and the growing size of corporate LANs are both driving the demand for increased network capacity. In addition to the need for raw bandwidth, new types of applications also require new levels of network control.

Historically, Ethernet at 10 Mbit/s and—recently—fast Ethernet at 100 Mbit/s have been deployed at the edges of networks for desktop and server connectivity. For building network backbones, however, greater speeds became necessary, giving rise to technologies like ATM. Gigabit Ethernet, however, now offers the option for network managers to scale their networks with Ethernet-only solutions. Using gigabit Ethernet, administrators can build scalable networks by incorporating Ethernet links with different speeds at appropriate places in their networks.

This chapter defines an Ethernet bandwidth hierarchy. It examines the various speeds of Ethernet available, as well as the technologies that underlie them, and considers where these different bandwidth levels may be deployed in a network. It also explains how to assign the dif-

---

* Bob Gohn is a product line manager in 3Com's Switching Systems Division, and 3Com-wide gigabit Ethernet program manager. Working in the networking industry since 1982, Bob has held a number of development and marketing positions involving technologies ranging from early packet radio systems to internetworking and LAN switching. Prior to 3Com, he worked for SMC, Raycom Systems, Netways, and Hazeltine. Bob holds an MSEE from Polytechnic University and a BEEE from the State University of New York at Stony Brook.

† Paul Sherer is vice president of technology development for 3Com Corporation. In this position he is responsible for bringing new networking technologies from conceptualization into product development.

Prior to this position, Sherer was director of technology development and director of systems architecture for 3Com's Network Adapter Division. He joined 3Com in 1984, and

ferent levels of bandwidth to different parts of the network. Next the chapter discusses some procedures net managers can use to categorize both their applications and users and make it easier to apply the various Ethernet options in networks of all sizes. Finally, the last section of the chapter provides a process, or formula, net managers can use to address design decisions when building fully scaled Ethernet networks.

19.2   Defining the Bandwidth Hierarchy

The advent of gigabit Ethernet technology is a response to new classes of applications that demand greater bandwidth, and it reflects the clear desire of network designers and managers to simplify network infrastructures. Running in parallel with the development of higher-speed Ethernet technologies is the widespread acceptance of LAN switching technologies.

But deciding which of the multiple-speed Ethernet solutions should go where in a network (not to mention whether a shared or a switched implementation should be used) is tricky. One way to make things easier is to use a concept called the Ethernet bandwidth hierarchy (see Fig. 19.1). This bandwidth hierarchy groups the various speed and switching options into four levels:

Level 1—shared 10 Mbit/s

Level 2—switched 10 Mbit/s and shared 100 Mbit/s

Level 3—switched 100 Mbit/s and shared 1000 Mbit/s

Level 4—switched 1000 Mbit/s

The bandwidth available at each level is best used for optimizing different parts of the network. Note that at levels 2 and 3 the modalities of Ethernet provide comparable—though not identical—levels of performance; this issue is discussed later in this section.

Before examining these options in detail, it is useful to review some of the basic media options and limitations for each of the Ethernet speeds and how this affects their use. It's also important to consider the

---

has been involved with the architecture and definition of several 3Com technologies including the Network Driver Interface Specification (NDIS), Demand Protocol Architecture (DPA), Heterogeneous LAN Management (IEEE 802. 1b), Softhub/linkWatch, Priority Access Control Enabled (PACE) technology for multimedia over Ethernet, the Parallel Tasking Architecture (used in the EtherLink III product family), and fast Ethernet (IEEE 802.3u). He helped found the Fast Ethernet Alliance and served as chairman. An inventor, he holds several U.S. patents related to network technology and has several pending U.S. and international patents.

Sherer holds a bachelor of science degree in electrical engineering from the University of Alabama.

Switched 1000 Mbit/s	Level 4
Switched 100 Mbit/s Shared 1000 Mbit/s	Level 3
Switched 10 Mbit/s Shared 100 Mbit/s	Level 2
Switched 10 Mbit/s	Level 1

Figure 19.1   Ethernet bandwidth hierarchy.

relative merits of shared and switched implementations at each of the Ethernet speeds (see Fig. 19.2).

19.2.1   10-Mbit/s Ethernet

Ten-Mbit/s Ethernet, which emerged in the late 1970s, was originally designed to run on two kinds of coaxial cable: IEEE 10Base5 or "Thick Ethernet," and IEEE 10Base2 or "Thin Ethernet." Using carrier-sense multiple access with collision detection (CSMA/CD) to resolve contention, a single cable was shared among various stations.

When the networking revolution took off in the late 1980s, unshielded twisted-pair copper media emerged as another option. The IEEE 10BaseT standard defines a device called a repeater that allows the use of Ethernet in a star-wired topology, where the repeater acts as a hub that links individual stations. This approach generally allows the use of voice-grade (Category 3) unshielded twisted-pair (UTP) cabling.

The advent of the star-wired hub essentially recast the deployment of Ethernet and spawned much more structured networks. The concept was also extended for fiber media via the 10BaseF standard, which allows for connections of up to 2 km in length.

	Ethernet	Fast Ethernet	Gigabit Ethernet design goals
Data rate	10 Mbit/s	100 Mbit/s	1000 Mbit/s
Cat 5 UTP	100 m (min)	100 m	100 m
STP/coax	500 m	100 m	25–100 m
Multimode fiber	2 km	412 m (hd) 2 km (fd)	220–500 m
Single-mode fiber	25 km	20 km	5 km

Figure 19.2   Ethernet technology comparison.

19.2.2   100-Mbit/s Ethernet

In the early 1990s, fast Ethernet arrived to provide Ethernet capabilities that, at 100 Mbit/s, were 10 times faster than the previous Ethernet standard. Because it provides higher bandwidth at low costs—particularly when compared with FDDI, the only other serious 100-Mbit/s LAN technology—fast Ethernet was quickly adopted for both switch and server interconnect applications. Fast Ethernet is usually run over high-quality Category 5 UTP cable. The IEEE 100BaseTX standard, which defines how 100-Mbit/s Ethernet runs over Cat 5, is similar to the 10BaseT standard in that it defines a topology in which repeaters share bandwidth in a star-wired topology. Fast Ethernet also can run over fiber—an application defined in the IEEE 100BaseFX standard. Recently the 100BaseT4 standard has also been defined, allowing fast Ethernet to run over Category 3 UTP cabling.

In today's networks, dedicated (or point-to-point) fast Ethernet connections are often used to link slower Ethernet segments. This has led to the development of full-duplex technology, which disables the CSMA/CD protocol to allow traffic to flow in both directions on a link at the same time, thus doubling the throughput of the connection.

19.2.3   Gigabit Ethernet

Like fast Ethernet, the latest Ethernet technology, gigabit Ethernet, is designed for point-to-point connections between a switch and a server and between different switches in a network. As such, most of the early implementations of gigabit Ethernet will be in full-duplex mode only. Because CSMA/CD is quite difficult to employ at gigabit speeds and limits the topologies and distances possible, it will most likely not be used or needed with gigabit Ethernet.

In another departure from slower Ethernet technologies, gigabit Ethernet, at least initially, will rely exclusively on fiber-optic cabling rather than unshielded twisted-pair copper, which has trouble achieving gigabit data rates. The limitations of copper may decrease over time, but it is not yet a viable option for use in gigabit Ethernet networks.

19.2.4   Switched connections or shared?

The evolution of Ethernet bandwidth has run parallel to the ongoing migration from shared to switched networks. Switched configurations, unlike shared ones, guarantee a dedicated Ethernet connection for each network user. Segments, and their users, can also be switched at various speeds. For the network designer, the question is: should a network be completely switched, or should it have some mix of switched and shared links? Are switched 10-Mbit/s connections to each desktop in a segment better than a single shared 100-Mbit/s connection?

In a segment consisting of 24 stations that share a fast Ethernet connection, each station can access to up to 100 Mbit/s of bandwidth. Hence, if station 1 is talking to station 24 and no other stations are communicating, stations 1 and 24 can communicate at a rate of 100 Mbit/s per second. However, if all stations are communicating at a relatively even rate, they must share bandwidth. This means each station gets only 1/24th of a 100-Mbit/s connection, or roughly 4 Mbit/s. In effect, though the peak bandwidth may be 100 Mbit/s, average bandwidth might be as low as 4 Mbit/s.

In contrast, for almost the equivalent costs, a switched 10-Mbit/s connection can be installed for that segment. Although peak bandwidth for each station is limited to 10 Mbit/s, every station is assured of a dedicated 10-Mbit/s link. In this instance, the safest course for the designer is to assume the worst-case scenario of a fully utilized network and compare the average bandwidth of 4 Mbit/s per user in a shared system to the average of 10 Mbit/s per user in a switched system.

A key advantage of the switched approach is its greater level of bandwidth control and management at each desktop. Since each port is dedicated to an end station, filters can be applied to every port. The degree of control offered by switching is much more difficult to achieve in a shared network. Perhaps the key consideration when deciding between switched or shared approaches is the applications used in a network; it is often the case that the applications determine which technology is preferable.

Clearly, while switched and shared modalities of levels 2 and 3 of the Ethernet bandwidth hierarchy may provide equal amounts of aggregate bandwidth, they offer different capabilities. In Sec. 19.3, this chapter addresses how each level of the hierarchy can be used to optimize various parts of the network and meet different needs.

19.2.5   Devices at each level

Just as Ethernet speeds can vary by a factor of 100, there is a wide range of devices and applications that operate best at various niches on the Ethernet bandwidth hierarchy. Devices that switch at gigabit speeds, for instance, will obviously be more costly but more suitable in network backbones than devices operating at shared 10 Mbit/s. At the desktop, network interface cards (NICs) offer either 10- or 100-Mbit/s connections and therefore would operate at levels 1 through perhaps 3 in the bandwidth hierarchy. Servers are likely to operate on either switched 10 Mbit/s, shared 100 Mbit/s, switched 100 Mbit/s (levels 2 and 3), or, increasingly, switched gigabit speeds (level 4). Repeaters could operate at levels 1, 2, or 3, depending on the technologies in use. Edge switches would provide performance at levels 2 through 4, while core switching would be used primarily at levels 3 and 4. These are the indicators to consider when constructing a migration path for higher-speed technologies.

19.3   Matching Networking Needs to the Ethernet Bandwidth Hierarchy

Each level of the Ethernet bandwidth hierarchy meets particular needs within the typical network infrastructure. No single level is appropriate for all places in the network. For example, level 1 is widely used for desktop applications, while levels 2, 3, and 4 provide the bandwidth required for data center aggregation. To help address the issue of how to deploy available Ethernet technologies, this section breaks down typical network architectures into five basic areas and examines the levels of Ethernet most appropriate for each area (see Fig. 19.3).

19.3.1   Client access

Almost all data in a network originates at or is destined for a desktop. The desktop is physically connected to the network with a NIC installed inside the desktop system. Therefore, the NIC is essential in determining the performance of the desktop. The criteria for selecting NICs are their performance, their cost, the type of bus within the desktop computer, and, increasingly, the system software that supports the NIC's participation in the network.

Depending on the performance needs, it is appropriate to use levels 1, 2, or 3 of the bandwidth hierarchy at the desktop. Most desktops today have shared or switched 10-Mbit/s connections, though 100-Mbit/s links

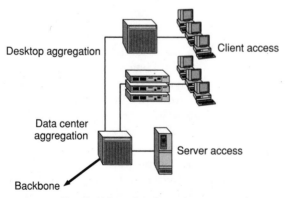

Figure 19.3   Bandwidth evaluation points.

are on the rise. Gigabit-speed NICs will be available, but they will be needed only for rare and extremely sophisticated applications.

To create a scalable Ethernet network, client systems should be equipped with 10/100-Mbit/s NICs. A 10/100-Mbit/s NIC can automatically detect and function at the fastest network speed available. Recently, the cost difference between 10-Mbit/s and 10/100-Mbit/s NICs has become so small that the latter is a prudent buy, even if an impending need for 100-Mbit/s operation is not immediately anticipated.

Since the purpose of installing a 10/100 NIC is to allow for future performance increases at the desktop, it makes sense to connect the NIC to a high-performance input/output (I/O) bus. The most common high-speed I/O bus in today's desktops is PCI. Ten Mbit/s is sufficient for desktops with ISA buses and 486-class CPUs. Desktops with peripheral connect interface buses and Pentium®-class or higher CPUs can benefit from 100-Mbit/s connections.

19.3.2   Desktop aggregation

In most network configurations, devices at the network edge might aggregate as many as 200 or as few as 8 desktops. The choice of technology here is critical—not only because it has one of the most direct impacts on network performance as perceived by end users, but also because reasonably large networks have many of these devices and they constitute a significant acquisition cost.

There are many choices for this part of the network within the bandwidth hierarchy, starting with level 1.

The traditional approach using level 1 is to have repeaters distribute shared 10-Mbit/s bandwidth to desktops. This solution offers the lowest cost, at about $50 per desktop connection, and is available with or without embedded management. Of course, shared 10 Mbit/s at level 1 of the hierarchy also provides the least bandwidth per user. Even for networks with low bandwidth demands it is unlikely that any more than 50 users per Ethernet segment would be desirable; thus, in a reasonably large network, an amount of switching will still be needed.

At level 2, switched 10 Mbit/s has been considered the state-of-the-art technology for desktop aggregation for some time. This solution uses edge switches to ensure that each user gets a 10-Mbit/s link. With or without embedded management capabilities, such as statistics gathering and RMON, switched 10 Mbit/s features a low cost of less than $80 per port.

Shared 100 Mbit/s, also at level 2, can be ideal for high-speed workgroups of 5 to 20 connected desktops. Available at costs approximately equal to those of switched 10 Mbit/s, it also is cost effective. Shared 100 Mbit/s offers a higher peak bandwidth than switched 10-Mbit/s technology, but may provide less aggregate bandwidth depending on the number of stations on the segment. Like shared 10 Mbit/s, shared 100 Mbit/s relies on repeaters.

For level 3 of the hierarchy, switched 100 Mbit/s is the emerging state-of-the-art technology. It is priced under $300 per port at this writing, and costs are expected to decrease over time. Switched 100 Mbit/s typically has embedded management capabilities and provides a dedicated 100-Mbit/s connection for each end station.

As for level 4, 1-Gbit/s Ethernet will not be a common desktop aggregation technology except in rare instances to provide connectivity for clusters of high-performance workstations running very bandwidth-intensive applications.

19.3.3   Server access

Server access is another key network area that can impact overall performance, and there is more involved with evaluating network server performance than the speed of the network interface. At a glance, it may seem that computers are getting faster at the same rate at which networks are getting faster, so that delivery of data to the application should not be a problem. However, it is essential to remember that computers' I/O buses and associated system memory have not always kept pace. In fact, many problems associated with turning good network throughput into equally good application-to-application throughput can be traced to the servers' I/O bus and memory bandwidth.

The overhead imposed on the server's performance by the handling of data to and from the server's NIC is another major issue. It is critical to ascertain that the NIC will not constrain performance between the network and server. Here are some guidelines for doing so.

Generally, application servers that provide significant services beyond file and print servers can sustain much higher levels of bandwidth usage. To take maximum advantage of such a server, it is reasonable to consider a gigabit Ethernet connection.

In 1998, new PCI systems will be able to double the width of the data path from 32 to 64 bits and increase the maximum clock rate from 33 to 66 MHz. Either of these changes would double the burst bandwidth of the bus from slightly over 1 Gbit/s to slightly over 2 Gbit/s. Applying both speed enhancements improves burst bandwidth to more than 4 Gbit/s. Over time, this indicates that gigabit Ethernet will increasingly become the server connection of choice.

Currently, the PCI bus operates at a burst bandwidth slightly over 1 Gbit. Generally, bus burst bandwidth should be at least twice the network bandwidth and preferably much higher. This apparent excess bandwidth compensates for other devices accessing the I/O bus and for overhead and inefficiencies in using the bus. For matching network bandwidth to server performance, the following guidelines are based on the architecture of the server:

- ISA—1–10 Mbit/s
- EISA—10 or 100 Mbit/s
- PCI (32-bit)—100 or 1000 Mbit/s
- PCI (64-bit)—1000 Mbit/s

Today, most new servers are installed with at least 100-Mbit/s connections. In addition, servers typically are no further than 100 m from the network access point.

19.3.4   Data center aggregation

The data center is where server connections and the aggregated desktop meet. "Downlinks" from the desktop aggregation devices located in wiring closets are fed to the data center, where servers are located. Since they reside at key junctions in the network, underpowered data center aggregation devices can have a negative impact on the performance of an entire enterprise. As a result, they require another order of magnitude of bandwidth if they are to operate without bottlenecks.

Shared 100-Mbit/s repeaters are rarely used as data center aggregators, except in relatively small networks. Fast Ethernet (100-Mbit/s) switches are most common, but shared gigabit Ethernet buffered repeaters—also known as *buffered distributors*—will provide low-cost gigabit solutions. As with repeaters, the nodes attached to a buffered repeater share the same bandwidth, and packets are sent to all attached nodes. Unlike with repeaters, all connections to the buffered distributor are full-duplex, in conformance with the IEEE 802.3x standard (in a traditional repeater, the connections are half-duplex). The distributor uses the flow-control mechanism of 802.3x to control when an attached node is able to transmit packets. For maximum performance, switched 1000-Mbit/s switches will provide the highest bandwidth to core switches, but at the highest cost.

19.3.5   The backbone

The backbone, of course, aggregates all data centers and is where the pressure for bandwidth is greatest in most networks. Only switched 100 Mbit/s (level 3) or switched 1000 Mbit/s (level 4) provide the necessary bandwidth to ensure scalability and performance for bandwidth-hungry applications like multimedia. Networks with shared 100-Mbit/s or switched FDDI backbones must make the migration to adequately deliver today's services. Generally, backbone connections are limited to less than 2 km.

19.4   Scaled Ethernet Design Rules

Network architects should keep this simple maxim in mind: *network needs determine network design.* Thus the first task in designing a network is to examine the network's goals. That can be easier said than done, however, and requires a strong knowledge of applications and the workgroups that will constitute the network.

19.4.1   Application classes

A useful way to identify available applications and their specific networking requirements is to divide applications into four classes. Not all real-world applications fit cleanly into these groups, but they offer valid guidelines for satisfying most networking needs and can be extremely helpful.

1. *Office automation.* Office automation programs include word processing, spreadsheet and financial applications, and integrated packages like Microsoft Office. They are common in networks primarily used as media for sharing such applications as e-mail, electronic messaging, and file and print services. Bandwidth demands on this type of network have traditionally not been very high, but they are now rising due to the increasing graphical content of the data. Multimedia, for example, is creeping into office automation applications; graphical images and presentations are routine in the workplace. Graphics are appearing even in e-mail.

    These bandwidth-hungry applications and their large files are straining the capacities of many networks from the desktop to the backbone.

2. *Applications run by technical users.* Technical users typically run high-end, UNIX-type workstations in environments, such as software development, where bandwidth demands are much greater than in traditional automated offices. Unlike users of office automation networks, technical users are less likely to obtain all their data from centralized servers and more likely to share data among specialized groups of peers.

    The desktops of technical users tend to be much more powerful than typical desktops. Many of these machines can drive 100-Mbit/s or even gigabit connections; therefore, the bandwidth needs for these users are growing at a faster rate than in the office automation world.

3. *Multimedia / multicast applications.* Certain user communities, such as those in the financial industry, require an emerging application called *multicasting*. Multicasting allows one station to broadcast data, usually multimedia-like video streams, to any number of desktops dispersed throughout the network. On the trading floors of financial institutions, for example, traders need ongoing CNN broadcasts delivered to their stations to keep abreast of breaking news stories that may affect their markets. Another example is video-based training applications, where an instructor at one station addresses users at desktops scattered throughout the network.

Multicasting traffic differs from the point-to-point connections of classical network traffic patterns because traffic is a constant stream originating from one location and is not constrained to any one region of the network.

4. *Integrated voice / video.* Integrated voice and video applications enable users to integrate even more sophisticated levels of multimedia services into a network. These applications include telephony and desktop videoconferencing—which is touted as an advanced office productivity tool.

Integrated voice/video applications, however, place new demands on the delay sensitivity of data traversing the network and introduce new traffic patterns. The increasing popularity of the World Wide Web and the integration of Web browsers into desktop applications and operating systems, for example, means that even the most remote network station might require access to advanced multimedia services.

Due to the applications noted in these four groups, design issues have escalated beyond bandwidth and now include the speed at which the network infrastructure can process delay-sensitive traffic.

19.4.2 Network user models

Having examined the four classes of applications, we will now briefly review the range of users these applications serve.

Light user. The light user creates roughly 10 Mbits of total network traffic an hour. The light user accesses smaller data items, which tend to create an average of 100 Kbits of total network traffic per access.

Medium user. The medium user is an order of magnitude more demanding than the light user. The medium user creates 100 Mbits of total network traffic in an hour and accesses data items that create an average of 500 Kbits of total network traffic per access.

Heavy user. The heavy user is an order of magnitude more demanding than the medium user. The heavy user's activities create 1 Gbit of total network traffic an hour. A heavy user will access data items that generate an average of 2.5 Mbits of total network traffic per access.

19.5   Designing a Fully Scaled Ethernet Network

One of the most critical steps when designing any network is to determine the future needs the network will be expected to meet. Questions such as how long the network is expected to last must also be asked and answered. The need to future-proof the network for new applications must also be weighed and appropriately woven into the network design.

Below is a process that network managers can use to make design decisions when building fully scaled Ethernet networks. The sequence is as follows: Determine the speeds that client systems require; choose server access; identify how data will flow between desktops and servers; and design the appropriate interconnect infrastructure.

1. *Determine the client access model.* Desktops are generally the starting and end point of all transmissions. Therefore, anyone designing a fully scaled Ethernet network must first decide which technology desktops require. To accomplish this, network designers must thoroughly evaluate desktop users' needs, including the kinds of applications they will be using and where on the network their data is located.

For office automation applications, shared or switched 10-Mbit/s Ethernet links (levels 1 and 2) will most likely suffice. Power users, such as a technical documentation group that generates graphics, might require shared 100-Mbit/s or even switched 100-Mbit/s connections.

When it comes to multimedia, the available aggregate bandwidth must be considered. In these cases, even relatively modest users will probably need switched connections of either 10 or 100 Mbit/s. Switching is important because it will provide capabilities like IGMP snooping, which, for IP multicast protocols, filters out the multicast stream from desktops that are not intended recipients. As a result of this capability, switched 10 Mbit/s is a better solution than shared 100 Mbit/s within level 2 of the hierarchy because switches will feed the IP multicast packets only to desktops participating in that session. CNN broadcasts, for example, will be distributed only to users who want them, thus reducing traffic elsewhere in the network and conserving bandwidth.

In environments that rely on integrated voice/video applications at the desktop, 100-Mbit/s switched devices are the preferred solution. This high-speed switching provides users with both the raw bandwidth and the control to ensure that transmission delays are within the tolerance of video and/or voice traffic.

Another key consideration when configuring desktop connections is the NICs used in each end station. Since the price difference between 10- and 100-Mbit/s NICs is small and shrinking, 10/100-Mbit/s NICs should be installed because they provide the most scalable solution in switched environments. 10/100 NICs enable workgroups with desktops of various capabilities or with limitations in the existing cabling infrastructure to migrate to 100-Mbit/s links over time. If parts of the workgroup are using Category 3 cabling, which typically cannot run 100 Mbit/s, switches will deliver 10-Mbit/s links to those desktops while providing 100-Mbit/s speeds to the others. This approach is ideal for gradual migrations. It also provides investment protection in existing equipment as well as future-proofing to satisfy tomorrow's networking demands.

2. *Server placement.* Server configurations depend less on applications—though they influence the size of selected servers—and more on ensuring that server access is bottleneck free. Servers are expensive, highly optimized machines designed to process applications and move data on and off disks very quickly. The investment in a server is undermined whenever its network connection is slower than the speed of the server itself. To avoid congestion, the network speed should be carefully matched to the server's capabilities. Therefore, a main criterion for determining the network connection to the server is the server itself.

Thirty-two-bit PCI servers, as well as slower machines, operate well with 100-Mbit/s connections. At this time, however, the newest generation of servers features 64-bit PCI buses that can support gigabit speeds. For these devices, gigabit Ethernet connections would be ideal.

Another consideration is whether servers are distributed throughout the network or centralized in a server farm. When servers are concentrated in one location, the issue becomes whether the bandwidth feeding the servers is peak or aggregated. In addition, server utilizations will impact the decision to provide switched or shared connections, even at gigabit rates.

If a server is utilized to back up data, then a shared connection makes sense because peak, not aggregate, bandwidth is all that is needed. However, if bandwidth demands are driven by client access to servers, then dedicated switched connections, whether 100 or 1000 Mbit/s, are most effective. Because fast Ethernet is now relatively inexpensive, 10-Mbit/s connections to servers, while still common, are not recommended for future installations.

3. *Data locality.* The next step in creating the network architecture is to determine the locality of the data. Designing networks is easier when the traffic patterns are known. For example, if it is known that users access data local to their workgroup or that a workgroup routinely obtains data from a specific server, the network can be designed so that users are close to their data and dedicated paths are provided. When the

locality of data is established, the network can be configured so a high percentage of the traffic will stay local within workgroups. Since little traffic will travel outside individual workgroups, little traffic needs to traverse the backbone. Therefore, the backbone's bandwidth can be a relatively small percentage of the aggregate desktop bandwidth.

On the other hand, if the locality of data is uncertain and users need routine access to information anywhere on the network, then the backbone must accommodate more traffic and its bandwidth must be a significant percentage of the aggregate desktop bandwidth.

4. *Interconnect infrastructure.* Once the issues of client access, servers, and data locality are resolved, interconnect infrastructures must be established between these devices. The first consideration is data center aggregation. If only 100 Mbit/s is required to the servers and client access is shared 10 Mbit/s or limited switched 10 Mbit/s, then 10-Mbit/s switches with 100-Mbit/s downlinks at the data center are perfectly adequate. However, if there are multiple 100-Mbit/s server links and switched 10 Mbit/s or shared 100 Mbit/s at the desktops, then the data center aggregation needs to be fast Ethernet switching, with perhaps the ability to downlink with gigabit Ethernet.

Finally, if there are some extremely high-speed servers in a network with switched 10/100 Mbit/s to each desktop, then the data center aggregation point needs to provide gigabit Ethernet switching.

Once the data center aggregation points have been resolved, the final issue is backbone bandwidth. This is the next step up in the bandwidth hierarchy. To achieve it, the network backbone needs to aggregate the various data centers throughout the enterprise. Typically, this will be accomplished with fast Ethernet or gigabit Ethernet, depending on the locality of the data and the technology used.

19.6 Conclusion

With the advent of gigabit Ethernet, network administrators can now meet the demands of today's bandwidth-intensive applications with an all-Ethernet solution. Spanning the bandwidth hierarchy from shared 10 Mbit/s to switched 1000 Mbit/s, Ethernet offers migration paths that simplify management and deployment due to the widespread familiarity with the technology. The key to scaling Ethernet networks is understanding how bandwidth needs differ at each part of the network, from the desktop to the backbone, and knowing that the best implementation depends on what users will actually need.

6

Gigabit Ethernet Design Tutorials

Purpose

To provide design tutorials that describe how to optimize the performance of either the gigabit Ethernet network itself or the equipment attached to it.

What Is Covered

The most dangerous myth about gigabit Ethernet is that it is a plug-and-play technology. It isn't. In order to obtain the performance they are looking for, network managers will have to do a lot more than plug this kit together and turn it on. As with any new, high-speed LAN technology, managers will have to tweak and tune these networks. How to do that is the subject of this section of the book.

Chapter 20 provides a straightforward lesson on how to build a gigabit Ethernet network. Chapter 21 takes a slightly different tack, concentrating not on the network but the servers attached to it. This tutorial starts from the premise that today's Intel CPU-based servers cannot keep up with or fill a gigabit connection. It goes on to explain how, by carefully choosing the right server components, net managers can squeeze the maximum throughput possible out of today's Intel processor technology.

Chapter 22 also focuses on the attached equipment rather than the network proper. In this case, it discusses ways to optimize the performance of users' end stations—in particular by choosing the right types of adapters.

Chapter 23 also has a specific purpose; it describes how to build a network that is designed to carry time-sensitive traffic.

Chapter 24, on the other hand, goes to the opposite extreme, describing how to build a data utility—simply, a dreadnought of a data network with so much spare capacity that it makes the bandwidth rationing of today's local area networks a thing of the past.

Finally, Chapter 25 provides essential guidance on how to successfully manage the gigabit Ethernet network.

Contributors

Bay Networks Inc.; Informed Technology Inc.; Alteon Networks Inc.; Extreme Networks; Plaintree Systems Corp.

How to Build a Gigabit Ethernet Network

Basil Alwan*
Vice President of Product Management, Bay Networks Inc.
Bob O'Hara†
President, Informed Technology, Inc.

20.1 Introduction

Gigabit Ethernet promises to bring a fast lane to the information superhighway that is both simpler and less costly than any other high-speed networking alternative. This chapter explains how to measure performance problems in existing networks and use that knowledge to determine where gigabit Ethernet should be installed to increase throughput. Four potential locations for gigabit Ethernet in enterprise networks are described: as desktop connections, as uplinks, in server farms, and as the foundation for a collapsed backbone.

20.2 Where to Begin

Gigabit Ethernet is without doubt one of the most promising remedies now available for performance problems. But in order to determine

---

* Basil Alwan is vice president of product management at Bay Networks. He came to Bay Networks via Rapid City Communications, a pioneer in gigabit Ethernet routing switches. Alwan joined Rapid City as VP of marketing after 12 years with AMD. At AMD he was director of technical marketing for AMD's Network Products Division with responsibility for AMD's FDDI, Ethernet hub, and Ethernet switching products. Alwan, 34, holds a B.S. in electrical engineering from the University of Illinois in Champaign-Urbana.

† Bob O'Hara is founder and president of Informed Technology, Inc., a network consulting practice. Bob has worked in engineering and engineering management since 1978, when he graduated from the University of Maryland's Electrical Engineering program. Bob's areas of expertise are network and communication protocols and their implementation, operating systems, system specification and integration, standards development, cryptography and its application, strategy development, and product definition.

where to add capacity and how much capacity to add, network managers first need to work out where the problem is on their existing networks.

That is done in two ways: first, by monitoring the installed network and analyzing the data gathered, and second, by examining the applications in use and identifying growth trends.

20.3   Monitoring the Network

Monitoring a network is second nature for net managers at most large network installations; most would rather plan their network upgrades and outages than react to the outraged calls of frenzied users who are unable to accomplish the profit-making activities that the network supports. Gathering the statistics for individual segment utilization rates, per-station data rates, and forwarding and filtering rates for each LAN segment and subnet provides the basis for deciding which segments and subnets need to be upgraded. Setting realistic targets for utilization that allow for future growth will ensure that newly upgraded segments and subnets are not in need of additional upgrades too soon.

Many tools can be used to monitor the network. If the current network equipment has extensive network-monitoring capabilities built in (RMON capability, for example) then a management application capable of reading the statistics from the network devices may be all that is needed. If the current network equipment does not include this capability, a stand-alone RMON probe can be connected to a network port. This method is much more labor intensive, however, because the probe must be moved from point to point in the network to determine the utilization of each key link.

20.4   Analyzing Traffic Patterns

Once the monitoring method has been selected, the next question is what to measure and why. In addition to the aggregate utilization data, a network manager must analyze the traffic patterns and address questions such as:

- Does most of the traffic stay on the LAN segment, destined for the local server?

- Is a large amount of traffic exchanged with stations further removed in the network?

- Is much of the traffic on the LAN entering the segment from outside, destined for a few servers?

- Does the traffic ebb and flow at various times throughout the day?

The answers to these questions can identify the key endpoints of traffic in the network and provide clues about where further segmentation of the LAN and the application of greater bandwidth will be beneficial.

20.5 Analyzing Applications

Analyzing the applications that are generating the traffic in the LAN will also provide important data that will affect the decisions on where to apply gigabit Ethernet. After a time of increasing usage of client-server applications, several new classes of the peer-to-peer variety of applications are gaining popularity. These include both shared whiteboard desktop conferencing and desktop videoconferencing. Given their potential for increasing productivity while eliminating travel time and costs, conferencing applications are likely to grow in popularity and usage.

20.6 Forecasting the Future

Once the data on the traffic and the applications have been gathered, it is necessary to determine how the amount of network traffic will grow. It is important to note that the traffic growth will be driven by a number of different forces. The first force driving growth is simply additional nodes.

The second force driving traffic growth is the replacement of existing nodes with more capable nodes with faster CPUs and more powerful internal architectures. This will provide a hidden component to traffic growth, because the absolute number of nodes remains the same, but their ability to generate traffic improves.

The final force driving traffic growth is the use of new applications. This often provides an unexpected component to traffic growth. Most network managers remember when the first videoconferencing applications were introduced or the departmental Web server went online.

20.7 Prescribing a Cure

Stage one in the process of building a gigabit Ethernet LAN is to monitor and analyze the installed network, examine the applications in use, and identify growth. Stage two is to use all of this information to decide where gigabit Ethernet can be best applied to solve performance problems in the most efficient and economical way possible.

There are four ways gigabit Ethernet can be deployed.

- As a desktop connection
- As an uplink from a standard hub or switch

- In the server room
- As a collapsed backbone

20.8 Desktop Connections

Applications with large data sets (sometimes called *big data applications*) can slow to a crawl if network bandwidth cannot support them adequately. In this environment faster link speeds are critical. If network monitoring reveals congestion caused by big data applications, gigabit Ethernet desktop links can be used to help solve the problem.

As an example, large-scale modeling is now done with number-crunching supercomputers that deliver a constantly updated stream of data to one or more visualization stations. The visualization stations provide real-time display as the user manipulates and commands the model. Visualization applications include mechanical modeling, chemical modeling, drug discovery, crystallography, and financial analysis.

Generally, the user interface is provided on a high-end Silicon Graphics, Sun Microsystems, or Hewlett-Packard workstation. These workstations display images of 1280×1024 pixels refreshed at rates of 15–30 images per second. This generates a tremendous load on the network, often in excess of 10 Mbit/s for a single visualization station.

The latest advance in visualization is to present the images stereoscopically, using LCD shutter glasses to present each eye with its own unique image adjusted for parallax. While the frame rates are currently equal to those for flat images, it is easy to see that there will soon be a doubling of the data rate to maintain the same degree of interactivity as with the flat image.

Another big data application is digital prepress operations, in which entire magazine pages are pasted up on a workstation, necessitating the moving of images scanned at 2400 dpi between the workstation and file server. Without sufficient bandwidth, an editor's productivity can suffer drastically due to the time needed for the images to load. Movie and television production, particularly the creation of computer-generated effects, requires an enormous amount of bandwidth. Again, the movement of images, this time simply the sheer number of them required for the 24- or 30-frames-per-second (fps) media, drives the bandwidth demand. Other big data applications are seismic analysis used in oil and gas exploration, medical imaging, and insurance claim analysis utilizing online, stored images.

In these applications, and others that are similar, the use of gigabit Ethernet links to the desktop can easily be justified. The installation of a gigabit Ethernet switch between the servers and the final link to the user provides the bandwidth necessary to satisfy the application's need for raw bandwidth (see Fig. 20.1).

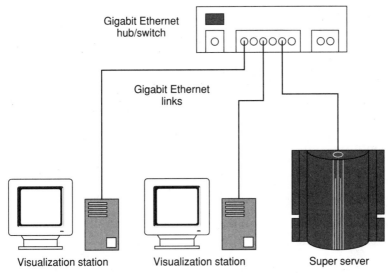

Gigabit Ethernet
hub/switch

Gigabit Ethernet
links

Visualization station   Visualization station   Super server

Figure 20.1   Gigabit Ethernet in big data applications.

20.9   Uplinks

Gigabit Ethernet can also be used as an uplink from a traditional hub
or switch. As networks migrate from a shared medium to segment
switching to fully switched configurations, it is important to have an
uplink that can aggregate bandwidth and maintain low latencies. This
is particularly true if servers are centralized and the uplink is highly
utilized.

In the past, uplink choices for network managers were limited. Only
ATM offered data rates greater than 100 Mbit/s. While ATM supports
data rates of 622 Mbit/s and beyond, it is not trivial to map standard
protocols (like IP) onto an ATM infrastructure. Use of ATM results in
additional complexity that should be avoided unless there is a particular current application that requires ATM. If you believe you do have
an application that requires ATM, it may be worthwhile to revisit
Ethernet switching. Ethernet switching has come a long way; since the
nondeterministic access mechanism used in shared Ethernet networks, CSMA/CD, is not utilized in full-duplex Ethernet switches, it is
now common to find Ethernet switches that offer consistent and low
latencies that once were strictly the domain of ATM. Gigabit Ethernet
offers a very straightforward uplink for switched Ethernet networks.
When gigabit Ethernet is combined with well-designed Ethernet
switches, it is possible to build a very high-performance infrastructure.

Some networks have several subnets present in each building (or
wiring closet!). In this case, the extra bandwidth provided by gigabit

Routing Switches

One of the benefits afforded by gigabit Ethernet is the reduced latency between the initial transmission of a frame and the delivery of that frame to its destination. This benefit is provided by the increased bit rate of gigabit Ethernet over either 10- or 100-Mbit Ethernet. This can only be guaranteed between two stations connected directly with gigabit Ethernet or connected by a gigabit Ethernet MAC (Layer 2) switch.

Once a frame must be processed above Layer 2, much of the benefit gained by using gigabit Ethernet can be lost to the serialization, the processing latencies, and the throughput limitations of classical routers. Recognizing this fact, a number of vendors have introduced what is called a *routing switch*.

The routing switch performs as both a Layer 2 switch and a Layer 3 router. It moves much of the routing algorithm from software to hardware. Thus, most of the latency suffered in traditional routers is eliminated as the MAC address substitution from the routing tables is performed by hardware added to the Layer 2 switch and performed at wire speeds. Where traditional routers might have latencies in the millisecond range, routing switches have latencies of less than 10 μs.

Conceptually, a routing switch will examine incoming frames from each port and determine the MAC destination address it contains. If this MAC destination address is directly connected to another port of the switch, the routing switch will perform exactly as any other Layer 2 switch and send the frame directly to the appropriate port.

If the MAC destination address is the address of the routing function of the routing switch, the Layer 3 destination address is extracted and used to find the correct MAC destination address for the next hop from the routing table. That new MAC destination address is inserted in the frame and the frame is sent to the appropriate port, the identification of which can also be contained in the routing table. Because the process of matching the Layer 3 address to the destination MAC address is performed in hardware, the latency of routing decisions performed in software is avoided. It is also possible to perform this address lookup and substitution process at each port of the routing switch, thus performing the routing function in parallel and eliminating the problem of serialization in traditional routers.

Routing switches may not be able to handle the routing of all of the protocols in use in a network. However, their use can be a significant boon to throughput, even when traditional routers must be used as well. The reason for this is that the load on the traditional router—the number of frames for which forwarding decisions must be made—is reduced by having the routing switch handle all of the frames of the protocols it supports. Thus, even if routing switches will not handle all of the protocols in a network, it is very often beneficial to install them to offload the traditional routers.

This routing switch is fundamentally different from an earlier technique that "switched" based on Layer 3 information. These earlier switches utilized Layer 3 information to build a virtual circuit over an intervening ATM network between two routers. While this may be a viable solution in some cases, it is really converting Ethernet frames into ATM cells and providing the additional information to transport them efficiently through an ATM network. This approach was used in Ipsilon Networks' IP Switching.

Ethernet is helpful but the router will still throttle intersubnet traffic. Several vendors of gigabit Ethernet switches have recognized this problem and are providing routing switches that perform the frame forwarding decisions extremely quickly. By using hardware to make the MAC address substitutions from the routing table, the forwarding portion of the router can keep up with the raw frame rate and be integrated with the MAC level switching, removing much of the latency and bottlenecking previously imposed by traditional routers. See the routing switch sidebar for more details.

Routing switches have several advantages over traditional Layer 2 switches. For example, routing switches allow routing functionality to be deployed wherever needed. Previously, much of the throughput gained by using a Layer 2 switch would be lost when a frame arrived at a backbone router. When this occurs, the router's large latency takes its toll and performance plummets. With routing switches, it is up to the network designer to decide where to deploy routing elements in the network—it is reasonable to consider enabling intersubnet routing right in the wiring closet (see Fig. 20-2).

20.10 Server Room

A third application for gigabit Ethernet is found in the server room, where a large number of high-capacity servers are clustered for security, administrative, and other reasons (see Fig. 20-3). Previously, FDDI and switched 100-Mbit Ethernet have been used to interconnect the servers and the backbone. The concentration of traffic in this location is tremendous, as nearly all of an enterprise's clients communicate with the central file, e-mail, Web, database, and computer servers. As the traffic demand increases throughout the remainder of the enterprise, that demand is multiplied by the concentration of resources in

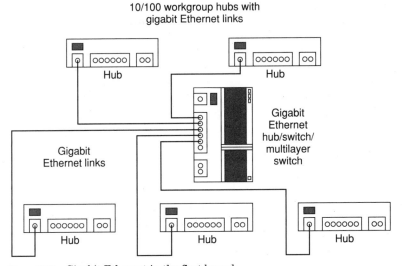

Figure 20.2 Gigabit Ethernet in the first branch.

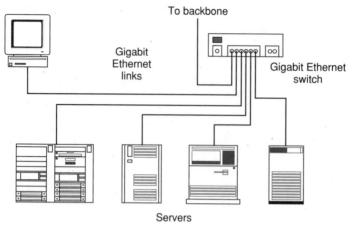

Figure 20.3 Gigabit Ethernet in the server room.

the server room. Thus, as processing power, capability, and raw numbers of nodes in the enterprise increase, driving increasing amounts of traffic, a very large portion of that traffic increase also is seen in the server room.

In addition to the external forces driving the increase in traffic, the server room has a number of internal forces that come into play. Shared file systems that allow a set of libraries or fixed data to reside in only one location while being used by many nodes create an amount of traffic that can be viewed as a minimum level before any external traffic is introduced. Similarly, file system mirroring also creates traffic in the server room to provide redundancy for high-value, dynamic data sets. Both of these internal forces are also affected by the increasing external traffic, in a heterodyning effect.

Gigabit Ethernet can easily replace older FDDI or fast Ethernet server room connections. In addition to providing more bandwidth for the current set of servers, the great headroom offered by gigabit Ethernet allows more servers to be added in the future. The use of a gigabit Ethernet switch instead of an FDDI concentrator or straight 10- or 100-Mbit Ethernet can increase the available bandwidth linearly for each link simultaneously in use.

20.11 Collapsed Backbone

The fourth application of gigabit Ethernet is in the collapsed backbone (see Fig. 20.4). In the backbone, there is an emerging trend toward moving the routing function out to the periphery of the network rather

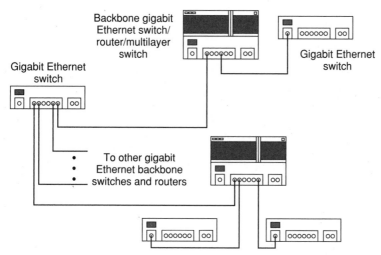

Figure 20.4   Gigabit Ethernet in the collapsed backbone.

than spreading it throughout the backbone. This trend was begun with the introduction of ATM into the backbone and continues with other switched technologies, such as switched 100-Mbit/s Ethernet and switched gigabit Ethernet. The reason this configuration is becoming so attractive is that when the routing decision is moved out to the periphery, the backbone can be built almost entirely with high-speed, low-latency switches. In many cases, this leaves only two high-latency routing decisions to be made, one on entry to the switched backbone network and the other on exit from the backbone network. And even the latency of these two routing decisions can be reduced through the use of routing switches. The new breed of routing switches provides flexibility in the network design by removing performance bottlenecks (serialization and latency) that exist in traditional routers.

   While this view may seem simplistic, it is a viable network architecture for many medium and large enterprises, as long as the physical limitations of the medium are not exceeded. The end nodes are connected to a 10/100/1000-Mbit/s routing switch, which is then connected to a gigabit Ethernet switch in the backbone. The first gigabit Ethernet switch is, in turn, connected to a second-level gigabit Ethernet switch. This layering of switches continues until the network is fully connected. Thus, the entire backbone is constructed with a hierarchy of gigabit Ethernet switches. In cases where additional routing is required in the backbone, the use of gigabit Ethernet routing switches satisfies that need and introduces little additional latency.

20.12   Implementing Gigabit
Ethernet to the Desktop

Before installing gigabit Ethernet to carry big data to the desktop, it is necessary to examine the requirements of three items: the end station, the gigabit Ethernet switch/router, and the cable plant. In this discussion, the end station represents both the user's workstation and the server feeding the workstation, which are assumed to be connected to the same switch/router (see Fig. 20.5).

The most important item to examine in the end station is the bus to which the network interface controller (NIC) will be attached. With the exception of PCI and S-Bus, the bandwidth available from the I/O buses in most end stations falls far short of being able to support the data rates of the applications described above, let alone the burst rate of gigabit Ethernet itself.

The data rates generated by the applications that would require gigabit Ethernet are in excess of 10 Mbyte/s. The burst rate of gigabit Ethernet is 125 Mbyte/s. The I/O bus in the end station must be able to transfer this data while also updating the display and transferring data to and from the disk. This will almost certainly result in periods of time in which the bandwidth to transfer the gigabit Ethernet data is not available or there is latency between the request to transfer data and the actual start of the data transfer.

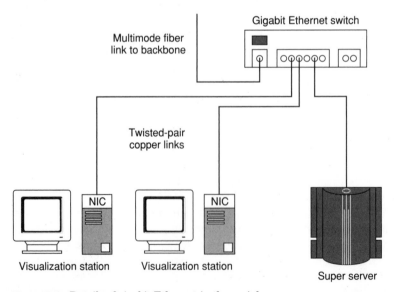

Figure 20.5   Details of gigabit Ethernet in the periphery.

One possible way to deal with the bandwidth availability problem is to use multiple I/O buses. In this case it is a simple matter to allocate the I/O controllers to each of the buses to apportion the load so that the needs of each controller are met. However, if there is only a single I/O bus available in the end station, this is not an option.

In the case of a single I/O bus, the NIC will need to provide data buffering in order to perform adequately in high-latency environments. The question then becomes, "How much buffering is enough?" The exact answer to that question is highly dependent on the individual application and I/O environment. In most cases, buffers of 32 Kbytes or more are needed to sufficiently handle incoming data.

The gigabit Ethernet switch/router supporting the connection to the workstations must be able to continuously deliver the data stream from either the upstream port on the backbone to the port connected to the downstream workstation or from one of the downstream ports to another. The switch/router must be able to accomplish this for each of the data streams that might be ongoing, while minimizing the possibility of internal blocking and dropped frames. This can be accomplished in a number of ways, including store and forward, back-pressure applied via the collision signal, or a PAUSE flow-control protocol now included in the IEEE 802.3 standard that will cause a node to cease transmitting for a short, predetermined amount of time.

The first of these methods attempts to deal with congestion in the switch by temporarily storing frames in the expectation that they will be sent within a short space of time. Trading the latency of storing the frame for the probability of being able to deliver it in the future can be very effective in many cases. However, since this method relies on a finite amount of buffer space in the switch, there is a small chance that the short-term network traffic will fill the buffer and some frames will be dropped due to the lack further buffer space.

The second method relies on the back-off mechanism built into Ethernet and the fact that the transmitting station will store a frame until it is either successfully transmitted or abandoned. This may reduce the cost and complexity of the switch somewhat, but may also result in some frames being dropped after a large number of retries during extreme congestion. In addition, this method may result in longer latencies than the store-and-forward method.

The third method, standardized in the IEEE 802.3x Full Duplex/ Flow Control spec, provides a flow-control mechanism for the switch to tell a station to stop transmitting for a short period of time. This method may provide the best mechanism for controlling congestion in the switch, as the switch can apply selective back-pressure to those stations responsible for the congestion. The flow-control mechanism makes use of a reserved multicast address and type field to add a thin MAC layer to a full-duplex Ethernet implementation. At present the only MAC type defined is PAUSE, used to implement flow control. However, it is easy to see how, once the MAC control sublayer is standardized, other control functions can be added in the future.

The final item to examine is the cable plant. While work continues in the 802.3z gigabit Ethernet workgroup to develop and standardize a twisted-pair copper physical layer (1000BaseT), the first physical layer for gigabit Ethernet relies on the fiber-optic specifications from the ANSI Fibre Channel standard. In particular, the gigabit Ethernet committee adopted the FC-0 and FC-1 layers of the X3.T320 Fibre Channel standard as a basis for the 1000BaseLX (long-wave laser) and 1000BaseSX (short-wave laser) physical layers.

The requirement of a fiber-optic physical layer would have been a significant detriment to a network standard only a few years ago. This is not necessarily so today. There have been significant advances in terminating and connecting fiber in recent years. This has led to a significant decline in the cost of installing a fiber-optic cable plant. In fact, recent data from commercial and government network installations indicate that the installation of fiber may cost only 5 to 20 percent more than copper for both material and labor.

When installing a cable plant for gigabit Ethernet in the periphery of the network, it is important to adhere to the requirements of EIA/TIA 568-A for structured wiring. This will ensure that the individual cable runs will not exceed the maximum length supported by the gigabit Ethernet physical layer. For the cable run using 1000BaseLX from the gigabit Ethernet switch to the workstation or server, the maximum length for both 50-mm and 62.5-mm multimode fiber is 550 m. If 1000BaseSX is used, the maximum length is 500 m for 50-mm fiber and 220 m for 62.5-mm fiber.

For the anticipated Category 5 UTP physical layers, the maximum run is 100 m. Be very conservative about installing a copper cable plant for gigabit Ethernet at this time. The standardization of the copper physical layers is at a very early stage, and the industry has yet to see a copper gigabit Ethernet physical layer implemented in silicon.

If the cable runs between the workstations, switch, and servers are shorter than 100 m, the use of Category 5 cable should definitely be considered. This is particularly true of existing installations, since the cabling can be reused with no added cost. As the cable runs begin to exceed 100 m, more consideration should be given to the use of multimode fiber.

20.13 Implementing Gigabit Ethernet as an Uplink

The most prevalent use of gigabit Ethernet is likely to be in a switch that has 10- and 100-Mbit/s ports on the downstream side and a gigabit Ethernet port on the upstream side connecting to the backbone (see Fig. 20.6). In this application of gigabit Ethernet, the requirements that must be examined are for the switch itself and for the cable plant on the gigabit Ethernet side of the switch.

The requirements for the workgroup switch with an uplink are somewhat more demanding than those for the switch in the desktop example, described previously. In this case, the network connections to the workstations and servers are using 10- or 100-Mbit/s Ethernet and only the backbone connection of the switch implements gigabit Ethernet. In this case, the potential for congestion in the workgroup switch due to a large number of frames arriving from the backbone, destined for a single workstation or server, is significantly greater than when all network connections are operating at the same data rate.

To minimize the potential for dropped frames due to switch congestion, the switch must implement a mechanism for applying backpressure to the gigabit Ethernet link to the backbone. As stated earlier, the two candidates for this mechanism are (1) the application of the collision signal to the gigabit Ethernet link when congestion

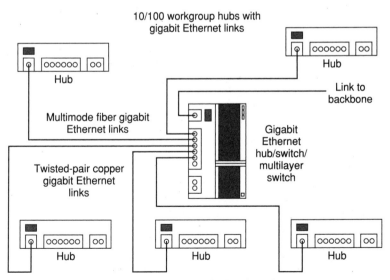

Figure 20.6 Details of gigabit Ethernet in the first branch.

occurs and (2) the use of the 802.3x flow control mechanism to apply selective back-pressure on the gigabit Ethernet.

The use of the first method—application of the collision signal—may result in an increased number of frames being discarded at the source, due to transmission being abandoned after the maximum number of retries is reached. This can be very detrimental to throughput as upper layer protocols are forced to wait for time-outs before attempting to recover. It is suggested that extensive simulation data and, if available, actual operational test data be examined before a product that uses this method of congestion control is selected.

The second method of congestion control—the 802.3x flow-control mechanism—is likely to be much more successful at reducing switch congestion than the collision mechanism. Stopping the source of a transmission before it begins subsequent transmissions is preferable because there is no risk of frames being abandoned or of bandwidth being consumed with frames that begin transmission and then experience collision.

In addition to the problem of congestion, the switch in this application is aggregating traffic from many stations on 10- and 100-Mbit links to the backbone on gigabit Ethernet. The use of gigabit Ethernet may provide little or no benefit if the throughput gained from the use of gigabit Ethernet is lost again in the added latency and low forwarding rates of a traditional router.

It is important to examine the protocol mix being exchanged with the backbone. If one or more of the protocols identified, e.g., IP, can be handled by a routing switch, then employing a routing switch will return significant benefits in throughput for the protocols handled in the routing switch as well as those that must be handled in a traditional router. See the sidebar on routing switches for more detail.

The cable plant for the gigabit Ethernet in this case is most likely to be either multimode or single-mode fiber, though the use of copper media is not ruled out. For planning purposes, Category 5 cable should only be considered for runs less than 100 m until standardization is complete. Once standardization of the copper medium for gigabit Ethernet is complete, it will be known whether cable runs in excess of 100 m will be supported. For cases where a very small number of runs exceed 100 m, particularly in an existing cable plant, some consideration should be given to using a buffered repeater, sometimes known as a *gigabuffer*. The insertion of a buffered repeater will nearly double the reach from one gigabit Ethernet node to another. However, the buffered repeater does introduce latency into the link, which may not be desirable.

If copper media cannot be used, fiber is the alternative. The selection of one type of fiber over the other will depend on the length of the cable run from the gigabit Ethernet switch to the gigabit Ethernet port on the next component of the backbone. It is also beneficial to think about long-range planning when selecting the fiber type to install in this uplink switch. If the current network architecture does not take the form of a collapsed backbone, then multimode fiber may well be sufficient for most small and medium networks. 1000BaseLX supports fiber runs up to 550 m in length on both 50-mm and 62.5-mm fiber. 1000BaseSX supports fiber runs up to 220 m on 62.5-mm fiber and 500 m on 50-mm fiber.

However, for large networks, for preserving the investment in network equipment should the backbone architecture change, and for collapsed backbones, single-mode fiber will be the best choice. The increased reach of the single-mode fiber physical layer, up to 5 km for 1000BaseLX, offers the most flexibility to the network planner. Single-mode fiber is not available with 1000BaseSX.

20.14 Implementing Gigabit Ethernet in the Server Room

In the server room, gigabit Ethernet is likely to be used to interconnect all of the servers and to connect the servers to the backbone (see Fig. 20.7). The requirements here are similar to those for using gigabit Ethernet in the other locations previously mentioned (server capabilities, switch capabilities, and cable plant).

The requirements of the server and its NIC are very similar to those of the desktop gigabit Ethernet application. The server must have sufficient bandwidth in its I/O subsystem to handle the 125-Mbyte/s peak data rates, as well as the average data rates that may approach a very significant portion of the peak rate. In the recent past, server comput-

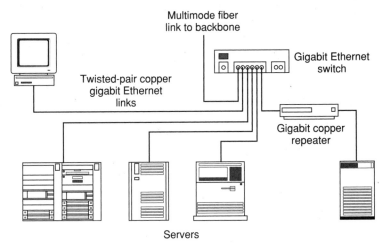

Figure 20.7 Details of gigabit Ethernet in the server room.

ing capacity has been able to stay ahead of the demands of the client workstations in large part because the network limited the number of client requests through the limited bandwidth available. With gigabit Ethernet making significantly more bandwidth available, this limitation is being removed.

With a large amount of bandwidth now available to the clients to make demands on the servers, the efficient connection of the NIC to the memory and computing resources of the server is critical. Sharing an I/O bus with other peripherals where the gigabit Ethernet interface would use more than 50 to 60 percent of the bus bandwidth is the same as installing a bottleneck that will limit the effective capacity of the server. This results in significant internal transfer latencies for the gigabit Ethernet interface as well as for the other peripherals sharing the I/O bus.

Even if the gigabit Ethernet interface is not sharing a bus with other peripherals, there will be a certain amount of latency between the request to begin a transfer and the actual start of that transfer; the interface will have to adapt to this latency. To meet this requirement, the interface must include some on-board frame buffering. In the server, as in the workstation, 32 Kbytes should be sufficient. The buffer should be evenly divided between transmissions and receptions.

Another critical item in the server is the network protocol stack. If the protocol stack is unable to process frames at a rate sufficiently high to keep pace with the network hardware, no amount of buffering will be sufficient. Thus, it is important to determine the frame processing rate supported by the protocol stack. In many cases, this rate will not support the rate at which frames may arrive from the gigabit Ethernet and the protocol stack will need to be upgraded.

The server room switch will likely use gigabit Ethernet on each port. For migration purposes the ports may also support 10- and 100-Mbit/s operation as well. Of course, this multiple data rate capability will only be possible if a copper physical layer is used, as the vast majority of installed 10- and 100-Mbit/s Ethernet adapters operate over a twisted-pair copper medium. The migration path may be more difficult if the server room requires a fiber cable plant. The prudent path, in this case, is to install a flexible gigabit Ethernet switch that also supports 10/100-Mbit/s connections. Servers can start out on the 10/100 mbit/s connections and migrate to the gigabit Ethernet ports as time and finances permit.

The switch in the server room must be capable of minimizing congestion and maximizing the exchange of frames among the servers and between the servers and the backbone. To minimize congestion, it must implement a flow-control algorithm, preferably the one described by IEEE 802.3x that applies selective back-pressure to the sources of congestion. Maximizing the exchange of frames among the servers can be handled by a high-performance gigabit Ethernet switch, since all of the servers in a single server room are likely to be on the same subnet and not require routing to communicate among themselves.

To maximize the exchange of frames between the servers and the backbone, a routing switch will provide a significant throughput boost over a gigabit Ethernet switch with a traditional router. As described in the sidebar, the routing switch moves part of the routing algorithm into dedicated hardware. A routing switch performs the frame-forwarding decisions and MAC address substitutions at the link rate, allowing the forwarding of routed frames at the same rate frames are switched and eliminating the router as a potential bottleneck at the entry and exit of the server room.

The final item to consider when installing gigabit Ethernet in the server room is the cable plant. Given the more restricted confines of most server rooms, it may be quite possible to use an existing Category 5 twisted-pair copper cable plant for gigabit Ethernet. Care should be taken to measure the lengths of each link from switch to server in order to ensure that the link meets all of the physical-layer requirements. Certainly, the money saved by using an existing Category 5 cable plant rather than installing a new fiber plant might make it cost-effective to relocate the switch from a distant wiring closet to a central location in the server room.

If the links are longer than those supported by gigabit Ethernet on copper media, selectively replacing copper links with fiber as individual servers are upgraded will provide a cost-effective migration path from 10- and 100-Mbit Ethernet to gigabit Ethernet. Another option to consider, if there are only a small number of cable runs that exceed the length supported by the standard but do not exceed twice that length, is the use of a buffered repeater in the middle of the run. Particularly in the server room, this may not be preferable to using fiber for these runs because of the additional latency a repeater adds to the link. However, there are conditions that may make this more desirable than installing new fiber-optic cable.

20.15   Implementing Gigabit Ethernet in the Collapsed Backbone

Installing gigabit Ethernet in the backbone may provide the most significant initial benefits. Gigabit Ethernet's large bandwidth can immediately eliminate the backbone congestion typically seen in today's networks. The items to examine when installing gigabit Ethernet in the backbone are the cable plant, the switches, and the other equipment on the backbone (see Fig. 20.8).

The cable plant is the first item that must be examined. For the backbone, very strong consideration should be given to the use of single-

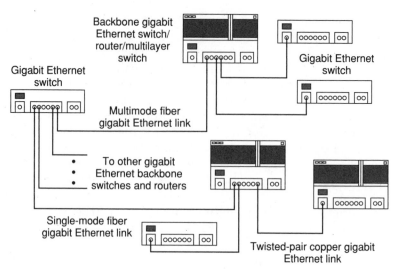

Figure 20.8 Detail of gigabit Ethernet in the collapsed backbone.

mode fiber. Installing this fiber from the outset will allow the most flexibility in the backbone architecture and provide the most options for future reconfiguration and upgrades. It may be the only choice available for some backbone links, due to their length. While multimode fiber may be applicable in some cases, if there is a mix of fiber types in the backbone, any cost savings available through the use of common test equipment, spare and replacement fiber, termination equipment, and nonduplication of training may be lost. It is not anticipated that copper media will be useful in the backbone application of gigabit Ethernet, except in a very small fraction of networks. The deciding factor in the choice of multimode or single-mode fiber may be the length of the cable run. Multimode fiber runs are limited to 550 m. Single-mode fiber runs may be up to 5 km in length.

The switches used in the network backbone must be capable of sustained nonblocking operation in order to provide the maximum benefit of the gigabit-per-second data rate available from gigabit Ethernet. This will require that some flow-control algorithm be included. The flow-control algorithm being standardized by the IEEE 802.3x workgroup will be satisfactory and should be implemented by all switches in the backbone if flow control is to be at all effective.

Again, routing switches should be considered to replace traditional routers for those protocols for which routing switches exist. The need for minimizing the latency and maximizing the throughput in the backbone should be obvious. (Again, see the sidebar on routing switches for details.)

The migration of an existing network to gigabit Ethernet may also be done in stages, upgrading the most heavily utilized links first, adding routing switches where needed, and proceeding to the next most heavily utilized links and routers in turn. This strategy for installing gigabit Ethernet in the backbone minimizes the initial capital outlay and will return immediate benefits in additional bandwidth and reduced latency in the backbone. As more segments of the backbone are upgraded to gigabit Ethernet, the benefits will spread to more of the network's user population.

20.16 Summary

Installing gigabit Ethernet is a task that makes use of extensions of the existing knowledge of the network manager. Because gigabit Ethernet is based on 10- and 100-Mbit/s Ethernet principles, many of the analysis tools and techniques for determining the need to upgrade network segments are already in use. With careful analysis of the network and methodical application of the design requirements for gigabit Ethernet, the installation of gigabit Ethernet is readily accomplished.

Choosing Server Components for a Gigabit Ethernet Network

Sean Riley* and Stephen Thorne†

Intel Corp.

21.1 Introduction

Gigabit Ethernet may sound like an ideal solution to the increasing bandwidth demands placed on today's servers—but there's a big problem: today's servers can't support gigabit throughput. Installing a gigabit Ethernet LAN adapter in an existing server generally won't yield the 10-fold performance boost over a fast Ethernet adapter that network managers might expect. How can these managers exploit gigabit Ethernet to get closer to 1-Gbit/s throughput today—and what can they expect tomorrow?

This chapter provides some context by describing the bottleneck problem on today's server hardware that gigabit Ethernet seeks to address—and the reasons a gigabit Ethernet adapter won't necessarily provide full gigabit-per-second throughput. It offers guidelines as to how to make hardware and software choices to maximize server throughput. Finally, it provides a road map to the key server technolo-

* Sean Riley, an Intel employee since 1992, is the director of marketing for Intel's Network Products Division in the Europe, Middle East & Africa (EMEA) market. NPD develops products consistent with Intel's high-performance workgroup strategy—delivering high-bandwidth solutions that take full advantage of next-generation PCs and servers and give more control to LAN managers and users. In his current position, Riley is responsible for managing Intel's networking and small business products and channels in the European, Middle Eastern and African markets.

Prior to his current position, Riley served as a fast Ethernet architect and was instrumental in delivering the EtherExpress™ PRO/100 LAN Adapter. The fast Ethernet adapter was introduced in October 1994 as the first 10/100-Mbit/s network interface card and is the market segment leader.

Riley has been an active fast Ethernet proponent from his involvement as a founding member of the original seven-company Fast Ethernet Alliance through to establishing

gies network managers can expect to see arriving in the next few years—technologies that will enable servers to run at true gigabit-per-second speeds.

This discussion focuses on the most popular server architecture in use today—Intel Pentium, Pentium Pro, and Pentium II processor-based servers. It does not address other types of servers, such as DEC Alpha or Sun SparcStation systems.

21.2   Gigabit Ethernet in Today's Servers

Gigabit Ethernet is a remarkable networking technology that stresses today's servers beyond their capacity. It builds on recent trends that are increasing network speed faster than the speed of server processors. For instance, 10 years ago, a system was high-tech if it had a 286-processor-based server running an early version of Novell's NetWare. This server was likely connected to a few other networked PCs via an early Ethernet card and a thick coax cable. Back then, 10 Mbit/s was a huge amount of bandwidth—similarly, a 10-Mbyte hard drive was a huge amount of server disk space. In the last 10 years, processors have increased exponentially in performance and hard disks have increased exponentially in size and speed; the network must increase in speed to keep up (see Fig. 21.1).

However, as one makes the leap from fast Ethernet (100 Mbit/s) to gigabit Ethernet (1000 Mbit/s), network capacity will leapfrog server capability. This leaves network managers in the unfamiliar situation of having a network that is more powerful than the servers and PCs it connects. This amazing increase in the speed of network connections means that the host system—rather than the network—is the throughput bottleneck.

the IEEE 802.3u subcommittee on 100Base-T. Riley was also a signing member of the 100Base-T specification in June 1995.

Riley is quickly becoming an established expert in fast Ethernet technology and deployment. He has written several application notes and white papers and is a coauthor of the book *Switched and Fast Ethernet: How It Works and How to Use It.* Riley has also been a guest speaker for a variety of Fast Ethernet seminars across the United States. Riley holds a BSEE and a MSEE in electrical engineering from Arizona State University.

† Stephen Thorne, an Intel employee since 1995, is a product manager at Intel's Network Products Division in Portland, Oregon. In his current position, Thorne is responsible for product development and marketing of Intel's family of Express fast Ethernet hubs. Prior to his current position, he managed and developed Intel's EtherExpress Server Adapters.

Thorne has written a variety of white papers and application notes on local area networking technology. His research focused on the areas of network performance and new server technologies—including VLANs, port aggregation, and NOS offloading. Thorne has a degree in computer science from the Georgia Institute of Technology, and lives in the Portland area.

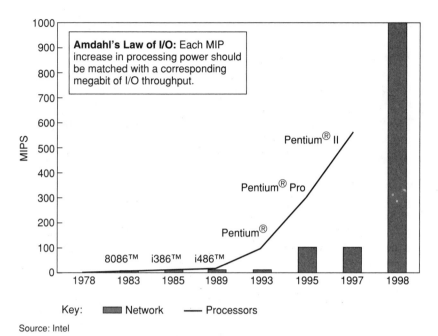

Key: ▬ Network   ── Processors

Source: Intel

Figure 21.1 Processor versus network evolution.

This is exemplified by gigabit Ethernet adapters in today's Pentium processor–based server systems. In studies at Intel labs, the basic server system was found to maintain 250–400 Mbit/s of sustained throughput. Network managers may find different results in their environments, depending on the type of server, operating system, and subsystem elements they use. No servers sustain gigabit-per-second speed today; however, future servers will be gigabit-per-second capable. The tips and guidelines described in this chapter can make all the difference to network managers seeking to gain greater throughput from the current generation of server hardware.

21.3 Understanding the Total Server System

To get the best performance from their servers, network managers need to understand the hardware and software components of those servers. Why? Because each one can affect throughput—for better or worse.

These components include microprocessors, memory subsystems, host bus interfaces (e.g., PCI chipsets), network adapters, disk subsystems, network operating system (NOS) software, and application software. A key principle of server design is that network managers need to use the highest-performing components throughout their servers in order to get maximum system performance—because one poorly chosen part can degrade overall performance. In other words, the system is only as strong as its weakest link. Network managers can readily improve some of their server components (e.g., add extra host processor chips), but, unfortunately, not others (e.g., network operating system code). This chapter discusses each major component and identifies which ones the network manager can adjust to boost throughput.

21.4   Performance Factors

Before discussing how to design a server for maximum performance, it's appropriate to spend a moment on the concept of *performance* itself. Performance is a difficult subject because the term means different things to different people. Performance can describe network latencies, network bandwidth, system efficiency or utilization, response time, and even more abstract concepts like uptime. For the purposes of this chapter, overall server performance is discussed in relation to network throughput and server CPU utilization. That is, performance here is the ability of the server to keep pace with a stream of data coming in and out of the system at 1 Gbit/s. The discussion that follows elaborates on these factors, looking first at factors related to the server's hardware subsystem, then at factors of the software subsystem, and finally at the interface between the two.

21.5   Hardware Subsystem

The hardware subsystem of a server contains many of the components that can affect total throughput. Think of the hardware subsystem as an engine. It can run fast or idle, run smooth or rough, run powerfully in a low gear or efficiently in a high gear. How its designer chooses and combines its components goes a long way in determining how it performs. Those components include processors, memory subsystems, host chipsets, bus types, and network adapters. It's not crucial to spend an extravagant sum for each component—a fast hard disk subsystem, top-end processor(s) and ample memory, along with judicious choices for the other components, will serve the network manager well.

21.6   Single-Processor Systems

If the processor is at the heart of the server, then the speed and type of processor that the network manager chooses is clearly at the heart of the ability to approach gigabit-per-second throughput. The processor frequency, for example, directly impacts the ability to crunch the networking data. The faster the server can execute code, the faster the bits will get on the wire, according to Amdahl's law of input/output.

Pentium and Pentium Pro processors bring a new class of performance to single-processor servers. With an older i386 or i486, it wasn't possible to begin to reach gigabit-per-second speeds. Amdahl's law states that 1 MIPS is needed for every megabit per second of bandwidth provided to the processor. That's about 1000 MIPS for gigabit-per-second throughput—requiring four Pentium Pro processors!

As processors scale upward in both architecture and frequency, they are less likely to create a bottleneck in a gigabit-per-second server. With the success of gigabit-per-second speeds, vendors will develop servers based on processors that can handle gigabit-per-second I/O.

21.7 Multiprocessor systems

Given that today's typical single-processor servers can't keep up with the amount of traffic in gigabit networks, it makes sense to use a multiprocessor architecture for better performance. There are two types of multiprocessor systems: symmetric and asymmetric. This distinction refers both to how the operating system deals with the multiprocessors and to how interrupts and tasks are handled in the server system.

In asymmetric multiprocessor systems, each of the CPUs is assigned to a different task. This is great for servers that perform embedded functions in the network. For instance, a set of distributed database application servers may keep one processor per server dedicated to the functions that reside within that server chassis, while the other processors are available to process requests from any other server on the network. Novell NetWare is a good example of a network operating system that makes good use of the asymmetric multiprocessor architecture. In symmetric multiprocessor systems, on the other hand, each of the multiple CPUs can handle any function. This means there is an intelligent interrupt arbiter within the server that sends tasks to different processors depending on which processor is most available. Windows NT uses this technique and masks the number of processors behind its DPC software layer.

It's not yet clear whether asymmetric or symmetric multiprocessing provides consistently better performance for near-gigabit-per-second throughput—and in fact, today there are still too many server bottlenecks to overcome before this becomes a significant issue. The best choice may turn out to be a hybrid of symmetric multiprocessors with one or more processors dedicated to networking. Multiprocessors, however, remain the clear choice over single-processor systems.

21.8 Memory Subsystem

Network designers and managers also need to keep in mind that the memory subsystem can affect overall system performance drastically.

Understanding the components of the memory subsystem, and how the characteristics of those components (wait states and latency, for instance) can affect server performance, can help network managers in getting closer to full gigabit-per-second throughput.

21.8.1 Memory type

It's easy for server memory subsystems to drag down throughput, because memory has had a hard time keeping up with advances in server processors. Most memories cannot move data at gigabit-per-second speeds, so servers that use these memories will not run at gigabit-per-second network speeds. Of course, even when network designers use the best memory types, they still don't see full gigabit-per-second throughput because of bottlenecks elsewhere in the server.

Network managers need to be aware of recent, dramatic changes in memory speed, type, and cost so they can make the best decisions for maximum server performance. In 1986, basic DRAM-type memory ran at about 20 MHz. Today, multiported EDO DRAM memory runs at in excess of 100 MHz. Multiport memories allow two bus masters to access the same memory chips simultaneously. Memory is expensive and can make up almost 25 percent of the server cost; but designers and managers shouldn't skimp on high-performance memory when they build gigabit-per-second servers, because more and faster memory will aid in the caching of data at the server level, reducing the need to access the hard disk and thereby boosting throughput significantly.

Accessing the server hard disk when operating at gigabit-per-second speeds on the network is like taking a tricycle out on the autobahn. To support faster throughput, memory will have to increase to compensate for slow hard disk subsystems. But the size of files transferred over the network is increasing too—putting yet more pressure on memory size and speed to support increased throughput. Happily, performance improvements in memory in just the past two years have been massive, so expectations for this area are high. The key for buyers is to look for server systems that are flexible enough to be upgraded to new memory types later.

21.8.2 Cache

Cache memory—which gives the processor faster access to frequently used code or data—can also influence the server's ability to approach gigabit-per-second speeds. There are two types of cache: L1 and L2. L1 is tightly coupled, and often integrated, with the host processor. L2 is slower, bigger, and coupled with the external processor bus and chipset. The two types have distinct advantages. Instructions that are used over and over again, such as the server OS kernel, will be in L1 cache a large percentage of the time and therefore accessed much faster. L2 cache typically holds less frequently accessed data, but can also be important for executing instructions.

The size and speed of L1 cache doesn't affect gigabit-per-second servers as much as L2 cache does. That's because L2 cache is typically much bigger and slower than L1 cache, and so can hold much bigger chunks of data and affect a larger portion of network accesses from the clients to the server. Since the difference in speed between L1 and L2 cache is minuscule compared to the difference between L2 cache and the hard disk, a larger investment in L2 cache, as opposed to the hard disk, pays off more in overall server performance.

However, running network data from cache may be a double-edged sword. That's because—although the processor can get to the information quickly—some server systems penalize PCI bus-mastering peripherals when they compete with cache for the same resources. Some servers hold off the host processor while the network card goes into the cache to get data. Overall, network managers won't be penalized for buying larger L2 cache for their servers—consider 256 Kbytes to 1 Mbyte or more. They should buy servers with which they can vary the size of L2 cache. Then they can use the L2 cache size that gives them the minimum CPU utilization. This is how they can achieve a good mix of cache size and cache efficiency. They also should be sure to discuss this issue—and how the server platform allows access to the cache—with their server vendors.

21.8.3 Latency and wait states

Latency and wait states are the two ways memory transactions are measured—and they are two more variables that influence the quest for near-gigabit-per-second throughput.

Wait states measure the speed with which consecutive entries can be read or written. This can also be thought of as how long it takes the system to go from one memory access to the next, a process known as *consecutive memory access*. Network managers want wait states to be as brief as possible—ideally zero—especially for large quantities of data for which the server will have to access memory several times.

Memory latency measures how long it takes to get the first data from memory. Memory is smart enough to know when it is accessed for the first time and to start fetching the next few entries automatically so they are ready for subsequent reading. Low latency—a shorter initial delay—is better for small amounts of data because it brings the data more quickly; low latency is generally associated with higher wait states (subsequent delays), but, if the amount of data is small, this isn't a problem since there isn't much data to retrieve on subsequent memory accesses.

Typically, low-wait-state memory results in higher up-front latency. However, low-wait-state memory is better for larger data loads because, while it takes a bit longer to deliver the initial portion of data, it takes less time to deliver the subsequent portions. Since most memory access in a server is sequential, low wait states increase performance. Gigabit-per-second network access requires 1 dword (4 bytes) to be fetched every 30 ns, so high-latency memory types with zero wait states (the fastest) will help the system get closer to full gigabit-per-second throughput, but this memory will be expensive. Even though it is expensive, network managers should buy memories with zero wait states and low access latencies. If budget is limited, they should choose memories with longer access latencies, but should not compromise on wait states. The cheap memory advertised in the back of many PC magazines probably isn't one of these two types.

21.8.4 Memory access

Another expensive but helpful choice for systems designers seeking near-gigabit-per-second throughput is dual-ported memory. This type of memory gives multiple processors a chance to read the same memory bank or chip. It's handy in symmetric multiprocessor server systems where any processor can go to any memory location at any time.

With gigabit-per-second access taking place on the I/O bus through the network adapter, dual-ported memory may be a wise investment for a high-performance multiprocessor server. Network managers who don't want to invest right away in dual-ported memory for their new servers should at least make sure that they can upgrade to this type of memory when the need becomes more pressing or cost effective. Network managers with single-processor servers don't need dual-ported memory, because their servers lack the second processor needed to exploit dual-processor access to memory resources.

21.9   Disk Subsystem

The hard disk subsystem could be the biggest roadblock to achieving near-gigabit-per-second server throughput. That's because hard disk access times are measured in milliseconds—a million times slower than the nanosecond during which each bit goes on a gigabit-per-second wire. If cache and memory utilization are stretched in near-gigabit-per-second systems—and, as discussed above, they are—then trying to read a file off the much slower hard disk is an even more serious impediment to maximum system performance. Even hard disks in the fastest SCSI-II and ultra-RAID systems can't keep up with the gigabit-per-second link to the network. So, in a gigabit-per-second server, it may be very economical to load up on main system memory and to force access to the server hard disk only when necessary.

Unfortunately, much of today's network traffic is made up of very large files, such as AVI video, that are generally too large to be stored in memory. So servers in the future will need newer and faster ways of accessing stored data in order to get closer to full gigabit-per-second performance.

Another factor for network managers to consider is the way their disk subsystems (especially bus master SCSI systems) will compete with network adapters and processors for PCI bandwidth—because that competition will also be a drag on throughput. The best course is for the network manager to choose these components, including the gigabit networking interface card (NIC), carefully and test them in various configurations to ensure they are not in constant competition for PCI bus access.

21.10 PCI Chipset

The network manager's choice of bus type and specific implementation also affects the ability to achieve near-gigabit-per-second performance. There are several bus types, each with its own place in computer history. They offer varying trade-offs of broad industry support, performance, flexibility, and cost. For server systems and adapter cards, Intel focuses on the PCI bus—and so will the discussion here. This is primarily due to the performance (bus mastering), flexibility (scalable bus width and clock speed), and industry support that PCI offers. Network managers may want to examine other bus types depending on their server strategies. Table 21.1 identifies the trade-offs among the bus types supported by Intel.

TABLE 21.1 Comparison of Bus Types Supported by Intel

	Bus width	Clock speed	Available bandwidth	Characteristics
PCI	32 or 64 bits	33 or 66 MHz	1–4 Gbit/s	Broad industry support; scalable bandwidth; can support Ethernet, fast Ethernet, and gigabit Ethernet.
EISA	8, 16, or 32 bits	8.33 MHz	264 Mbit/s	Introduced bus mastering; better performance than ISA; backward-compatible with ISA add-in cards, but limited band width capabilities.
ISA	8 or 16 bits	8 MHz	32 Mbit/s	Very broad industry support, but limited performance, no bus mastering, and no scalability. Cannot support high-speed Ethernet networking.

21.10.1   PCI bus width

PCI bus standards include both 32-bit and 64-bit widths. Network designers should choose the 64-bit width to get closer to full gigabit-per-second throughput. That's because all of the devices on a single 32-bit, 33-MHz PCI bus must share the 1 Gbit/s of available PCI bus bandwidth. A single gigabit-per-second adapter can tie up that bandwidth by itself, leaving no room for more PCI resources or other peripherals and thus degrading throughput.

Fortunately, 64-bit PCI standards are already in place so products are becoming more widely available that can take advantage of them. In addition, 32-bit cards work well in 64-bit slots, so there are few compatibility issues facing managers who want to go with 32-bit PCI today and upgrade to 64-bit PCI later. Designers and managers developing their strategies for future servers should be sure to include 64-bit PCI server slots and gigabit-per-second server adapters.

21.10.2   Multiple PCI bus systems

Network managers also should look for servers with multiple PCI buses. These servers can support more peripherals than single-bus servers can, increasing throughput. Also, managers should install the gigabit NIC on the primary PCI bus to ensure it takes priority when peripherals on different buses are contending for the same processor and memory resources. It might not be readily obvious which PCI slots are primary in the server, so network managers should consult the server user's manual or call the server vendor. This can make a big difference. As mentioned above, managers should test their NIC and storage peripherals in a variety of configurations to ensure they are not in constant competition.

21.10.3   PCI frequency

In addition to doubling PCI data bandwidth, the PCI bus can be scaled up in frequency, further increasing system throughput. When server manufacturers solve the electrical problems of routing a high frequency through a connector, network managers will start seeing PCI frequencies as high as 66 MHz (see Fig. 21.2). Because frequency can be increased without changing other parts of the PCI bus, the PCI specification is inherently scalable and backward-compatible. That means managers will find it relatively easy and cost-effective to take advantage of frequency enhancements. Buyers should look for server vendors that can deliver scalable PCI bus width (32 to 64 bits) and PCI frequency (33 to 66 MHz).

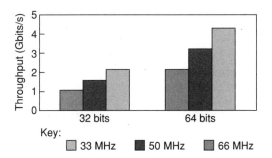

Figure 21.2 Relation of PCI frequency and throughput.

21.10.4 PCI latency

PCI latency affects performance just as memory latency does. It's measured by how long it takes to get the first amount of data in a transfer on the PCI bus. Typical PCI latencies range from 3 clocks to more than 30 clocks. Low PCI latency retrieves the first bits of data quickly but makes the user wait for subsequent data. Higher PCI latency typically delivers the first data more slowly, but delivers the rest of the data stream more quickly.

Which is best depends on whether users are working with small amounts of data, for which low latency is more effective, or with large amounts of data, for which high latency is better. Unfortunately, both situations are equally likely in gigabit networking. Excessively long latencies can translate into poor system performance. Network managers contemplating new servers should ask their vendors to explain what PCI latencies they can expect—and how this will affect performance on their particular systems. In general, managers should consider PCI bus designs that have short up-front latencies but can deliver a long stream of data at one time.

PCI access latency and interrupt latency will be increasingly hard to tolerate in a gigabit-per-second server. A latency of only a few microseconds could cause a server to completely miss an entire gigabit-per-second Ethernet frame. Unfortunately, as server architectures become more parallel (like multiprocessing), interrupt latency and PCI access latency will be driven up. It will be a major system design challenge to keep latencies down while driving up processor power.

One way to do this is to run the gigabit-per-second adapter in polling mode, in which interrupts are ignored and the network adapter is checked regularly for changes in status (e.g., for an incoming frame or a frame that needs to be sent). This would be much more efficient in a gigabit-per-second server. Although polling mode will allow server throughput to increase, the drawback is that CPU utilization also goes up. It's worth it. Network managers should check with their network server adapter vendors about their ability to support this feature. There is more information about adapter hardware and software later in this chapter.

21.10.5 PCI memory commands

PCI memory commands also affect overall system performance. Today, most PCI chipsets use 32-bit commands to move data to and from memory, but different commands have different characteristics. One set of commands is optimal for short data transfer, because the commands have a low up-front latency but longer wait states between the transfer of subsequent data bits. For instance, a short move of this type would use a "mem write" or "mem read" command. Another set of commands is optimal for longer data transfers, because these commands transmit a larger chunk of data quickly during the transfer, although they have a longer initial latency. Examples of these long-latency commands are the "mem read line" and "mem write and invalidate" commands.

Hardware vendors and designers, therefore, have a choice between low latency with small data chunks and high (or long) latency with large data chunks. This forces a trade-off similar to the one that shoppers face at the grocery store. Shoppers who want only a few things can get served quickly at the express aisle, but they can't buy more than 10 items. Shoppers who want a lot of items need to go to a standard aisle and wait a while longer up front, but they complete the purchase much faster than they would by returning to the express line over and over again.

Typically, network managers should want PCI chipsets that support both types of commands and that use each at the correct instance. Network managers will pay more for a server that supports both modes, but it's well worth it for faster data access and throughput. The same holds for the network adapter; that is, network managers should choose adapters that use the proper PCI commands for different types of memory access scenarios.

21.10.6 PCI cache/posting

PCI cache/posting enables the chipset to fetch additional, consecutive data bytes from memory before they are actually requested, on the likelihood that a user who asked for data in, for example, locations 1 through 12, will probably next ask for the data in location 13. Network managers seeking near-gigabit-per-second performance should definitely include PCI cache/posting on their feature checklists. It's crucial in servers where large chunks of data get transferred all the time. PCI cache/posting usually supports from 32 to 256 bytes of data.

The user can post in either the read or write direction. In read posting, the data is fetched in advance. In write posting, the user writes data to a buffer in the chipset, which then slowly changes the data in main memory while the user moves on to the next task.

21.11 Adapter Hardware

Network interface cards (NICs) are another server component that can boost or inhibit network throughput. Many companies produce and market NICs and, because these NICs are designed to varying specifications, they present network managers with yet another set of decisions in attempting to achieve near-gigabit-per-second throughput.

NIC buyers should closely inspect the advertised feature list of any network adapter they're considering, to make sure that it meets their basic requirements. For instance, buyers should make sure that their NIC hardware supports their server's bus interface (e.g., PCI 32, PCI 64, Sbus, and so on) and that the NIC supports the correct network topology and cabling (e.g., 1000BaseSX, 1000BaseCX, and so on).

Buyers should also make sure that the NIC vendor offers software device drivers for the intended network operating system (e.g., Windows NT 4.0, Novell NetWare 4.11). It's common for vendors not to support older or less popular operating systems (e.g., OS/2, Windows NT 3.51). One area of differentiation for some of the better-known network adapter vendors is the software driver sets they support. Typically, the larger brand-name vendors (e.g., HP, Intel, 3Com) support larger device driver suites and frequently update their device drivers with bug fixes, performance improvements, and support for new network operating systems.

After eliminating adapters that do not support their required features, buyers should look for high-performance adapters. Selecting the best network adapter is important in designing a high-performance gigabit server. That's because the architecture of the network adapter contributes greatly not only to network latencies, but also to the overall efficiency of the entire server system. What constitutes the "best" network adapter? The rest of this section describes some of the more important aspects of network adapter hardware (see Fig. 21.3 for a diagram of a typical adapter).

21.11.1 Bus master versus slave architecture

Network adapters traditionally have used either a bus master architecture or an I/O slave architecture. Bus mastering adapters are more efficient, because they take the responsibility of moving data across the

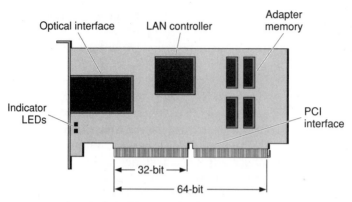

Figure 21.3   A typical gigabit network adapter.

system (PCI) bus to or from memory that resides on the network adapter. This architecture is beneficial because it allows the CPU to perform other tasks while the bus mastering adapter is using its DMA engine to transfer data.

Adapters based on a simpler slave architecture require software running on the host processor (device drivers) to move data across the bus. This is accomplished via memory mapping, or I/O mapping, memory space on the adapter into host system memory space. The slave architecture is inherently less efficient than the bus master architecture because the slave architecture requires the CPU to move all of the data. This is significant because host CPUs don't have enough bandwidth to move 1 Gbit/s of data (or more, at full-duplex) across the system bus.

Although slave architectures are less efficient, they can provide lower initial latencies in some cases. This is because there is a certain amount of latency involved in having the bus mastering adapter fetch descriptors (requests to send or receive data) out of memory, process those descriptors, and then fetch the data. However, because efficiency is generally more important than latency at gigabit speeds (because the system, not the network, is the bottleneck), bus mastering architectures are recommended over slave architectures.

21.11.2   Adapter memory size and receive overruns

All network adapters have some amount of memory (sometimes referred to as FIFO space) on the card. Sometimes this memory is incorporated directly into the network MAC controller; in other cases it is located in separate memory chips external to the MAC controller. In either case it is very important that the memory bandwidth on the adapter be greater than the theoretical bandwidth of the gigabit wire (2 Gbit/s at full-duplex).

When data is received from the network, the MAC controller places the data in its adapter memory. Subsequently, the data is transferred across the system bus (by the adapter if the adapter is a bus master) and into host memory. If the rate of transfer across the system bus is slower than the rate at which data arrives from the network (which is a maximum of 1 Gbit/s), there is a potential for receive overruns. Overruns occur when the adapter's memory is full and the adapter does not have room to store incoming data. In this case, the incoming packet is dropped. Dropped packets severely hurt performance, because the sender of the packet has no way of knowing that the receiver dropped it. Eventually, protocol software running on the sender will time-out and resend the dropped packet. However, it usually takes the protocol many milliseconds, if not seconds, to recover from the time-out. This degrades performance tremendously on a gigabit network.

The best way to avoid overruns is to use a high-efficiency bus mastering adapter that processes incoming packets with a minimal amount of overhead before transferring the packet data in long bursts across the system bus. The system bus can be a real bottleneck here. That's why—as discussed earlier—it's best to choose adapters and systems that support wide (64-bit as opposed to 32-bit) high-speed buses. At a minimum, network managers should use 64-bit, 33-MHz PCI adapters.

To avoid receive overruns, it also helps if a large amount of adapter memory (48 Kbytes or more) on board the NIC is dedicated to storing receive packets. Larger receive buffering helps when the system bus is heavily shared with other devices and initial access times to the bus may be quite high. This suggestion for more memory is different from, but analogous to, the earlier suggestion to maximize server memory.

The on-board NIC memory also stores incoming and outgoing packets before they are sent on the wire. However, not much memory is required for outgoing packets because the adapter can queue transmit packets in host memory until it has available resources on the adapter to store those packets.

21.11.3 Flow control

Flow control is another way to eliminate receive overruns and contribute to near-gigabit-per-second performance. Not all adapters support flow control, so network managers should be sure to put this feature—especially full-duplex flow control—on their checklists when considering a server system.

Flow control provides a way for a receiving station to warn the sending station that its buffers are nearing capacity. By sending pause frames, the receiving station can clear its buffers to eliminate dropped packets. Slowing transmission via flow control is much more efficient than incurring overruns, because it avoids protocol errors and time-outs. A short pause in transmission consumes much less time than a dropped packet and ensuing time-out. Only full-duplex connections have the ability to support flow control.

21.11.4 Hardware/software interface

The hardware/software adapter interface is another variable in the throughput equation. Adapters that have complicated interfaces will require more software interaction to manage the interfaces. The greater the software interaction, the greater the number of CPU cycles spent processing packets—and the lower the performance, because these CPU cycles may be needed elsewhere to run applications or handle I/O operations for other devices.

Superior software/hardware interfaces require the device driver to perform as few tasks as possible, thus sparing CPU resources. This can be accomplished by adapter hardware whose host interface mimics the lower end of common network driver interfaces (e.g. NDIS, ODI). Section 21.12 addresses these issues in greater detail.

21.11.5 Half-duplex versus full-duplex

Network managers facing the server design and purchase decision also have to choose between half-duplex and full-duplex adapter hardware. They should choose full-duplex hardware. Here's why.

Most early gigabit hardware (both adapters and switches) only supports full-duplex operation over fiber media (1000BaseSX or 1000BaseLX). Half-duplex products will be rare (no vendors are currently shipping products of this type) until the 1000BaseT standard for Cat 5 copper is complete. And half-duplex operation over fiber wiring doesn't make sense, because fiber wiring includes dedicated send-and-receive strands (making it inherently full-duplex).

Also, gigabit adapters running at half-duplex incur performance penalties. Short packets under half-duplex require a carrier extension to ensure that they occupy at least 512 byte times on the wire. The carrier extension is required in order to maintain a 200-m collision diameter at gigabit-per-second speeds. The graph in Fig. 21.4 shows theoretical maximum throughput at half-duplex, with and without carrier extension.

Unless there are extenuating circumstances, network managers should plan to purchase full-duplex-capable adapters and to configure gigabit servers to run at full-duplex.

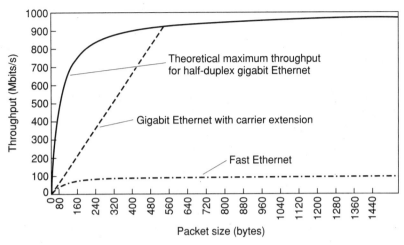

Figure 21.4 Half-duplex throughput with and without carrier extension.

21.11.6 The cost of interrupts

Network adapters generate system interrupts to inform the CPU that they have either received a packet that the CPU needs to have processed or that they have completed sending a packet. Unfortunately, these interrupts, if generated frequently, can stifle system performance, especially under Windows operating systems.

The problem is that each time an interrupt is generated, the processor needs to save its current context and then execute the interrupt service routine (ISR) of the software driver for the device that generated the interrupt. Finally, the processor needs to switch back to its original context. The cost of an interrupt varies from one operating system to another. Generally, the interrupt latency (i.e., the time from when the interrupt is asserted to the time that the ISR is called) is several microseconds. The context switch alone usually consumes hundreds of CPU cycles.

Since gigabit adapters are very high-speed devices, they can generate an extraordinary number of interrupts. For instance, if an adapter generates one interrupt per received packet, and it receives a stream of minimum-size Ethernet packets at full gigabit speed (1,953,125 packets per second for 64-byte packets at 1 billion bits per second), it could attempt to generate more than a million interrupts per second. However, if it takes the system 10 µs to process each interrupt, the system can handle only a maximum of 100,000 interrupts per second. This is a problem.

The solution is for the adapter and its software driver to attempt to coalesce interrupts. That is, software should handle multiple events each time it processes an interrupt. Further, modifications can be made to the adapter hardware so that it generates fewer interrupts and gives the software an opportunity to process multiple events each time an interrupt occurs. One hardware implementation would be to delay the generation of receive interrupts in the hope that multiple packets will have been received by the time the interrupt is asserted.

However, there's a drawback to relying on the hardware to delay the interrupts: the delay increases latency. That is, the first received packet is not delivered to the network operating system as quickly, because the indication to the system is that the received packet was delayed. In general, it's beneficial to reduce the number of interrupts (thus increasing the system efficiency), but this needs to be done in moderation to ensure that latencies are not increased dramatically.

21.11.7   Improving CPU utilization

Adapters with the highest level of integration and efficiency can boost server throughput in another way: by reducing host CPU utilization. This is important because CPU utilization is, and will continue to be, a crucial factor in server performance. Network managers can expect their first gigabit servers to run continuously at 100 percent CPU utilization, because the network is overpowering the server. Reducing CPU utilization gives the host processor more cycles to process tasks other than networking activities. Hardware vendors also facilitate low CPU utilization in their servers by focusing on optimal PCI bus efficiency, interrupt reduction (i.e., polling mode), and offloading software tasks such as checksumming to hardware.

21.12   Software Subsystem

Earlier, the server's hardware subsystem was compared to an automobile engine. If the hardware is the engine, then the server's software subsystem can be compared to the automobile's driver. Just as the driver directs the engine and the car, the server software directs the hardware. In an automobile race, the winner is determined both by the capabilities of the automobile and by the skill and competence of the driver. An unskilled driver could have a great car but still finish last in a race because he or she didn't know how to coax the highest performance out of the vehicle.

This analogy is useful in pointing out how crucial the software components are to the performance of the server system. Sometimes software is overlooked, especially in discussions regarding performance. Too often, performance is equated only to MIPs or some other hardware-centric metric. In reality, software plays an important role in both the functionality and the performance of the system. For instance, inefficient code requires more processing power to perform a given task in a timely manner. Performance can often be increased either by optimizing code or, in some cases, by offloading to hardware tasks that are normally performed in software.

The following section describes some of the software subsystem components that affect server throughput, how those components relate to each other, and how they relate to the hardware components. Unfortunately, network designers and managers can't directly address many of the throughput problems related to these software components. That task can only be solved by the software and hardware vendors themselves.

However, network managers can query their adapter and server vendors to gain helpful specifics about the design of their software device drivers. They can implement aggressive beta tests for new operating systems. And they can provide more performance-related feedback to their operating system vendors. Network managers can also impress on software vendors the importance of addressing the software issues that delay the achievement of full gigabit-per-second throughput. Network managers should also be flexible, frequently testing and implementing updates to software operating systems and device drivers.

21.12.1   Core operating system

The core operating system defines the architecture of the individual software components in the system and is the glue that holds everything together. Most network operating systems (NOSs) have a modular architecture. This is good because it promotes portability and extensibility. On the other hand, it's bad because it can increase software overhead and make the system less efficient. For example, Fig. 21.5 shows the block architecture of the networking elements for Windows NT 4.0. As the figure shows, the distinct components of the networking stack in Windows NT are modularized. Each of these layers has APIs that communicate with other layers. The transport driver interface (TDI) is important because it specifies a higher-level interface for protocols, so those protocols can communicate with higher-level software in a standardized fashion. Network Driver Interface Specification (NDIS) acts as a lower-level interface between protocols and network device drivers. It provides a standard way of abstracting communication between protocols and device drivers, so that each protocol does not need to have any hardware-specific code and each network device driver does not need to be aware of which protocols are running. This allows multiple protocols to be used simultaneously with different network devices.

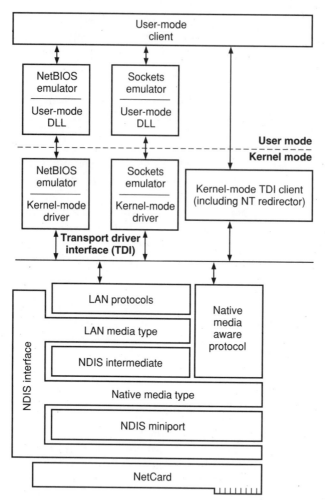

Figure 21.5 Block architecture of networking elements for Windows NT 4.0.

21.12.2 The cost of copying data

Unfortunately, one of the side effects of a modular approach is unnecessary data copying. For instance, when data is received over the network, it is initially placed into buffers owned by the device driver for the network adapter. Later, that same data is often copied into protocol buffers so that the protocol can store, examine, and manipulate the data. After that, the data may be copied again by one or more upper layers (applications, user mode DLLs, TDI clients, etc.). Data copying is a particular problem at gigabit-per-second speeds; at much slower 10-Mbit/s Ethernet speeds, the memory copy is much faster than the network transmission.

The copying of data is a very expensive operation, because the time it takes is generally limited by the speed and latency of the memory subsystem rather than by the instruction execution times of the processor. Caching does not always help reduce the copy time, because the cache locality of network data is not high. For example, when a packet is received over the network, it is usually not stored in the processor's data cache. When the processor goes to examine or copy the packet, the packet data must be retrieved from main memory and placed into the cache.

21.12.3 Protocol stacks

Most NOSs support the use of multiple LAN protocols. Windows NT ships with transport drivers for IPX, TCP/IP, and Netbeui. The modular architecture of Windows NT also allows for the use of third-party protocols. These protocols are implemented as transport drivers, and support either a TDI interface or a private interface at their upper end and an NDIS interface at their lower end.

Transport drivers under Windows NT implement the functionality of multiple OSI* network layers. The transport layer of the OSI model (Layer 4) ensures that messages are delivered error-free between end points. It establishes and terminates connections between end points through the use of naming and addressing. The network layer of the OSI model (Layer 3) controls the operation of the subnet. It determines the physical path that data should take on the network. This includes the routing of packets through the network and the accounting of those packets. This layer completely masks the network topology from the upper layers. The data link layer of the OSI model enables an error-free delivery of frames between nodes. It is responsible for the sequencing of frames, frame flow control, frame error correction, and the delivery of frames to the physical media (Layer 1). Generally, Windows NT transport drivers implement Layers 4 and 3 and parts of Layer 2 of the OSI model.

21.12.4 The TCP/IP protocol

An example of this implementation would be Microsoft's implementation of the TCP/IP protocol in a single transport driver: TCPIP.SYS. For those not familiar with TCP/IP, Table 21.2 provides an overview.

The two most common Internet Protocols are UDP and TCP. UDP is very simple. Basically, it's an unreliable protocol that relies on lower network layers to deliver UDP segments to their destination. TCP is much more sophisticated. It's a reliable protocol that guarantees delivery by requiring the receiver to acknowledge received data. If a TCP

---

* To learn more about the OSI model, see Andrew S. Tanenbaum, *Computer Networks* (Englewood Cliffs, NJ: Prentice-Hall, 1988).

TABLE 21.2   Examples of TCP/IP Layers

TCP/IP layer	Example
Application	Telnet, FTP, e-mail, etc.
Transport	TCP, UDP
Network	IP
Link	NIC and device driver

segment is not acknowledged within a certain amount of time (a time-out period), then TCP resends the data to the desired destination.

It's important to note that the complexity of handling the addressing, sequencing, acknowledging, framing, and retransmitting of data—to ensure reliable, error-free delivery of information between to end points—leads to a lot of overhead in code that must be executed by a system running in a TCP/IP environment. Experiments performed in Intel Labs show that, when using TCP/IP under Windows NT 4.0, it is difficult to sustain more than 270 Mbit/s of throughput through a server system. This is largely due to the tremendous amount of processing required to execute Microsoft's implementation of the TCP/IP protocol.

But there are ways to increase throughput under TCP/IP. The standard TCP/IP configuration allows a small number of frames to be transmitted before an acknowledgment is required. This is known as the *transmit window*. Some adapter vendors mitigate the overhead by allowing the adapter to be configured to transmit a larger number of TCP/IP frames before requiring an acknowledgement. These larger transmit windows can help to improve performance because fewer acknowledgments produce less overhead.

On the downside, larger windows usually mean longer recovery times (time-outs) when a packet is dropped. Network managers should check to see whether the adapter vendor supports larger TCP/IP windows. Microsoft's implementation of the TCP/IP protocol stack enables sliding windows that adjust dynamically, allowing a balance between window size (less overhead) and the cost of time-outs.

21.12.5   Network adapter device drivers

Windows NT, NetWare, and many forms of UNIX all have modular device driver models. A good example of a modular device driver model is the NDIS miniport model used by Windows NT.

In its current form (NDIS 4), the NDIS miniport model is not very efficient. NDIS abstracts the NT kernel and HAL from the NDIS miniport device driver and requires the driver to go through NDIS to communicate with protocols or the hardware. While this has the benefit of simplifying device driver development—making it easier to support many varieties of network hardware under Windows NT—it has the drawback of adding overhead for each packet transmitted or received. In addition, NDIS serializes communication with the device driver, which hurts performance in multiprocessor environments because multiple processors are not allowed to execute miniport device driver code at the same time.

Further, NDIS presents data to the network drivers in a way that is not necessarily conducive to high performance. That's because the driver sometimes must manipulate that data and present it in a different form to the networking device. For instance, when NDIS asks a miniport driver to send a packet, the data in the packet is presented as a linked list of multiple packet fragments. For bus mastering hardware, the device driver is required to sort through the fragment list and build a scatter-gather descriptor list for the hardware from which to DMA the data. Further, the driver must get the physical address for each fragment and lock that fragment down in memory. This process can be fairly time consuming.

In comparison, the ODI interface used by NetWare presents data to the device driver in a slightly friendlier way. For bus mastering devices, the packet fragments may be given to the driver as an array of physical address and fragment length pairs. If network hardware was designed to use the same descriptor format as the fragment address/length pairs, the hardware could simply use NetWare's fragment list as its descriptor list, eliminating the process of building a new list. Intel adapters utilize this feature under Novell NetWare. Microsoft Windows NT needs to evolve to this model in new OS revisions.

21.13   The Interaction of Hardware and Software

As mentioned earlier, the interface between server subsystem hardware and software is often not ideal for high-performance networking. Examples described in the preceding text include the cost of interrupts and frequent cache misses with network data. Unlike many hardware problems, these interface problems do not have ready solutions and network managers must factor them into their performance projections. However, one important issue remains to be discussed: how fragmented network traffic can drastically reduce bus bandwidth.

21.13.1   Practical versus theoretical bus bandwidth

Most network packets are relatively small. A rule of thumb is that 80 percent of Ethernet packets are small (less than 200 bytes in length), yet 80 percent of network data is contained in larger packets. Because packet data from multiple packets is not contiguous in host memory, bus mastering devices must transfer packets one at a time over the host bus.

Here's what happens with a typical host bus such as PCI. Shorter packets result in shorter bursts over the bus. Shorter bursts in turn result in more overhead consumed in setting up the initial PCI transaction. If data is read out of host memory, it takes the memory controller a certain amount of time (latency) to fetch that data. This latency reduces the amount of bandwidth that is available to a given device.

In addition, software usually organizes transmit packets into multiple fragments in host memory. It's common for one fragment to contain the Ethernet header, one fragment to contain the protocol header, and another fragment or two to contain the data portion (or payload) of the packet. This packet fragmentation results in the bus mastering adapter having to fetch data from noncontiguous memory locations, which increases the number of PCI bursts and reduces the possible burst length. With each burst there is more overhead.

One way to increase performance in this scenario is for software to coalesce (or copy) small fragments into a single larger fragment. For example, Intel adapters coalesce small fragments. Coalescing may increase available bus bandwidth, but it places additional stress on the processor, which then must copy packets from one location to another. A good trade-off would be for software to coalesce only small fragments and to map the larger fragments. To implement this trade-off, network managers should check to see whether the NIC vendor's software drivers are written to coalesce fragments and align on larger byte boundaries.

Software often makes no attempt to align data on any boundary, let alone a large one. This can lead to shorter bursts on the bus (the chipset may disconnect during unaligned transactions) and stalls due to partial cache line writes. PCI chipsets, in contrast, are often optimized to handle data aligned on larger (e.g., 4, 8, 16, 32) byte boundaries.

What can network managers do about these software challenges? There is considerable room for improvement in the protocol stack implementation under Microsoft Windows NT. Unfortunately, much of the improvement will only occur in the longer term, after gigabit-per-second servers begin to be deployed. Network managers should stay abreast of operating system developments and look to software updates and new releases. In the short term, they should consider implementing operating systems other than Microsoft NT, such as Novell NetWare, SunSoft Solaris X86, or other UNIX implementations. Also, they should be sure to check to find out whether the adapter vendor's device drivers implement efficient features such as descriptor coalescence. Finally, they should be sure to configure their servers with plenty of L2 cache memory to help mitigate software inefficiencies.

Over the longer term, many tasks performed in software may eventually migrate to the silicon on the adapter, where they can be performed much more rapidly than by the host processor. This integration will allow repetitive tasks to be performed directly on the adapter at much greater speeds, reducing latency and host interrupts. For example, NOS vendors are providing greater access to portions of the NOS code for the adapter. It is possible to load portions of the protocol stack directly onto the Intel EtherExpress PRO/100 Server Adapter under Novell NetWare. Look for more offloading in future NOS releases, such as Microsoft Windows NT 5.0.

Checksumming is one example of a software task that can be offloaded to hardware to boost throughput. A checksum is typically performed on a protocol header to ensure data integrity before a packet is passed up the protocol stack. Performing checksumming on a packet before it is passed up the stack reduces latency and increases performance. Today, the protocol handler and the host processor perform the checksum. In the future, the adapter will perform this activity to improve throughput. Pertinent protocols include offloading TCP, UDP, or IP v4 checksums. Future NOS releases, such as Windows NT 5.0, will support this feature.

21.14   Looking Ahead

In future servers delivering full gigabit-per-second throughput, hardware must help the CPU and not get in the way. At gigabit-per-second speeds, the hardware will be moving very fast and needs to enable the CPU to execute without waiting. The interface between the server hardware and server software must also remain simple and fast in order to run at gigabit-per-second speeds. How will the industry bring about these changes so servers can readily support gigabit-per-second network performance? Here's a look at some of the likely advances over the next few years.

21.14.1   Hardware subsystem

Faster and expanded processing power.   The key constraint in a gigabit-per-second server system will likely be code execution time and memory bandwidth. Raw MIPS will address this problem tremendously, which is why most gigabit-per-second servers that push the edge of performance will be multiprocessor. Also, by doubling clock speed, system designers can almost double the MIPS of the processor and therefore, by Amdahl's law, the ability of the server to handle I/O. Servers with 1000 MIPS will need to be developed to support a half-duplex gigabit-per-second server adapter. Servers with 2000 MIPS will be needed for full-duplex gigabit-per-second service.

Faster bus bandwidth. Even at 64 bits, PCI only offers a maximum bandwidth of 2 Gbit/s. Therefore a "PCI-2" may need to be invented for truly high-performance servers with gigabit-per-second adapters. This is a natural next step from today's PCI and there are several industry leaders already working on it.

Improvements in storage subsystem. Improvements in storage subsystem architecture will help improve gigabit server performance. The storage subsystem may benefit from intelligent coprocessing through the I_2O Specification. The goal of I_2O is to reduce CPU utilization by offloading activity from the host processor to an intelligent coprocessor through an industry standard interface. Some manufacturers are already supporting I_2O with fast Ethernet storage components and device drivers. Microsoft is expected to support I_2O for disk drives in the next release of Windows NT.

Fibre Channel is another storage technology for gigabit-per-second data transfer. It maps common transport protocols such as SCSI and IP, merging networking, and high-speed I/O in a single connectivity technology. It is an open standard, defined by ANSI and OSI, that operates over copper and fiber-optic cabling at distances of up to 10 km.

21.14.2 Software subsystem

Low overhead. Often, stacks are optimized for low-speed network media such as modems. Stacks need to be more efficient and to support a greater level of parallelism. In low-speed networks, the stack has the luxury of longer idle time, allowing more leeway for acknowledgments and copying data from one layer to the next. Either the number of layers must be reduced or tasks should be performed in parallel.

New hardware/software interface. In the future, one way to provide gigabit-per-second throughput will be through a cluster of multiple servers. To address this need, Compaq, Microsoft, and Intel released Version 1.0 of the Virtual Interface Architecture (VIA) Specification in December 1997. It describes a new software/hardware interface for server area networking (SAN).

The purpose behind VIA is to greatly reduce the latency and overhead involved in the processing of network traffic for clusters of servers that are used in distributed computing environments. Robust, high-performance building blocks, combined with industry standard high-performance clustering techniques (and the right kind of cluster-enabled database software), will allow the construction of high-performance and high-capacity server systems that are relatively low in cost when compared to proprietary alternatives. Expect to see VIA products in 1998 and beyond.

Do away with highly layered stacks. Highly layered stacks lead to inefficiency in the data flow, as the cost of copying data is expensive. Improvements to TCP/IP are becoming necessary as more and more users gravitate to this protocol. Future stack implementations should reduce the number of layers and abstractions, improving performance. Vendors such as Microsoft need to focus on this area as they release updates and new versions of operating systems.

21.15   Gigabit Ethernet Will Be the Server Connection of Choice

The available processing power on desktops is continuing to increase. More and more networks are migrating these desktops to fast Ethernet to take advantage of this processing power. As a result, network managers can expect to see bottlenecks at their servers. Today's servers will not fully support gigabit Ethernet; however, improvements in server architecture, network architecture, and operating systems will make gigabit Ethernet the preeminent server network connection over time. In summary, network managers should keep the following factors in mind when making purchase decisions:

- Scalable multiprocessor servers
- The amount and type of memory
- The number of PCI buses
- PCI chipsets with 64-bit/66-MHz PCI slots
- The type of disk drive
- Updates and new releases to operating system software
- Sixty-four-bit Gigabit NICs with support for a variety of network operating systems
- Efficient NIC device drivers
- Support for full-duplex and flow control
- Updates and new releases to NIC device drivers
- TCP/IP configuration parameters

22

How to Optimize the Performance of Gigabit Ethernet End Stations

Selina Lo
Alteon Networks

22.1 Introduction

"I don't know if I have any network bottlenecks. All I know is that my application response time slowly degrades. When it becomes ridiculous, we add a server and split the user load. I know I'm merely putting a Band-Aid on the problem, but I don't have time to figure out what the problem is."

This is a common complaint from information systems managers. Performance optimization is a complex art. To realize maximum performance from a client/server infrastructure, IS managers are faced with difficult multidimensional design questions.

For example, what is the appropriate speed and capacity of the clients and servers? And where should upgrades be undertaken in order to obtain the best end-to-end performance?

Also, what is the appropriate speed and capacity of the network infrastructure that binds together clients and servers? What is the impact of client technology, server technology, the applications supported, and the network topology on overall network/system performance?

---

* Selina Lo is vice president of marketing and products at Alteon Networks, and is responsible for product definition, positioning, market development, and corporate communications. Prior to joining Alteon Networks, Lo served as vice president of product management and marketing for the enterprise backbone switching business at Bay Networks after it acquired Centillion Networks, a LAN and ATM switching company that Lo cofounded in September 1993. Before that, Lo spent three years as senior director of product marketing at Network Equipment Technology, where she was responsible for NET's interworking, frame relay, and ATM product lines. Before NET, Lo held a variety of positions with Hewlett-Packard Information Networks Group.

There are no simple, universal answers to these questions. However, one thing is clear. Any approach that improves only one specific technology area (client horsepower, network speed, or server capacity) generally does not significantly increase overall system performance.

Instead, the discrete components in a communications network must be capacity-balanced to achieve the highest possible performance at the lowest possible cost.

What is capacity balancing? Simply put, it means sizing the capacity of each element of the system to handle the maximum possible load from its input sources. The output load can then be metered so as to allow the next element in the end-to-end path to adequately handle the traffic.

There are a number of key elements in a client/server network whose interaction with each other can affect overall performance. They are: effective CPU power (in the server and client), protocol stacks (in the server and client), network adapter and driver (in the server and client), and the intervening network (between the client and the server). (See Fig. 22.1.)

Studying the progression of client/server performance issues over the years best illustrates this concept.

In the beginning, clients and servers connected to the same Ethernet or Token Ring LAN. In general, these LANs had much more capacity (remember when 10 Mbit/s seemed infinite?) than the attached sys-

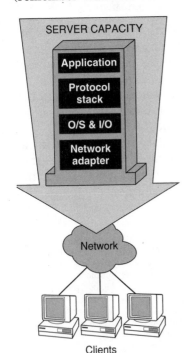

Figure 22.1   There are a number of key elements in a client/ server network whose interaction with each other can affect overall performance.

tems. Therefore performance bottlenecks generally were related to the adapter and driver solutions in the servers.

When wire-speed adapters became available, servers quickly became the bottleneck. Exacerbating the server bottlenecks, these fast adapters "stole" server CPU cycles to support faster drivers and protocol stacks. Meanwhile servers were increasingly being consolidated and centralized, helping to justify further investment in server capacity.

The increase in server and client horsepower in recent years has outstripped the capacity of what are now considered to be low-speed LANs like 10-Mbit/s Ethernet. This has driven the development of higher-speed LAN technologies such as fast Ethernet as well as other switching technologies. Switching increased performance by enabling highly segmented shared-media networks, thereby driving the trend toward installing dedicated bandwidth for each desktop and one or more fast Ethernet connections per server.

As clients have been upgraded to Pentium PCs equipped with switched 10-Mbit/s or 100-Mbit/s network connections, the bottlenecks have once again returned to the network and server connections. Now gigabit Ethernet has emerged as a viable solution to increase the capacity of the backbone and the servers to cope with the dramatic bandwidth increase at the edge of the network. But the introduction of gigabit Ethernet to the network requires a rebalancing of the capacity and load on many network elements (see Fig. 22.2).

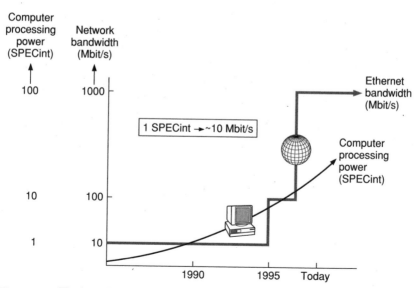

Figure 22.2 The introduction of gigabit Ethernet to the network requires a rebalancing of the capacity and load on many network elements.

A perfectly balanced network system should ideally exhibit four characteristics:

1. Client network connections equal to or greater than the peak capacity of the client computers

2. Backbone network capacity sufficient to handle the peak aggregate capacity of all simultaneous network traffic

3. Adequate server horsepower to handle the maximum potential client access

4. Server connections equal to or greater than the capacity of the server

This chapter focuses on the topic of networked computer (client and server) performance optimization, a key facet in designing such a balanced or tuned network.

22.2 Pinning Down the Problem

The way to start the process of optimizing system performance is to work out where the performance problem lies—with the end station or the network. When a networked PC or server exhibits poor performance, the cause can typically be found in one of five places: memory, disk, CPU, adapter, or network.

The easiest way to find the culprit is to examine statistics from the network adapter that indicate the percentage of received packets dropped and the number of back-pressure events, if available. A high percentage of dropped packets (over 1 percent) or back-pressure events indicates that the receiving system is the bottleneck.

There, the bottleneck can be the adapter, the CPU, memory, or (with applications that are disk I/O-heavy) the disk. The bottleneck can be further isolated by viewing the system statistics on memory, CPU, and disk utilization that are readily available on most PCs and servers.

If none of the system statistics indicate abnormally high utilization, then the bottleneck is most likely to be on the network connection. The simplest solution to the problem is to upgrade the speed of the connection (both adapter and network switch) and in the process ensure that the end-to-end path is capacity-balanced.

22.3 Network Adapter Performance Metrics

But which adapter to install? Some adapters are designed to maximize throughput by stealing host CPU cycles for network processing, while others are designed to offload the host CPU at all costs. Using a "dumb" adapter that relies heavily on the host processor in a congested end system can actually make performance worse by upsetting the balance between the volume of incoming traffic and the available processing cycles.

Some network adapters alleviate this problem by offloading processing from the CPU on the end station. This is particularly important when the adapter is installed in a server. Thus, when host cycles are expensive, as in the case of most enterprise-class servers, a key criterion for evaluating high-speed networking adapters is the amount of host CPU they consume. This number is usually measured as a percentage of the server CPU utilized to process traffic from the adapter. It should be recorded when the CPU utilization parameter on the host system reaches its steady state in a file transfer or equivalent process.

There are three other performance metrics that network designers should use when selecting a network adapter:

- *Throughput*—the user data throughput, measured in megabytes per second. This value is typically measured by the speed at which a large file can be transferred from one system to another. Measured at the user level, it takes into account the Ethernet, IP, TCP/UDP, and application processing overhead.

- *CPU efficiency*—the CPU utilization per megabyte per second of throughput, measured as a percentage. In general, CPU utilization grows linearly with throughput. This measurement is generally used to compare adapters of different speeds or technologies.

- *Latency*—the time it takes the driver in the adapter to receive and transmit a packet; last bit in to first bit out. Although a relevant metric for delay-sensitive applications, latency measurement on a particular system is only meaningful as a component of the end-to-end latency between client and server.

As a rule of thumb, server adapters must be designed to maximize host CPU efficiency, since server cycles are expensive both in absolute costs and in their impact on an organization's productivity. Desktop adapters, on the other hand, may be developed intentionally to optimize throughput at the expense of host CPU utilization, in order to reduce adapter cost.

22.4 Stealing CPU Cycles

Designed to minimize costs, PC desktop adapters that "steal" host CPU cycles to maximize performance are typically unintelligent, relying on the host CPU to perform as much of the frame processing as possible (see Fig. 22.3).

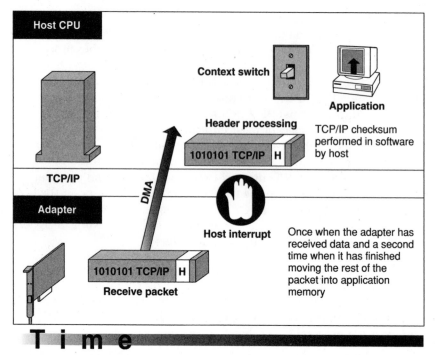

Figure 22.3 PC desktop adapters that "steal" host CPU cycles to maximize performance typically rely on the host CPU to perform as much of the frame processing as possible.

The advent of faster applications is driving the need for 10- and 100-Mbit/s Ethernet adapters that can move data in and out of the host more quickly. Some high-performance adapters implement a concept known as *look-ahead buffers* to increase performance. On an unintelligent adapter, this approach will consume even more host CPU cycles. Here's how it works.

Step 1. The adapter moves the first 64 bytes of received data to the host protocol stack through a buffer pre-allocated by the host. The first 64 bytes include header information and some data.

Step 2. The adapter notifies the host that it has received data by issuing an interrupt.

Step 3. The host protocol stack analyzes the headers and tells the adapter where in application memory to put the remaining data from the packet.

Step 4. The protocol stack moves any data (minus the headers) from the initial 64 bytes to application memory.

Step 5. The adapter moves the rest of the data into application memory.

Step 6. The adapter tells the host protocol stack that it has finished moving the rest of the data packet into application memory by issuing an interrupt.

Step 7. The host protocol stack performs a checksum of the packet in the application memory space.

Step 8. The protocol stack informs the application that data has arrived. The operating system then suspends itself after saving its state information, and invokes the application.

Transferring only the first 64 bytes of received data, instead of the entire packet, to the protocol stack allows the protocol stack to identify the appropriate application and memory locations into which the adapter should deposit the data directly for the target application. Meanwhile, the adapter can receive the rest of the packet in parallel. This reduces latency and minimizes host copy operations. Without this step, the adapter would have to move the entire packet into host memory so that the protocol stack could examine the protocol header and determine the target application. Then the protocol stack would have to copy the data into application memory. Copy operations consume CPU cycles and slow down the receive process, resulting in reduced network throughput.

The problem with this look-ahead adapter implementation is that it generates multiple host interrupts per packet. A costly function in terms of CPU utilization, an interrupt causes two context-switching operations (when a system's CPU must switch from processing one application to another) on a host.

In the first context switch, the host operating system has to stop whatever it is doing, save the current state information so that it can resume processing afterward, and then handle the interrupt. Once interrupt handling is completed, a second context switch occurs during which the operating system recovers its previous state and resumes processing.

With its interrupt overhead, the look-ahead method is only suitable when network throughput is valued over preserving host CPU cycles. For this reason it should be avoided on very high-speed network connections such as gigabit Ethernet, where the network outpaces the majority of computers attached to it.

22.5 Host-Optimized Network Adapter Design

The adapter design in the preceding example is unsuitable for strategic systems such as servers and mission-critical workstations. In these systems, maximizing overall performance (the response times for the entire client population of a server, for example) is more important than optimizing any single component in the network. In this case, the right adapter to use is one that optimizes host performance by maximizing both throughput and host CPU efficiency.

Three activities in network processing consume a lot of host CPU time:

1. Interacting with protocol layers—adding protocol headers, removing protocol headers, generating checksums, and so on

2. Moving data within the host memory system

3. Network interrupt handling

An intelligent adapter can minimize host CPU overhead in three ways. The first is by using an embedded, or on-board, RISC processor or custom packet-processing silicon. With this feature, the adapter can offload from the host CPU-intensive tasks such as removing and adding packet headers, checksum computation, and byte swapping—all of which are more efficiently handled in adapter hardware than in host software.

Second, to minimize expensive data copies, the adapter should support scatter/gather Direct Memory Access (DMA)—the ability to take host data from multiple memory locations and deposit network data into multiple memory locations. This eliminates the need for the host to coalesce data and packet headers via a copy operation for adapter transfers. In addition, when a packet is deposited from the adapter to host memory, the adapter can place the header and data in separate memory locations so the host does not have to divide the header from the data for the target application. The adapter can also make sure that data is deposited into memory locations aligned for the application, a function commonly known as *virtual page swapping*. This eliminates another copy operation when the host operating system prepares to execute the application.

Last, an intelligent adapter should only issue host interrupts when necessary. As discussed in the previous section, each interrupt causes two context switches by the host operating system. Typical adapters are designed to interrupt the host immediately on every packet received to pass the packet to the protocol stack. For high-speed adapters, this design only works well when traffic is light. A burst on a fast Ethernet LAN, for example, can generate over 100,000 packets in a second; on a gigabit Ethernet, over a million packets per second are generated. Interrupting the host at such rates can quickly consume all the available cycles on a system. Thus, when the host is busy and received packets will simply be queued at the protocol stack to await processing, an intelligent adapter should issue a single interrupt for multiple data packets.

Specifically, when packets arrive continuously, the adapter should "coalesce" or collect interrupts by receiving a number of packets before alerting the host. This minimizes interrupt handling overhead, allowing the host to spend its cycles on processing the queued packets instead of constant context switching.

A smart adapter should be able to automatically adapt its interrupt frequency based on traffic conditions (see Fig. 22.4).

Here is an example of the data receive process of such an adapter:

1. The adapter extracts the header from the received packet and puts it into a buffer space provided by the host operating system. At the same time, it begins to perform a TCP/IP checksum.

2. The adapter sets an indicator that informs the protocol stack of the checksum offload and puts the packet data into a separate buffer space provided by the operating system, aligning it on the page boundary (making it ready for application access later). This process (steps 1 and 2) will continue as long as there are still incoming packets and/or adequate buffer space and no preset timer has expired.

3. If there are no more incoming packets or available buffer space, or if a timer has expired, the adapter issues an interrupt telling the protocol stack how many packets have been copied to the buffer.

4. The protocol stack determines the target application for each packet. If target application memory is aligned on a page boundary also, the protocol stack and application memory simply swap pointers, obviating any copies. Otherwise the protocol stack copies the data to application memory.

5. The protocol stack informs the application that data has arrived (i.e., a context switch occurs).

When the network is not busy, the adapter reverts to interrupting the host as soon as a packet is received to minimize network latency. This approach offers both performance and host CPU conservation at the same time.

22.6 Standards and Emerging Standards Impacting Network Adapter Designs

Besides performance optimization, some emerging standards will fuel the need for a new generation of adapter designs. Next-generation Ethernet adapters should readily support these standards:

22.6.1 IEEE 802.3x

An optional flow-control mechanism defined by IEEE 802.3x is available for full-duplex Ethernet operations. It works similarly to XON/XOFF flow control.

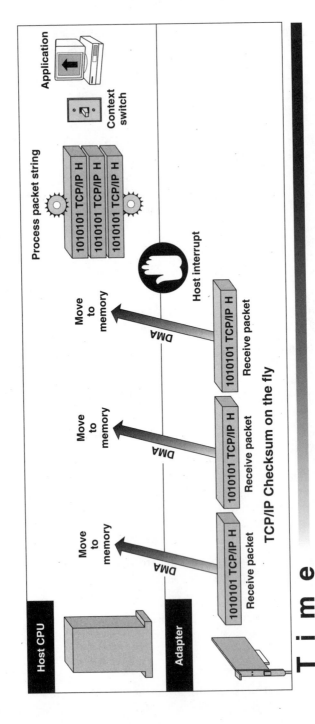

Figure 22.4 A smart adapter should be able to automatically adapt its interrupt frequency based on traffic conditions.

A receiving station at one end of the point-to-point connection can transmit a frame to the sending station at the opposite end of the connection instructing the sending station to stop sending frames for a specified period of time. The sending station ceases to transmit frames until the period of time has passed or until it receives a new frame from the receiving station with a time of zero, indicating that it is okay to resume transmission. This flow-control mechanism is designed to accommodate occasional, but not sustained, bursts.

The 802.3x standard is not specific to any particular speed for Ethernet. It can be used for Ethernet, fast Ethernet, and gigabit Ethernet. An intelligent adapter should provide support for this standard in hardware.

22.6.2   IEEE 802.1p/Q

The IEEE 802.1p specification is a supplement to IEEE 802.1D, and extends the concept of Filtering Services introduced in that standard. It defines additional capabilities in bridged LANs that allow expedited traffic capabilities (priority), end-to-end signaling of user priorities regardless of the underlying media, and a dynamic method of defining and establishing group membership.

IEEE 802.1Q leverages the mechanisms of LAN bridging defined in 802.1D and 802.1p and defines additional mechanisms that allow the implementation of virtual bridged LANs. VLANs can be partitioned based on criteria of which Layer 2 devices are normally unaware, such as IP subnets, protocols, and applications. 802.1Q-compliant devices utilize a VLAN tag in addition to the destination address in frame forwarding. 802.1Q specifies the format and the rules for VLAN tagging. 802.1p/Q-compliant adapters allow a connected host to be a member of multiple VLANs over a single physical connection. Intelligent adapters should have the provision to support these standards when they are finalized, through either custom hardware or software executed by an on-board processor.

23

How to Prioritize Time-Sensitive Traffic on a Gigabit Ethernet LAN

Gordon Stitt*
President and CEO, Extreme Networks

23.1 Introduction

Data traffic on local area networks has historically been text- or graphics-related. "Time-sensitive" in this environment has meant the time necessary for the user to read the screen. Since the advent of the Web browser, and with the emergence of digital video sources, *real-time* takes on a very different meaning. For example, an MPEG compressed video stream requires 3–5 Mbit/s per stream.

On today's networks, time-sensitive data is that which is disrupted or becomes significantly less valuable if it incurs delays as it passes across the network to the user.

A classic example of delay-sensitive traffic is networked video. This is sent as a continuous stream of data. MPEG, the most common form of stored video, is in fact continuously bursty. In other words, it is a continuous steam of data with frequent higher-bandwidth bursts. For the display of video to be smooth, video frames must be received in time for display. A delay causes the screen to freeze until the next frame

---

* Gordon Stitt is president and CEO of Extreme Networks, which he cofounded in 1996 along with Herb Schneider and Steve Haddock. Extreme Networks develops and manufactures third-wave LAN switches based on gigabit Ethernet technology. Prior to the founding of Extreme, Stitt was cofounder of FDDI and LAN switching pioneer Network Peripherals, where he held the position of vice president, marketing and vice president and general manager of the OEM Business Unit. Prior to NPI, Stitt spent 2 years at Sun Microsystems and 11 years in both the modem business and the early PC business, in both R&D and marketing roles. Stitt lives in the heart of Silicon Valley with his wife, two children, and beagle. He holds an MBA with an emphasis in marketing and finance from Haas School of Business, University of California, Berkeley, as well as a BSEE/CS from Santa Clara University, Santa Clara, California.

arrives. This freezing makes the video very difficult to view, and therefore much less useful. Thus video poses a difficult problem for network designers: it represents a continuous and bursty flow, it uses a fair amount of bandwidth, and delays are very visible to the user.

There are other types of delay-sensitive data, of course. Consider a bank that needs to process the day's transactions during the evening and complete them before the next day's opening. Although the time involved is several hours, a massive amount of data must be transferred across the network in a fixed amount of time or there will be serious problems for the company. Network delays could delay the bank's opening in the morning.

But compared to the demands of video, other types of time-sensitive traffic are relatively easy to handle. That is why this chapter deals with solutions for prioritizing video. However, the reader may apply the same rules to other types of traffic as well.

This chapter looks at some of the sources of delay in Ethernet networks and the solutions. The solutions come from careful network design, the correct choice of switches and routers, and the implementation of emerging quality of service (QOS) technologies.

23.2 Where Network Delays Come From

The ideal network would deliver infinite bandwidth with no delay to any number of users. It would look like a single switch/router that connected all of the organization's desktops, servers, and WAN connections. Each server connection would be many times the speed of the desktop links (so as to serve many desktops simultaneously) and the single switch/router core would have enough bandwidth to handle all servers and desktops at full speed. If this were available, network planning and quality of service (QOS) wouldn't be necessary. A nice idea. But products that handle 1000 client connections and 100 servers in a single switch/router aren't available—and are not likely to be anytime soon.

In reality, today's networks are a collection of network segments (operating at different speeds) and various internetworking devices (some switches, some routers) with varying performance levels. In these real-world networks, packets may traverse several network segments and be processed by half a dozen routers on their journey from server to desktop. Along this path delays occur. Typical causes of these delays are:

- Bandwidth-constrained network segments
- Congested network segments
- Congested internetworking devices
- Transitions as traffic is sent between network segments running at different speeds

The easiest way to understand these problems is to examine them one at a time.

23.2.1   Problem: bandwidth-constrained network segments

Gigabit Ethernet is one of the fastest networks available. Yet it is unlikely that any corporation will be lucky enough to own a pure gigabit network. Within any network there will be slower segments. The location of these 10- or 100-Mbit/s segments can have a significant impact on the overall performance, and can induce latency as the data waits to cross the slow segments.

Bandwidth constraint on segments is a slightly different problem than pure congestion on the segment. A bandwidth-constrained segment may not have collisions, but may nevertheless still be a limiting factor in network performance. For example, a direct link between routers may run out of bandwidth, causing the routers to utilize a different (and possibly slower) link to communicate.

23.2.2   Problem: congested network segments

When Ethernet is used in a hub or repeater form, it operates in a shared mode and uses collisions based on the CSMA/CD protocol to produce fairness in access to the network. When an Ethernet station has information to transmit, it listens to the network and, if there is no one using it, begins transmitting. If another station begins transmitting at the same time, a collision occurs. Both stations then back off and wait a random (but short) period of time and then try again. This wait adds latency to the network. When many stations are using the network, or loads are very high, there are many collisions and the latency adds up and slows operation of the network.

As the number of collisions increases, the actual efficiency of the network decreases due to the wasted network bandwidth consumed by collisions and back off times. The higher the utilization of the network, the greater the potential for latency introduced by retransmissions.

23.2.3   Problem: congested internetworking devices

All switches and routers have latency. Latency is the time required from the moment the first bit of a packet is received to the moment the first bit appears on the output port. Even so-called cut-through devices incur some latency; it is inherent in a switch/router. There is, however, a large difference in latency between different devices. For example, a router typically has much longer latency than a LAN switch. Routers typically use a processor that performs complex calculations before forwarding a packet, whereas a switch forwards in hardware based on a simple address. As the traffic through a router increases, latency typically increases as the forwarding processor becomes busier. Router latency—often several milliseconds—can contribute significantly to the latency of a network.

Congestion in routers or LAN switches can come from many sources (see Fig. 23.1). A device may have limited backplane bandwidth that limits the speed at which it can handle data. For example, a router with 20 full-duplex fast Ethernet ports may have a 1-Gbit/s backplane. If all ports were active at full bandwidth, 4 Gbit/s of backplane—four times what is available—would be required to prevent the router from acting as a bottleneck. When congestion like this occurs, data is held up while the congestion clears. In addition to backplane limits, a router's processor may have a maximum packet per second (pps) capability that could be exceeded in high-traffic-load situations. Also, the port configuration of the device may be such that too many segments are routing to a single segment, or a high-speed input link may also be routing to a lower-speed output link, causing congestion.

23.2.4 Problem: traffic sent between segments of different speeds

Anytime there is a transition from a high-speed LAN to a lower-speed network (for example, going from gigabit Ethernet to 100-Mbit/s fast Ethernet), there is the potential for the network to slow down and for latency to increase. This happens when a burst of high-speed data

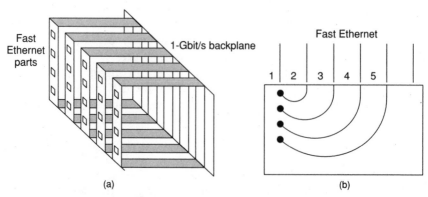

Figure 23.1 Congestion within routers and switches. (*a*) A router with a 1-Gbit/s backplane has five slots, each holding a module with four fast Ethernet ports. These 20 total ports—with maximum bandwidth of 4 Gbit/s—compete for the only 1 Gbit/s that is available. (*b*) Congestion caused in router when four fast Ethernet ports are sending data to a single fast Ethernet port.

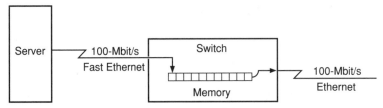

Figure 23.2 Speed mismatch can cause delays. Whenever a high-speed port sends information to a lower-speed port, data is stored in memory.

waits for the slower link to send that same burst. This situation can create backups that cause all traffic to be delayed while the slower link empties out (see Fig. 23.2).

23.3 Solutions

Having examined how latency can be introduced into the network, it's now time to look at the solutions for avoiding latency and building a high-performance network for delay-sensitive data. Of course gigabit Ethernet itself provides high speed and high bandwidth, and in themselves these can solve many latency problems. Unfortunately, however, not many users will have pure gigabit networks anytime soon: it would be prohibitively expensive and impractical from a cabling standpoint. Gigabit Ethernet networks will have many segments at mixed speeds. A typical gigabit Ethernet network is likely to have dozens (and possibly hundreds) of 100-Mbit/s fast Ethernet and 10-Mbit/s Ethernet segments.

It is this mixed-speed network that network managers work with every day. It is in the careful design of these networks that managers can solve the end-to-end latency problem and build networks that can support multimedia now and in the future.

Before exploring the solutions, it is worth making a comment regarding bandwidth. A common solution proposed by some systems integrators and vendors is to simply throw bandwidth at the problem. As can be seen in the following sections, bandwidth is only part of the solution; a network that handles video and provides scalability into the future requires a balance between speed, bandwidth, and congestion control.

To achieve the goal of transporting delay-sensitive traffic, several things need to be considered:

- Network topology
- Internetworking devices
- Congestion avoidance protocols (both network Layer 3 and MAC Layer 2)

23.4   Solutions: Network Topology Design

The topology or layout of the network is the most important first step in designing a gigabit Ethernet network for video and delay-sensitive data. What follows is a list of guidelines, or rules, to use when designing the network topology.

Keep in mind, however, that the design rules in the following section should also be used in conjunction with knowledge of the traffic patterns on the network. Many network managers have already analyzed network traffic on their networks and have mapped common traffic patterns. Consider the e-mail server—a device to which all users must be connected. Connections to the e-mail server do not need to deliver delay-sensitive data, but they do need to transport large amounts of data.

Further, the topology design for the network must also take into account unknown traffic patterns. Why? Because networks are changing rapidly and new, emerging technologies could have a dramatic impact on traffic patterns. Again, using the example of an e-mail server, what would happen if video e-mail became widely deployed? What would happen if the vendor of the e-mail server used it for desktop video conferencing as well?

Thus, network topology design must take into account current patterns, but must also be designed to be scalable. It must be able to handle higher-speed applications, higher bandwidth, larger networks, and more delay-sensitive traffic. The following design sections should be applied with an eye to future applications as well as dealing with today's problems.

23.4.1   Low-latency backbones

Much of the traffic will traverse the backbone, making the design of the backbone key. Many market analysts are predicting that the Internet, and deployment of Intranet applications, will cause backbone traffic to increase from 20 percent of LAN traffic a couple of years ago to 80 percent by the year 2000. This dramatic increase is one of the drivers for deployment of gigabit Ethernet.

The backbone must be the highest-speed, highest-bandwidth part of the network. This may seem obvious, but it is often overlooked. Historically, the backbone has either been a large router, often referred to as a *collapsed backbone,* or a small group of FDDI connected routers (referred to as a *distributed backbone*). Both of these designs have severe bottlenecks. The collapsed backbone produces bottlenecks within the router. The distributed backbone, in contrast, suffers bottlenecks on the shared FDDI links. In a gigabit network these problems can be avoided by building the backbone using a single high-performance, high-bandwidth switch or a series of interconnected switches.

The new generation of gigabit Ethernet switches have 16 or more gigabits of internal bandwidth—much higher than the 1-Gbit internal capacity typical of today's routers. These devices also have forwarding rates—some in excess of 10 million pps—much higher than the several hundred thousand pps typical with routers. A collapsed backbone with these specifications provides in excess of an order of magnitude greater performance and scalability than the typical current collapsed backbone.

For users with a distributed backbone, gigabit Ethernet provides great scalability. For one thing, the links between switches (or routers) are switched, rather than shared. Further, the speed of gigabit Ethernet—10 times that of FDDI—combined with the bandwidth of a dedicated link (versus the shared FDDI ring) also provides scalability. In addition, most gigabit Ethernet products allow *trunking*, where several gigabit Ethernet links are connected in parallel, providing further increased bandwidth. Shared-media backbones should be avoided. (see Fig. 23.3).

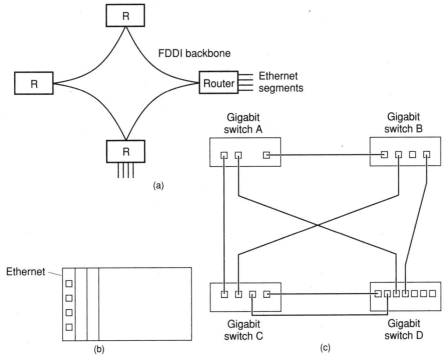

Figure 23.3 Legacy LAN backbones. (*a*) Typical distributed backbone using FDDI to connect routers together. Performance is limited by FDDI shared bandwidth to 100 Mbit/s for the entire backbone. (*b*) Typical collapsed backbone is a single router. Performance is limited by backplane of router. (*c*) Gigabit Ethernet backbone. All links are switched for very high bandwidth. Routers C and D are connected with trunked links, doubling bandwidth to 4 Gbit/s full-duplex—40 times the performance available with FDDI.

23.4.2   Interconnection of servers

Servers should be connected as close to the backbone as possible and at the highest available speed. In this sense, *close* means the least number of router/switch hops away. In a gigabit Ethernet network that means servers are attached to the backbone directly, using gigabit links. The direct backbone connection is likely to be a switch/router port on a device that forms the collapsed or distributed backbone. Given the continued improvement in server performance, it is important that servers be connected to the highest-bandwidth network. Although it may be tempting to put servers on shared gigabit, it is important for servers—as the source of much data—to have deterministic access to the network, which means a dedicated switched link. Better still, servers should be connected to full-duplex switched links.

The new Intranet model encourages the use of workgroup connected servers by making it very simple to convert a desktop into a server. In cases where servers are located away from the backbone, gigabit Ethernet should be used to connect the servers to a workgroup switch, which then connects to the backbone through further gigabit links (see Fig. 23.4).

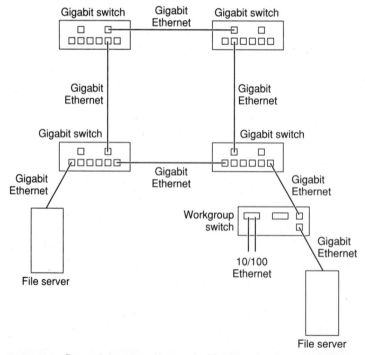

Figure 23.4   Server interconnection to gigabit networks. Servers should be connected directly to backbone switch ports or to workgroup switches that have Gigabit connections to the backbone.

23.4.3   Dedicated client/desktop connection for low latency

The desktop recipient of data has bandwidth needs less than those of a server, but the same need for uncongested access. The cost of desktop switching has come down substantially with the migration from 10-Mbit/s Ethernet to 100-Mbit/s Ethernet well under way. The network design for client networks is simple: minimize latency to the backbone and provided dedicated switched links.

There are two steps a network designer can take to minimize latency to the backbone. The first is to minimize the number of switch/router hops between desktop and backbone and ensure that those switches/routers are high performance. The links that aggregate desktops should also increase in performance as they get closer to the backbone. For example, consider a situation with 24 desktops in a workgroup. Each of these desktops is currently connected to a 10-Mbit/s switch. That switch has a 100-Mbit/s uplink that connects it to another switch, which ultimately connects to the backbone. To futureproof this client network, the aggregating switch—the 100/100 switch that connects multiple workgroups—needs to have a higher-speed connection to the backbone. This is the place where gigabit Ethernet is introduced to the network as a method to aggregate data and futureproof the network by providing additional bandwidth to the backbone.

The final step to providing low latency is to connect desktops to switched segments—eliminating any congestion caused by network access to shared media. The cost of 10-Mbit/s switching is very low, and the price of 100-Mbit/s switching is coming down fast. Users should eliminate shared Ethernet—even at the desktop—and run dedicated switched links. This ensures that each desktop will have dedicated bandwidth to handle video and new applications that could be a problem on shared networks.

23.4.4   Minimize network hops

As networks grow, the number of tiers, or levels, they contain tends to increase. The backbone is the highest speed; the next level is floors, the next is the wiring closets. At each level, switching or routing is used to deliver data to the next level—ultimately to a desktop connection (see Fig. 23.5).

It is unavoidable that every time a packet traverses a switch or router some latency will be introduced. In the Intranet world, packets may come from a desktop, go all the way to the backbone, and traverse it. A response follows the same path. Minimizing the number of switches or routers between the backbone and the desktop minimizes latency. This is known as keeping the hop count down.

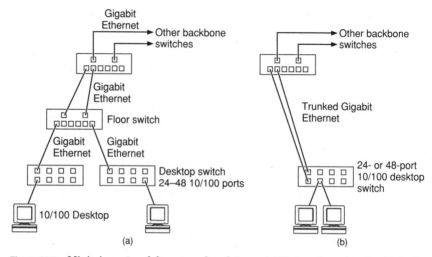

Figure 23.5   Minimize network hops to reduce latency. (*a*) Enterprise network with backbone, floor, and desktop/wiring closet switches. (*b*) Higher-density desktop/wiring closet switches allow for simpler, faster network design.

One way to do this is to have a single, flat network. This is rarely advisable; it keeps latency low, but other issues come up, such as single point of failure (what happens if a big router fails?), and congestion due to broadcast and multicast packets also occurs. The size of a switched network today is largely limited by the size of the broadcast domain or IP subnet. The larger the switched LAN, the greater the percentage of broadcast traffic—and the greater the chance for congestion. Further, many organizations have already assigned IP addresses, and introducing a large flat network would require major reconfiguration.

Perhaps most importantly, no one has been able to build a really big router/switch that can form a single network. The sheer number of connections required would cause congestion in the box as packets competed for resources. And then there is the matter of cabling. Buildings, particularly high-rises, are often precabled, and this may limit the ability to aggregate all networking into one big switch.

The solution is to compromise. The hop count should be minimal, but at the same time the size of network segments and the congestion within the switches/routers should be taken into account.

This is done by starting at the desktop and working toward the backbone. The first-level switch (the desktop switch) would be selected first. The choice is made based on the number of ports, the bandwidth, and over-/undersubscription. For port count, more is generally better (because of lower cost), but the number of ports is limited by cabling distances (all users must be within 100 m of the desktop switch) and bandwidth. Desktop switches typically have limited bandwidth. The number of ports is typically 16, 24, or 48. If these are 10-Mbit/s Ethernet ports, then a fully subscribed switch would need 320, 480, or 960 Mbit/s of backplane bandwidth. For cost reasons, desktop switches are typically oversubscribed by 1.5:1 or 2.5:1. If smaller desktop switches (8 or 12 ports) are used, then another level of switch may be required to aggregate the desktop switches. This would add a level to the network and increase the hop count by 1. In general, then, it's better to have higher density on a desktop switch as it tends to reduce hop count.

The uplink on the desktop switch would then connect to the next-level switch. In the case of a large organization, this would be a floor switch. In a smaller organization the desktop switch may connect directly to the backbone. The number of ports on a floor switch is typically smaller. A secondary floor switch may be required in the case of a very large building. The floor switches must be nonblocking and under-subscribed/fully subscribed.

Finally the floor-level switches connect to the backbone. In this example, traffic crosses four hops maximum.

23.4.5 Summary of topology

In summary, the topology of a network should follow these rules:

- Backbones need bandwidth—lots of it. This is provided within switches/routers and between them on a distributed backbone.

- Servers should be connected at the highest speed possible, as close to the backbone as possible.

- The backbone itself, whether distributed or within a single box, must provide very high bandwidth and sustain high speeds.

- Flatten the network. Minimize hops.

- Switch rather than share. When costs permit, the model of a single station per link is better than that of traditional shared Ethernet.

23.5 Solutions: Dealing With Congestion

There are many sources of congestion in a network. Some forms of congestion can be avoided by utilizing different equipment or through network design. But these solutions can't prevent other types of congestion. Take the case of two servers sending data to a single client. No matter how fast the network, the client network can't handle twice the traffic and the result will be congestion and delays. In cases such as this, quality of service protocols such as RSVP and 802.1p are utilized to resolve the congestion. In many cases, congestion is still unavoidable.

23.5.1   Congestion on links

Congestion on network segments can be caused by too many stations sharing a link. Ethernet becomes less efficient, and the latency increases, when many workstations share it. The solution is to reduce the number of stations and—ideally—go to a fully switched network with a single station on each segment. With one station on each switch port, there is no competition for bandwidth on that segment. The other benefit of the single-station model is that full-duplex operation is possible. The full-duplex mode allows information to travel in both directions on the same link at the same time, effectively doubling throughput and reducing latency. This is currently cost effective for standard Ethernet, and will soon be cost effective for fast Ethernet.

Congestion can also occur on links between switches. As discussed earlier, sometimes a link is simply too slow to handle all of the information being sent across it. The best solution here is to upgrade that link to a higher speed. An alternative is to load share across multiple links. In other words, provide multiple links between switches. Load balancing is provided by router protocols, or it may be implemented in some vendor-specific features.

23.5.2   Solving congestion in switches/routers

Congestion in switches and routers is a common source of delay in networks. Some problems—such as oversubscription and blocking—can be solved by replacing existing routers. Others, such as delays due to speed mismatch and feature slowdown, can be handled by product or network reconfiguration.

A blocking switch/router has internal limitations that can cause congestion. A nonblocking device, on the other hand, has the ability to handle all ports at full wire speed simultaneously without any internal congestion. Blocking is caused by limiting the switch fabric performance or (as is often the case with a router) the performance of the processor.

Vendor specifications can be confusing, so here is a method to determine nonblocking. To be nonblocking, a device must handle all ports at once with no performance degradation. To work out whether a switch is nonblocking, take the number of ports, multiply by the speed, and then (if full-duplex) multiply by 2. The specified switch bandwidth should equal or exceed this number. For example, a switch with two gigabit Ethernet ports and 16 fast Ethernet ports would require 7.2 Gbit/s of bandwidth:

$$2 \times 1000 + 16 \times 100 = 3.6$$

$$3.6 \times 2 = 7.2 \text{ (for full-duplex)}$$

If the product being considered has less than 7.2 Gbit/s of bandwidth, then it is a blocking device.

Another issue with blocking architectures is packet throughput. This relates to the ability of the device to handle high-volume, minimum-size packets. Here the rule of one and one-half is used To calculate this, take the required bandwidth (not including full-duplex) and multiply by 1.5. Using the above example:

$$3.6 \times 1.5 = \text{required pps performance}$$
$$(\text{in millions of packets per second})$$

The reason for the 1.5 ratio is that Ethernet allows a minimum packet size of 64 bytes. Including headers, that calculates out to 14,880 pps for Ethernet, 148,800 pps for fast Ethernet, and 1,488,000 pps for gigabit Ethernet. For the sake of simplicity, 1.48 is rounded to 1.5.

If the router or switch has less than the required packet forwarding rate, then it is blocking.

Nonblocking is clearly superior to blocking. But nonblocking routers are rare. Sometimes reducing the number of ports on a router configuration can move it into the nonblocking class.

23.5.3 Congestion caused by oversubscription

Oversubscription in a switch is often confused with blocking. Oversubscription occurs when the port configuration and use of the switch are such that there is more data capacity on one side than on the other. This can be shown by looking at two examples.

A 24-port fast Ethernet switch has a single gigabit Ethernet uplink. This product can be oversubscribed 2.4:1 ($24 \times 100 = 2.4$ Gbit/s). In other words, all of the fast Ethernet ports can drive the gigabit port and overwhelm it.

A 16-port fast Ethernet switch has 2 gigabit Ethernet uplinks. This product is undersubscribed ($16 \times 100 = 1.6$, which is less than 2).

Oversubscription can cause congestion and therefore induce latency. The key word here is *can*. Oversubscription is perfectly appropriate for a desktop switch because it is highly unlikely that all desktops will be operating at full speed at the same time. So the 2.4-Gbit/s rate may never be realized. On the other hand, a switch connecting servers to a backbone should *never* be oversubscribed, and the same goes for a backbone switch.

In summary, users should be mindful when choosing routers or switches that all backbone or server connection products are nonblocking and undersubscribed (or fully subscribed) (see Fig. 23.6).

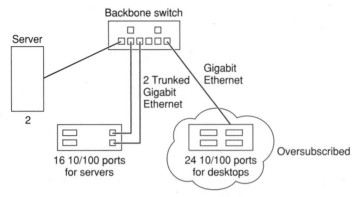

Figure 23.6 Network oversubscription occurs when there is more bandwidth required on the downlinks than is available on the uplinks. Modest oversubscription (3:1) is OK for desktops; servers and backbones should not be oversubscribed.

23.5.4 Speed mismatch

Wherever there is a speed mismatch, there is the potential for congestion. This is unavoidable as a network will always have segments of different speeds coming together. For example, if a server on gigabit Ethernet sends data to a desktop on a 100-Mbit/s Ethernet, there will be congestion. The buffers in the switch will store the data, but at a 10:1 speed differential even the largest buffer will fill up.

This problem is typically solved by upper-layer protocols (the second block isn't transmitted until the first is acknowledged), and by true quality of service protocols such as RSVP.

23.6 Solutions: Solving Congestion Using Protocols

Quality of service (QOS) is a general term that describes a group of technologies that provide different service levels for different types of traffic. These service levels, for instance, can be in the form of a latency guarantee. Typically, QOS classifies traffic into classes, each of which is given varying priority on a dedicated and limited resource.

QOS is generally implemented as a protocol that provides a way to request and grant levels of service. The device involved (router or switch) must then enforce the request and provide the service. These protocols can operate in routers at the network layer (Layer 3) and in switches at Layer 2. Some new switches may have the capability to handle Layer 3 QOS protocols.

23.6.1 RSVP

RSVP is an emerging IETF standard associated with providing quality of service in IP networks. This protocol, when implemented in switches and routers, allows users to request service levels from the network. Going back to the video example, if a user is going to run a live training video on a server, RSVP could provide special treatment for that session. RSVP itself is a protocol for passing reservation messages; IETF Integrated Services (IntServ) is a set of definitions that determine the actual type and quality of service that is provided. The two are always used in conjunction with each other.

There are three parts to RSVP: admission, enforcement, and policing. RSVP is a recipient-initiated protocol—that is, the recipient of the delay-sensitive data makes the request for service. This usually means that a user application makes the request to the network. The request is made to the network—which means to the routers that control the network traffic. In this scheme, different users of a server can request (and receive) different levels of service from the same source. The IETF IntServ is defined for router-to-router use. There is a subset of IntServ—Integrated Services over Specific Lower Layers (ISSLL)—that can work over Layer 2 switches. Although RSVP can be used to regulate non-IP traffic, in reality it is an IP-specific protocol.

23.6.2 RSVP in action

Before looking at what RSVP can do, it's a good idea to look at what happens in a router without RSVP. Without a protocol to prioritize traffic, the router provides a service level known as *best effort*—the router will attempt to get all of the packets where they want to go, with minimal delay, but without any guarantees. In a congested network, or with an overloaded, low-performance router, best effort may not be suitable for delay-sensitive traffic. As mentioned, RSVP itself doesn't provide services, it provides signaling to request and grant services. The services are provided by IntServ. The services defined fall into two basic areas:

- Controlled load
- Guaranteed delay

Controlled load allows the allocation of bandwidth for a particular session. For example, a user wanting to set up a one-way network video session may need 5 Mbit/s of bandwidth. The request would be made for this amount of bandwidth. When the request was received by the router(s) along the path, the router(s) would check to see whether 5 Mbit/s of bandwidth could be reserved for that session. If so, the request would be granted and the session could begin. This example has been simplified somewhat—the actual request can contain three parameters: minimum required bandwidth, peak bandwidth, and burst length. This allows the router to make a more intelligent decision in granting a specific request. Under the controlled-load scenario, a router can oversubscribe—or overpromise. The reason is the basic premise that not all users will utilize the full burst rate at once. This can, of course, be a problem in a heavily loaded network.

The other IntServ that can be requested is guaranteed delay. Guaranteed delay is just that—a guarantee. The router receiving a request cannot oversubscribe. Granting a guaranteed delay is much more difficult for the router than the guarantee of bandwidth, but it also provides a stronger guarantee to the user. This is difficult for routers, because they rarely have extra, or spare, performance. It is very difficult to predict future traffic and router loads and a burst of traffic coming in one port may consume a tremendous amount of router processor resources, causing the router to violate the guarantee.

If a router cannot provide the desired level of service to the user, the request is denied. This doesn't mean that the user cannot use the network; it just means that the level of service guarantee that was requested cannot be met. The user then receives best effort delivery. By default, if you bothered to go through the reservation process, you will get better performance than those that did not. In other words, asking and being denied is better than not asking at all.

RSVP holds great promise for providing guarantees in the delivery of delay-sensitive traffic. RSVP is relatively simple and can be implemented in a wide range of devices. There are some limitations, however. First of all, if the entire network (that is, all routers, switches, and end stations) doesn't support RSVP, its usefulness is limited. A guarantee of delay or reservation of bandwidth must be end to end: server to desktop. Second, the end stations must participate. That is, the application must be modified to make the RSVP request. It could take some time for applications to become RSVP aware. Another limitation of RSVP is the anarchy factor. RSVP operates on a first come, first served basis. What happens on a 100-user network when every user requests the full 100-Mbit/s bandwidth of the network? The first requests are granted, and the others fall into best effort. If a user with a mission-critical application requests late, he or she is out of luck.

23.6.3 MAC layer protocols

The 802.3 standard itself does not provide for traffic prioritization or quality of service. Actually, this is not a limitation, since it provides for flexibility by detaching the work on QOS from the work on Ethernet.

This partition also enables innovation: Ethernet is here to stay, QOS will evolve and become more powerful.

23.6.4   IEEE 802.1p

IEEE 802.1p is an emerging standard for prioritizing traffic over Ethernet (including gigabit Ethernet). 802.1p provides a scheme for assigning a priority to packets (rather than to specific sessions). Under 802.1p a tag is attached to each packet. The tag contains a priority field (3 bits for eight levels of priority) that is promulgated with the packet across the network. Each router or switch that handles the packet will recognize the tag and treat the packet with the appropriate priority.

802.1p has some advantages over other schemes. First of all, 802.1p is a Layer 2 protocol. It can be used in relatively dumb switches as well as in routers. It suffers some of the same limitations as RSVP: it must be implemented in all of the devices to be effective (end-to-end solution) and it suffers from the anarchy syndrome. In addition, 802.1p provides only a priority scheme. Traffic can be broken into categories, but there is no way to guarantee a particular bandwidth or delay (as can be done with RSVP). The assignment of categories is left to the user—which means that different devices may have different interpretations of what to do with a certain level of priority traffic. The strengths of 802.1p are its simple design and potential wide implementation in switches and routers.

24

Building a Data Utility

Paul Weiss*

Plaintree Systems

24.1 Introduction

Gigabit Ethernet has been the subject of great fanfare. What's going on here? Is there something magical about such a straightforward extension of technology that justifies all of the excitement and hype?

The answer is simple—Gigabit Ethernet has great significance because of the elegant simplicity of the next wave of local area networks that will use it—networks that approach the ideal data utility, requiring little more effort to manage than today's typical voice/PBX network.

A utility is something that is usually taken for granted. It serves many users with high reliability and high availability, independent of user location. It is perceived to do its job well because it meets expectations. It changes very slowly (at least from external appearances), so it is expected to work tomorrow as well as today. As capacity increases are needed, users assume that the increases will be provided for in a way that is invisible to them. A utility takes very little effort to administer because changes are infrequent and because new users or services can be added without disturbing the existing user community. The only times it is even thought of are those instances when something interrupts service, and they are noteworthy because they are so rare. A utility is just *there!*

This chapter addresses the background of why local data networks are in desperate need of simplification and utility-like behavior, and

* Paul Weiss has over 30 years of data communications experience in engineering and marketing positions with GTE Sylvania (Electronic Systems Division), Motorola-Codex, Network Systems-Bytex, and Plaintree Systems. He has also been principal of Wayside Consulting, where he interpreted technology for a number of firms in the data communications business. His work has involved LAN, WAN, and radio communications, and is currently focused on campus internetworking and gigabit Ethernet. Mr. Weiss has bachelor's and master's degrees in physics.

then outlines how gigabit Ethernet can help network managers bring systems to this level of performance.

24.2   Simplicity Is a Winner

When technology developments evolve to the point where they make people's lives simpler, they turn into real winners. Consider the many technology-based products that are used every day: personal computers, telephones, automobiles, and so on. The sophistication of their underlying technology is taken for granted, because no one has to worry about them; people just use them.

The Internet was churning along in a small growth pattern for years within the research and academic communities. It wasn't until HTTP and the World Wide Web made it simple enough for everyone to use that the Internet began its surge into the explosive growth pattern that is witnessed today.

Radio telephones were expensive, clumsy, and geographically restrictive (and therefore scarce) until cellular technology changed the rules. Now cellular telephones are a common sight everywhere from elevators to restaurants.

In all of these cases, simplicity on the surface was achieved with embedded sophisticated technology, but it was the simplicity that was the catalyst for dramatic growth.

24.3   Local Data Networks—
Missing the Boat

With all of the high-tech talent applied to the development of local area data communications over the years, one might assume that things would be getting simpler. On the contrary, information systems executives are faced with an increasingly bewildering array of choices—whiz-bang products and technologies that all seem to extract the toll of adding another layer of complexity as payment for whatever benefit they offer.

A lot of the complexity observed today is rooted in the history of local area networking. The first local area networks (LANs) tied together isolated groups and were administered by one of the group's members. Each group was free to choose among competing solutions such as Ethernet or Token Ring at the physical level and IP, IPX, Appletalk, or DECnet for their networking protocol.

The evolution of these early LANs into today's campus internetworks has followed a series of small steps, most of which just piled new improvements on top of previous ones. It started out with snaking cables or concentrators (later renamed *hubs*) tying small groups of users together. Bridges allowed these groups to communicate outside their closed environments, but it turned out that early bridging techniques didn't scale too well. The door was opened for routers to correct the bridging problems and establish dominance both in the local area backbone and across the wide area network (WAN).

Eventually, the high costs of adding more and more router ports resulted in the emergence of LAN switches. Initially, LAN switches were just bridges reduced to a hardware implementation, making them faster and cheaper than earlier programmed, or software-based, versions. LAN switches experienced instant acceptance in the market because they fit neatly into a middle ground between hubs and routers. In this intermediate position, they economically boosted the aggregate bandwidth of selected portions of the local network. Unfortunately, however, the bandwidth boost from LAN switches couldn't keep backbones from becoming the bottleneck of the late 1990s. Changing traffic patterns have forced ever more traffic into the backbone, where throughput at the core of the most networks is capped off by the backplane capacity of big routers.

Thus, most campus networks today look something like Fig. 24.1. Users are cabled to hubs, which are linked to backbone routers, with LAN switches sandwiched in between. This is extremely difficult to manage because, among other things, there are too many inflexible partitions and too many different types of equipment. The term *network management* has become an oxymoron. Despite the expenditures of billions of dollars on the management software, network complexity is advancing at a faster rate than the automation of management functions.

Corporate networkers are cringing at the thought of exacerbating this already-confusing scenario by accommodating the emergence of

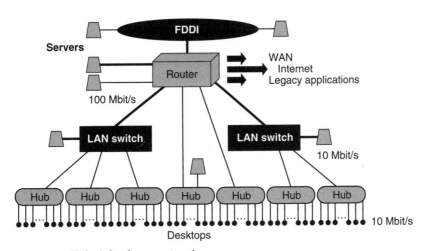

Figure 24.1 Today's local area networks.

switching hubs, routing switches, and switching routers while at the same time trying to jockey around the battles for high-speed supremacy among FDDI, fast Ethernet, ATM, and now gigabit Ethernet.

Let's face it! Campus networking has been on a road to nowhere, creating a money pit along the way that has consumed budgets and threatened careers at the same time it has denied organizations the ability to deploy new applications that might make a difference in profitability or even survival.

24.4   It's Time to Simplify

It's time to eliminate recurring network-related activities that contribute nothing to an organization's business objectives, yet consume valuable resources—"churning" activities such as:

- Administering moves, additions, and changes for network users (IP addresses, router/switch parameters, etc.)

- Segmenting, resegmenting, and regrouping LAN partitions to relieve congestion and to keep users and servers close to one another, in a futile attempt to divert traffic away from the backbone

- Resisting the deployment of applications that are likely to overwhelm the network, irrespective of their potential benefit to the organization

It's time to stop applying Band-Aids to gaping network wounds. The time for network *evolution* has given way to the time for *revolution*.

It's time for the campus data network to become a simple part of the building infrastructure—a utility, just as the power and phone networks are utilities.

24.5   The Data Utility— "If You Build It, They Will Come!"

What does it take to get a local data network into a position where it is thought of as a utility in this sense? Before answering that question, it is useful to confine the scope of discussion. First of all, only the campus environment—a building or a confined campus of buildings—will be examined, eliminating consideration of wide area network (WAN) issues. Second, the discussion will focus on connectivity, not applications, so anything above Layer 3 of the OSI model falls outside the scope of the data utility.

Is such a limited scope useful? Even if it were possible to achieve utility-like characteristics in this bounded space, would anyone care? The answer is "Yes!" and here's why.

With respect to limiting the scope to LAN versus WAN, the issue is economy of bandwidth. Local bandwidth is becoming so inexpensive that most organizations will find it within their means to have the capacity for gigabits or tens of gigabits per second of bandwidth in their campus backbones. On the other hand, WAN bandwidth continues to be so expensive that all but the biggest enterprises will pay for no more than megabits or tens of megabits per second—at least three orders of magnitude smaller than the local bandwidth. The economics of the LAN and WAN environments are therefore so fundamentally different that it makes good business sense to consider them separately. This posture allows organizations to apply techniques that can bring dramatic simplification and cost savings to the LAN portions of their enterprise networks that are handling more than 99.9 percent of the traffic.

For the application/connectivity issue, it is important to recognize that there will always be continuous growth and change in the application space, since this is where human productivity is enhanced. A new age is dawning in which network-based applications like interactive audio/video/imaging will create higher levels of productivity and a whole new way of working. However, such applications can't be deployed unless there is a high-powered communications infrastructure to support them.

Therefore, if enterprises want to benefit from the next wave of applications, they must capitalize on the fact that the data communications industry has finally grown to a place where a utility-like connectivity infrastructure is within reach. If the local network will support them, new applications will burst into use.

"If you build it, they will come!"

24.6   Characteristics of a Data Utility

Network managers setting out to design a data utility should aim for it to exhibit the following characteristics:

Wide accessibility. Wherever a network user goes within the campus, there should be uniform access to resources for which that user is authorized. Moving from one office (or classroom) to another should have no impact on service availability, and availability should not depend on actions by an administrative bureaucracy.

Continuously available service. The network must be continuously available, such that users can depend on network-based applications in the performance of their work. Quality of service must not be affected by the presence or activities of other subscribers to the utility.

Stable service offerings. Changes in procedures or service levels disrupt the flow of work and undermine the confidence of users. It follows that the data utility had better be based on very mature standards and on equipment that is unlikely to require continuing upgrades or retrofits.

Inexpensive service. Cost is forever a concern. A network infrastructure will not advance to utility status unless the costs per desktop/ server connection drop out of the budget-busting category.

Simple management. Network administrators should never have to get involved in the movements of users, should not be impacted by the constantly changing organizational structure of the enterprise, and should not be faced with constant concerns about whether a new application will choke the network.

24.7   Ethernet Dominates

The overwhelming majority of desktop connections today are Ethernet 10BaseT. Despite previously optimistic projections for the growth of competing technologies such as Token Ring, FDDI, 100VGAnyLAN, and ATM, good old Ethernet appears to have won the day. Ethernet's characteristics—maturity, low cost, and simplicity—are at the root of its victory, coupled with the fact that Ethernet's 10-Mbit/s bandwidth is more than sufficient to support virtually all end-user applications for the foreseeable future. The only thing that is likely to change at the desktop is that the shared LAN connections will be replaced with dedicated Ethernet switched connections.

The same characteristics that vaulted Ethernet into desktop dominance also favor the continuation of Ethernet technology into the backbone. The ability to simply extend Ethernet's characteristics into the backbone is the reason why gigabit Ethernet and an all-Ethernet infrastructure are favored to prevail for campus internetworking. Simply put, gigabit Ethernet and the high-capacity switches that use it are the one and only technology available today for constructing a homogeneous, utility-like LAN infrastructure that doesn't require replacement of all desktops.

24.8   Campus Data
Infrastructure—A New Model

Figure 24.2 shows one view of what a data utility campus infrastructure will look like. Ethernet frame switches exist at every fan-out point in the hierarchy of the building or campus wiring structure, from 10BaseT switches in the wiring closet to gigabit backbone switches at the top of the network.

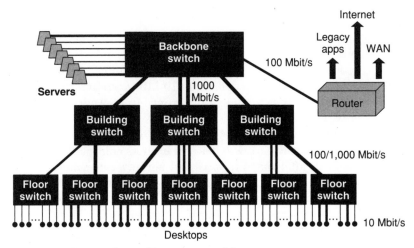

Figure 24.2 Campus data utility—a new model.

Most desktops link into the network at 10 Mbit/s via a dedicated 10BaseT Ethernet connection, although users with special needs may have a dedicated fast Ethernet (100-Mbit/s) link. All desktop connections are switched, not shared.

The switches are tied together with a combination of fast Ethernet and gigabit Ethernet pipes, depending on the population of users beneath. Where 1 Gbit/s is too much bandwidth, but 100 Mbit/s is not quite enough, multiple 100BaseT pipes are ganged together to form logical, load-sharing trunk groups of 200 Mbit/s or more, in increments of 100 Mbit/s. Similarly, very high-priority, high-capacity connections are formed out of multiple load-sharing gigabit Ethernet links—for redundancy, for capacity, or for both.

Note that almost all servers connect to the network at the top of the hierarchy, where they are equally accessible to every user on the campus and where they can most easily be consolidated, maintained, and protected (although nothing prevents servers from being positioned elsewhere, if it makes sense to do so). The top of the network is the most appropriate point to connect to the WAN, to the Internet, and to legacy applications through a router. Almost all of the intracampus traffic avoids the router entirely.

Good news! The cable plant of most organizations is already structured to support this model, and new buildings are almost exclusively being cabled in this hierarchical fashion—using unshielded twisted-pair (UTP) wiring from desktops to the wiring closet, primarily fiber from the wiring closet to the building's network center, and exclusively fiber between buildings in multibuilding campuses.

This new model exhibits a combination of power and simplicity.

24.9 Power

Powerful new switching products having gigabit Ethernet ports and tens of gigabits per second of bandwidth make possible, for the first time, unprecedented levels of performance—utility-like performance. Every user can have, in effect, a dedicated 10-Mbit/s virtual link to any and all services within the campus; interference from other users' traffic is practically impossible.

This level of performance is achieved when all of the switches are non-blocking and the bandwidth of switches and uplinks is more than ample for the maximum traffic that stations and servers can generate. For example, consider a switch at the top of the network that has, say, 20 Gbit/s of nonblocking throughput, with comparable subordinate switches and interconnecting links below it. At least 1000 10BaseT users at the bottom of the network can all simultaneously talk to servers and routers at the top, running full-duplex at their wire speed of 10 Mbit/s. Anything that a user or server pumps into this fully switched nonblocking network will emerge at its intended destination, regardless of what other users or servers are doing! (Of course, individual servers may become overloaded with requests from more users than they can handle instantaneously, but this is a server capacity issue, not a network one.)

If one is willing to back away slightly from absolute nonblocking, the same 20-Gbit/s backbone switch can serve campuses with thousands of users. Quality of service levels that users perceive to be extremely high can be maintained because, in any reasonable scenario, the probability is vanishingly small that thousands of users will consume their full 10 Mbit/s simultaneously. To further protect quality of service, particularly in the face of moderate levels of bandwidth oversubscription, switches with multiple priority buffers can utilize mechanisms such as RSVP/RTP and IEEE 802.1p to prioritize real-time jitter-sensitive traffic. With such protection, all real-time traffic such as voice or video continues to enjoy virtually guaranteed quality of service, while background activities such as file transfers are allowed to be subject to a tiny probability of infrequent slowdowns.

24.10 Simplicity

Having a network that can't get congested no matter what clients and servers throw at it changes the rules for network managers. Gone are the long cycles of planning for network upgrades, because the network is already provisioned for the future. Gone is the need to accumulate reams of traffic-monitoring data prior to deploying a new network application. Gone are the complaints from users about unreliable service or about the IS group being overly cautious when challenged to deploy new network-intensive applications.

Another major simplification is the elimination of IP address administration. This can occur in two ways: either by employing switches with automatic VLANs based on IP address (so that no matter where a user moves within the campus, his or her IP address goes along with the workstation and is automatically reassociated with the VLAN containing other members of the same IP subnet) or by eliminating IP subnetting altogether (so that IP addresses are programmed once and never changed, or are distributed dynamically via DHCP irrespective of location).

The option of eliminating IP subnetting altogether has only recently become available, made possible by switches that prevent the kinds of problems from broadcasts and multicasts that used to be associated with large single-partition networks. These switches read network layer information in every packet and apply Layer 3 intelligence to direct broadcasts and multicasts only to those stations needing the information. They also allow broadcast levels to be restricted below an operator-selectable threshold.

Continual reprogramming of the backbone router(s), a complex task usually reserved for a select minority of highly trained experts, joins some of the other regular management functions already mentioned as relics of the past. It is gone because the network changes that precipitated it are gone.

Maintenance and troubleshooting are greatly simplified by the fact that the entire campus network is implemented with one class of equipment and all users are treated identically. Furthermore, it is finally possible to clean up the wiring closets. The overall reliability of the network is enhanced when the rats' nests of patch cables and random collections of networking equipment are replaced with a neatly cabled, stable installation.

24.11   How to Build the Data Utility

Migrating a campus network to a position of utility status should begin at the very next equipment purchase. Here are some questions to ask a networking equipment vendor before making the next set of major network investments.

1. Are there clear, simple steps for migrating the network from wherever it is today to a data utility?

2. Does the recommended equipment have enough power (port capacity, bandwidth, priority buffering) to provision the network so that traffic flows are effectively nonblocking in all circumstances?

3. Does the equipment support all of the network interfaces that are needed today or that may be needed for future growth?

4. Does the equipment have Layer 3 support that will help simplify network topology without introducing the danger of broadcast/multicast interference?

5. Does the equipment have extensive redundancy and fault-tolerant features that will keep the network in continuous operation?

6. Does the new equipment protect existing investments in desktops and routers while relieving the need for further purchases of expensive routers and obsolete hubs?

7. Is it easy to understand how recurring operation and management costs will be greatly reduced?

8. Is it easy to justify that this is a good value relative to competing solutions?

In short, the guiding principle in building the data utility is to keep it simple, making sure that all future investments are in equipment with the raw power and intelligent features needed for a simple network.

24.12 Conclusion

It is homogeneous switched Ethernet throughout a campus, scaled from 10-Mbit/s to gigabit Ethernet, that promises to revolutionize local area networking, changing it from a perennial source of problems and wasted resources to an invisible and economical part of a building infrastructure—a data utility!

25

Managing Gigabit Ethernet Networks

Edward Chang*
Senior Product Manager, Bay Networks

25.1 Introduction

Gigabit Ethernet offers network designers dramatic capacity improvements for bandwidth-intensive applications. However, network managers face many challenges as gigabit Ethernet is adopted. Not the least of these is how to manage the new networks. The good news is that since gigabit Ethernet shares a similar frame format with existing Ethernet and fast Ethernet technologies, the currently available Simple Network Management Protocol (SNMP)-based management systems should only need minor upgrades to accommodate gigabit speeds. On the other hand, some gigabit Ethernet–specific network management standards and functions will also need to be developed and deployed. These will include:

- Performance and fault monitoring of gigabit Ethernet managed objects

- Performance monitoring of IETF Remote Monitoring (RMON) extensions for gigabit Ethernet

In addition, gigabit Ethernet will also be used to provide high-bandwidth pipes for Layer 2 and Layer 3 switches. As these switches

---

* Edward Chang is a senior product manager in Bay Networks' Network Management Division. Ed manages the Windows-based Optivity network management product line. Prior to joining Bay Networks, Ed was with AT&T in various areas, including Bell Laboratories, where he specialized in developing telecommunication services and in network management outsourcing. Ed holds a BSEE from Rensselaer Polytechnic Institute, an MSEE from the University of Southern California, and an MBA from NYU's Stern School of Business.

are developed and deployed, network managers will face configuration and management issues including:

- Virtual LAN administration as outlined in IEEE 802.1Q
- Quality of service (QOS) administration as outlined in IEEE 802.1p/Q

This chapter examines issues unique to managing gigabit Ethernet and discusses how gigabit Ethernet will fit into an existing network management infrastructure. But before specific gigabit Ethernet management issues are discussed, a brief summary of Ethernet management is presented to provide a better understanding of the impact of the new Ethernet speed.

25.2   The Evolution of Ethernet Management

When Ethernet was introduced in 1980, it was impossible to envision the market reception it would receive. Initial management tools were basic, because nobody fully understood the potential for network complexity. As Ethernet became more commonplace, networks grew larger every year, creating the need for more robust, feature-rich management tools and applications.

The first set of basic proprietary management tools evolved into standards-based tools via an industrywide agreement on the Simple Network Management Protocol (SNMP), which introduced a standard language for communicating with devices for the purpose of management. This first step resulted in the development of more sophisticated management tools that relayed more detailed information to the network manager. Standards-based MIBs—such as MIB2, BridgeMIB, and RepeaterMIB—gave managers the capability of gathering a set of basic performance statistics for specific devices on the network. Customer demand for traffic management inspired vendors to quickly develop and deliver proprietary extensions to these standards for their own products.

Although the extensions provided increased functionality, their interoperability was by definition limited. Solutions to this problem came in the form of multivendor management platforms such as Sun Microsystems Inc.'s SunNet Manager and Hewlett-Packard's Open-View applications, which have the ability to communicate with all types of vendor devices.

Next, the first step toward standards-based performance monitoring was taken with the development of the Remote Monitoring (RMON) standard. RMON offers remote monitoring to identify over-utilized segments, collision problems, and packet capture/decode for detailed diagnostics—all from a single management console. The initial version only collected data from the first two layers of the OSI reference model: the physical and data link/MAC layers. In 1997, RMON's successor, RMON2, took traffic monitoring to a new level by providing visibility from the network layer through the application layer.

Systems-based management applications that take advantage of RMON and RMON2 have become the standard. And the latest products to take advantage of these standards are designed to help the network manager analyze the data collected throughout the network and plan for growth, while at the same time making network management more convenient. New HTML-based management tools, for example, provide desktop Web browser access to these sophisticated management applications while also enabling individual device configuration and management, as well as providing corporate IS with easy access to network status data.

As users increase their reliance on the network to conduct business transactions, network complexity will increase. Applications such as Web browsing, e-mail, and file sharing create an even greater ability to exchange information, resulting in increased and unpredictable network traffic. Because of the increased number of users and traffic on the network, network managers are dealing with thousands of network devices. Management tools need to evolve with the networks. Keeping a complex network functional requires yet more sophisticated systemwide network management applications to monitor traffic patterns and device operation.

Today, network managers' new requirements include graphical user interfaces and powerful information collection tools capable of assimilating and analyzing information collected from devices throughout the network. Managers need to interact with connected devices anywhere on the LAN or WAN. Reconfiguring remote routers, downloading device software upgrades, managing logical VLANs, and implementing user-specific security have become absolutely essential. Effectively administering the network requires systemwide applications to manage mixed-media Ethernet and fast Ethernet from a single graphically rich screen.

Now, with the introduction of gigabit Ethernet, these same tools will need to be upgraded to manage gigabit Ethernet implementations.

25.3   Gigabit Ethernet
Specification Differences

There are few differences in the specification for gigabit Ethernet management versus the familiar Ethernet or fast Ethernet specifications. There are, however, minor exceptions within the 802.3z gigabit Ethernet specification that have implications for both the designers and users of network management systems. For instance, network management systems and tools must be upgraded to collect gigabit Ethernet specific managed objects. The following parts of the gigabit Ethernet specification require changes:

1. Repeater function

2. Burst mode counter

3. Collision algorithms

4. Counter length

5. Auto-negotiation

6. Physical layer interfaces

It's worth looking at these changes in more detail. Specifically, the following modifications were made by the IEEE 802.3z subcommittee to clause 30 of the 802.3u (100BaseT) specification.

25.3.1   Repeater function

The definitions of the repeater functions have been modified to accommodate 1000-Mbit/s operation. Seven repeater functions (activity timing, carrier event, collision event, CRC, framing, octet counting, and source address) are used to collect statistics on the activity received by the repeater port. The exact details of the new content required for gigabit operation are contained within the 802.3z technical draft.

There is only one type of repeater in gigabit Ethernet—the 1000-Mbit/s baseband type—as opposed to the two types of repeaters—100-Mbit/s baseband, Class I and 100-Mbit/s baseband, Class II—in fast Ethernet.

The repeater port autopartition, receive jabber, and carrier integrity functions are reinitialized whenever acPortAdminControl is enabled.

The 1000-Mbit/s repeater does not count or increment alignment errors.

25.3.2   Burst mode counter

A new repeater attribute—bursts—is introduced in 1000-Mbit/s operation. It is a generalized, nonresettable counter and has a maximum increment rate of 235,000 counts per second at 1000 Mbit/s. This counter increases by 1 for each carrier event with activity duration greater than or equal to slot time, during which the collision count increment state of the partition state diagram has not been entered. This attribute is only relevant to the half-duplex mode of operation.

25.3.3 Collision algorithms

The collision algorithm is essentially the same, except during burst modes. At the time of publication, the spec for counting collisions during burst mode had not been finalized.

25.3.4 Counter length

Counters specified for 10-Mbit/s operation are appropriate to 1000-Mbit/s operation, but have 100 times the stated maximum increment rate.

25.3.5 Autonegotiation

The autonegotiation, or link start-up protocol, has been altered to perform at gigabit rates. Additionally, the full-duplex PAUSE attribute has been expanded to include asymmetric and symmetric PAUSE operation for full-duplex operation.

25.3.6 Physical layer interface

There are several new physical layer (PHY) and Media Access Unit (MAU) types. Additions to the PHY entity attributes include enumerations for two new PHY types (1000BaseX and 1000BaseT) to accommodate 1000-Mbit/s operation. Additions to the MAU entity attributes include enumerations for full- and half-duplex, long- and short-wavelength laser physical media–dependent (PMD) entities. Full- and half-duplex 150-Ω balanced copper cable PMD entities have also been added.

In addition to the IEEE 802.3z gigabit Ethernet specification changes, network management systems will ultimately have to be able to collect and monitor any gigabit extensions to MIB2 or RMON/RMON2. Both MIB2 and RMON/RMON2 provide performance statistics for network interfaces, and it is anticipated that they will be augmented to support gigabit Ethernet.

In summary, the changes required to manage gigabit Ethernet rather than Ethernet or fast Ethernet are minimal. Network managers will see the addition of a few counters and manageable attributes, but overall they will be able to manage the gigabit Ethernet portions of their network as they do their installed, lower-speed LANs.

25.4 Benefits for Applications Today

So, while minor adjustments may be required during implementation, the overall modifications to the IEEE 802.3z Ethernet specification are not extensive. That is not coincidental; the standard was carefully developed with the idea of keeping changes to a minimum. In fact, the rationale behind the standard is to make it as easy as possible to fully integrate gigabit Ethernet into existing networks and to manage all three varieties of Ethernet as one.

That said, the changes that have been made are significant, providing, for instance, the ability to manage gigabit Ethernet devices using smart agents that can monitor the four main RMON groups (statistics, history, alarms, and events). Further, the changes will make it possible for vendors to add to their existing management application suites gigabit Ethernet tools that share the same look and feel as existing applications. This means network managers will be able to directly apply their experience with Ethernet and fast Ethernet management applications to gigabit Ethernet. And both users and vendors will be able to retain and apply practically all of their existing investment, training, procedures, and control methodologies.

But the benefits go beyond ease and simplicity to include new levels of management integration. As an example, a gigabit Ethernet uplink will often reside in a switch that also carries a mix of Ethernet and fast Ethernet, sometimes called an *aggregating switch*. The efforts to create a gigabit Ethernet specification based on the existing protocols allow this type of mixed Ethernet product to be managed by a single device.

This principle also applies to all current and future network devices—including buffered repeaters, gigabit Ethernet switches, and the aforementioned aggregating switches. Buffered repeaters are very similar to existing repeaters and it will not be necessary to develop new MIB management techniques to control and monitor them. Gigabit Ethernet switches will offer per-port RMON capability, and they will provide full visibility into switch ports and applications to help manage the flow of traffic throughout the network. And, meeting the demands of new network topologies, the gigabit Ethernet management specification is also capable of managing without significant changes those routers or Layer 3 switches that have gigabit Ethernet interfaces.

25.5   Gigabit Ethernet Management Tools

The goal of the Gigabit Ethernet Task Force (IEEE 802.3z) is to extend the functionality of Ethernet (IEEE 802.3) to 1000-Mbit/s operating speeds while maintaining compatibility with the installed base of Ethernet and fast Ethernet devices. Gigabit Ethernet is based on both the Ethernet and fast Ethernet standards and the high-speed Fibre Channel specification. This assures a high level of compatibility with existing Ethernet/fast Ethernet nodes and network equipment. That's significant because it means management vendors will be able to easily extend existing management functions and systems to standards-based gigabit Ethernet implementations.

The similarity of the gigabit Ethernet standard to Ethernet/fast Ethernet will allow management application vendors to retain and apply their investment in development tools and to offer a complete suite of gigabit Ethernet applications almost immediately—rather than starting with the basic tools that, historically, have been all that's available when a new standard is first finalized. Tools will be extended to provide network managers with the capability to troubleshoot and manage gigabit Ethernet networks. Specific areas where network managers will require gigabit Ethernet management systems and tools include troubleshooting, performance and fault monitoring, and configuration. Troubleshooting and performance/fault monitoring systems must incorporate the IEEE 802.3z managed objects and any Internet Engineering Task Force RMON or MIB2 gigabit Ethernet extensions. Configuration systems must incorporate 802.1p and 802.1Q VLAN and QOS specifications.

Troubleshooting tools for gigabit Ethernet will include protocol analyzers and testers that can operate at a 1000-Mbit/s line rate, decode and analyze packets, and perform link integrity tests. These tools will aid network managers in isolating gigabit Ethernet protocol problems and conducting inline performance testing. Gigabit Ethernet protocol analyzers and testers from Hewlett-Packard, Finisar, Netcom Systems, and others are starting to show up.

Performance and fault management systems will be extended with gigabit Ethernet capabilities as networks are deployed. These systems are provided by network device vendors such as Bay, 3Com, and Cisco, or by network management application vendors such as Concord Communications. Performance management systems will allow network managers to collect, monitor, and report gigabit Ethernet statistics including IEEE 802.3z managed objects and any RMON/MIB2 extensions. Fault management tools will provide network managers the ability to recognize and correct gigabit Ethernet specific fault-related managed objects. Performance and fault management systems will become generally available as gigabit Ethernet hardware implementations increase.

Configuration management systems will manage VLAN administration and traffic prioritization (QOS) on gigabit Ethernet devices. These tools will be supplied primarily by network device vendors such as Bay, 3Com, and Cisco. Gigabit Ethernet devices are also being married with Layer 2 and Layer 3 switching capabilities. As these devices are delivered, network managers will require (and will get) configuration tools to administer VLANs while managing traffic priority and performance on high-speed gigabit Ethernet trunks.

25.6   Implementation Issues

As gigabit Ethernet networks are deployed, network managers will face the issue of how best to deploy, or roll out, management. Network managers will require tools to help them with all aspects of a gigabit Ethernet implementation—including planning, design, implementation, monitoring, and trouble resolution. Since gigabit Ethernet offers the potential to aggregate traffic for high-speed links, network managers will need to plan and design network capacity carefully to avoid congestion or performance problems. A common approach to using gigabit Ethernet trunks is to utilize server or backbone trunks. Traffic patterns on these trunks need to be analyzed to determine the anticipated capacity. In this area, performance monitoring tools that feature capacity-planning facilities will be of aid to network managers trying to optimize their network designs to fully utilize gigabit Ethernet trunks.

Once gigabit Ethernet networks are deployed, network managers face ongoing configuration, fault isolation, and performance monitoring issues. The gigabit Ethernet management systems and tools described earlier will allow network managers to perform monitoring and troubleshooting. These tools will complement existing Ethernet and fast Ethernet management systems used by network managers. Network managers can troubleshoot protocol performance and errors using gigabit Ethernet protocol analyzers, just as with earlier versions of Ethernet.

Performance monitoring systems from network equipment and management vendors will allow network managers to collect gigabit Ethernet traffic statistics. As gigabit Ethernet trunks fill with mission-critical data, performance monitoring of the data and network flows within gigabit trunks will allow network managers to determine traffic prioritization and establish QOS levels. In addition, network managers will require traffic reporting tools to analyze traffic trends and predict where bottlenecks or congestion will occur.

25.7   Conclusion

Gigabit Ethernet offers network managers extra capacity. But it also brings with it unique management requirements. Fortunately, the standard's developers have set out to make the job of managing a gigabit Ethernet network as simple as possible, and also to ensure that network managers should largely be able to upgrade existing management tools to perform the job. Network managers would be well advised to invest in such tools as an aid in designing, deploying, and running their gigabit Ethernet networks.

Available Gigabit Ethernet Hardware

Purpose

To describe the key features to look for when choosing gigabit Ethernet products.

What Is Covered

When it comes to the first gigabit Ethernet adapters, hubs, and switches, it's a case of buyer beware. In order to find a product that matches the needs—and budget—of a particular network, net managers need to look beyond the marketing brochures and sweat the details of the products on offer. To that end, Chapter 26 provides guidelines on what to look for when assessing gigabit Ethernet products, and Chapter 27 reviews some of the first products on the market. To maintain this book's neutral stance, both chapters are written by Data Communications *magazine staff rather than by the vendors themselves.*

Contributors

The staff of Data Communications *magazine*

26

Selection Criteria for Gigabit Ethernet Switches and Hubs

Erica Roberts*
Senior editor, Data Communications *magazine*

26.1 Introduction

Looking for a little bandwidth boost? How about gigabit Ethernet and its 14.5 million frames per second? Think about it. One billion bits. Better yet, argue the vendors, it's tried-and-true Ethernet. Just drop a gigabit switch into the congested part of the network and sit back and wait for the performance to pick up. Clean. Simple. Plug and play.

But as corporate networkers already know, when it comes to speed, nothing's that simple—except maybe the math. Fast Ethernet tops out at 145,000 frames per second. Gigabit Ethernet is engineered to deliver 10 times that. And that opens the way to potentially catastrophic frame loss.

Think of it this way: tying a gigabit segment to a fast Ethernet segment is like hooking a fire hose to a lawn sprinkler.

Flow control is the obvious answer. The IEEE ratified 802.3x in March. "Too simple and too slow" is how Leonard Kleinrock, professor of computer science at UCLA, has characterized the IEEE effort. Worse, 802.3x can actually spread the effects of congestion to uncongested segments.

Faulty flow control may ring a bell with corporate networkers: it was one of the early problems plaguing ATM. Unfortunately, gigabit Ethernet shares another flaw with the high-speed switched transport: interoperability. Sure, gigabit Ethernet uses the same framing format as its 10- and 100-Mbit/s predecessors. But that doesn't mean products will automatically play together—particularly in the first year.

---

* This chapter first appeared as an article in *Data Communications* in May 1997. Vendors' product offerings may have changed since that time.

Then there's quality of service (QOS). Gigabit Ethernet is designed for speed, not service classes. Standards bodies are working on several prioritization schemes, but at best they're a force fit. Bottom line: don't expect gigabit Ethernet to furnish guaranteed delivery of delay-sensitive traffic.

26.2 Growing Pains

So is gigabit Ethernet in trouble? No. It's just going through the birth pangs that inevitably accompany any new networking technology. Gigabit has plenty going for it, starting with scalability. End-to-end Ethernet eliminates the patchwork approach to high-speed networking. And the price is right: 10 times the performance of fast Ethernet at 2 to 3 times the cost.

What's more, gigabit has garnered plenty of industry support, so net managers don't have to worry about it disappearing anytime soon. Many vendors are already shipping or readying switches—boxes that can deliver dedicated gigabit connections. Several have hubs—shared-media equipment that divvies up 1 Gbit among end stations and nodes—in the works. And other suppliers are selling adapters; some of these folks also make other gigabit gear, which could turn out to be a godsend given the likelihood of interoperability snafus.

The key switch consideration: internal architecture. Some vendors are retrofitting their gear, but the resultant blocking architecture can squelch performance. Nonblocking is the way to go, but only some vendors offer it. Almost all switches support slower Ethernet connections. Some also field ATM, FDDI, or token ring—which makes for easier migrations. Flow control is essential, and prioritization is a big concern. Virtual LANs (VLANs) are important for controlling broadcasts and segregating traffic. And the ability to route traffic at gigabit speeds is also critical. Management could turn out to be tricky, especially since Remote Monitoring (RMON) remains unproven at gigabit speeds. Things are simpler when it comes to hubs—mainly because they don't deal with advanced mechanisms like VLANs and multicast. That makes price the chief consideration.

26.3 What's in a Name?

As its name implies, gigabit Ethernet—officially known as IEEE 802.3z—is the 1-Gbit/s extension to the 802.3 standard already defined for 10- and 100-Mbit/s service. Gigabit Ethernet is specced for shared media and switched links over single-mode and multimode fiber, as well as for shorter runs over copper. But it's flow control rather than cabling that should concern net managers—as Ethernet and fast Ethernet switches have already shown. And it stands to reason that buffer overflows and junked frames will be even more of a problem with gigabit Ethernet.

Why the worry? Ethernet already has carrier-sense multiple access with collision detection (CSMA/CD), which acts as a simple, effective flow-control mechanism by preventing more than one station from transmitting at a time. But most vendors are producing gigabit Ethernet switches that can work in half- and full-duplex mode. In full-duplex, CSMA/CD is disabled and throughput is doubled—to 2 Gbit/s per connection.

The IEEE's answer: 802.3x. This XON/XOFF protocol works over any full-duplex Ethernet, fast Ethernet, or gigabit Ethernet link. When a switch buffer is close to capacity, the receiving device signals the sending station and tells it to stop transmitting until the buffer is freed up.

26.4   They Call Me the Wanderer

Nice and simple, right? But some warn that while 802.3x will work adequately on small nets, a more sophisticated scheme is needed for larger switched networks; otherwise, the result could be wandering congestion.

Net managers may not be familiar with the term, but it's something they want to avoid. Wandering congestion occurs when switches propagate jamming signals onto uncongested network segments, preventing users from sending data. Essentially, it creates the same effect as congestion—on segments that have plenty of capacity.

Corporate networkers who can't afford to have congestion wandering their nets need to contain 802.3x. The scheme works best on connections between switches and end stations. Using it on switch-to-switch links is borrowing trouble.

Here again, gigabit Ethernet echoes ATM. High-speed switched transport started life with rudimentary flow control. But early problems with dropped cells forced the ATM Forum to develop a more sophisticated spec—traffic management 4.0. And many in the gigabit community believe the IEEE will have to do the same. For example, some vendors now are pushing the institute to consider a credit-based scheme that can respond with different rates instead of just on and off.

26.5   Quality Control?

It doesn't matter how fast the network is if delay-sensitive traffic gets sent after coach-class data. And that's exactly what happens with

Ethernet's best-effort attempts. The IEEE is hammering out two specs that will help Ethernet provide QOS.802.1Q tags traffic for VLANs and for prioritization. 802.1p is a signaling scheme that lets end stations request priority and allows switches to pass these requests along the path. But even with these mechanisms in place, net managers shouldn't expect Gigabit Ethernet to furnish anything like ATM's guaranteed quality of service. For one thing, gigabit Ethernet deals with variable-length frames (not uniform cells). Thus, it's possible for a small delay-sensitive frame to get stuck in a buffer behind an enormous data frame. For another, ATM boasts very sophisticated flow control, which makes it possible to carve out a chunk of bandwidth for voice or video. Gigabit Ethernet can only push one frame ahead of another; it can't reserve bandwidth for an entire stream. Corporate networkers hoping that Resource Reservation Protocol (RSVP) from the Internet Engineering Task Force (IETF) will transform Ethernet into ATM may be disappointed. True, RSVP can reserve bandwidth (hence the name). But RSVP is a simple hop-by-hop signaling system. Each link is negotiated separately, so there's no way to guarantee service end to end. ATM, in contrast, sets up the entire connection and ensures a constant class of service for the duration of the session.

26.6 Stack and Stand

Switch vendors are eager to catch the gigabit wave (see Table 26.1). For the most part, the boxes are ready to be tied into Ethernet and fast Ethernet. Smaller switches are shipping or on the way from start-ups and old standbys, including Acacia Networks Inc. (Lowell, Massachusetts), Alteon Networks Inc. (San Jose, California), Bay Networks Inc. (Santa Clara, California), Extreme Networks Inc. (Cupertino, California), Foundry Networks Inc. (Sunnyvale, California), Nbase, and 3Com Corp. (Santa Clara, California).

Stackables and stand-alones tend to be fixed-configuration devices with a limited number of high-speed uplinks. Alteon's Aceswitch 110, for example, has two gigabit Ethernet ports. Its 160 offers six. Foundry's Fastiron and Netiron come with a pair of gigabit ports; 3Com's Superstack II 1000 and 3000 have one each. Two of the smaller switches are gigabit-only: Extreme's Summit1 features eight gigabit ports. The same goes for 3Com's Superstack II 9000. These boxes are intended to consolidate workgroup switch-to-switch or switch-to-server connections.

TABLE 26.1 Gigabit Switches

Vendor	Product	Configuration	Internal architecture	Maximum ports	Other switched connections	Flow control	Quality of service	Virtual LAN support	Layer 3 switching	Price
Acacia Networks Inc. Lowell, Massachusetts 508-458-7200 http://www.acacianet.com	Novaswitch 1600	Stackable	Blocking	16 Ethernet/6 fast Ethernet/1 gigabit	ATM, FDDI	802.3x	802.1p and 802.1Q planned	802.1Q planned; port (proprietary); Layer 3	None	$7,490 for 16 Ethernet ports and 1 gigabit port
	Novaswitch 2400	Stackable	Blocking	24/6/1	FDDI	802.3x	802.1p and 802.1Q planned	802.1Q planned; port (proprietary); Layer 3	None	$8,990 for 24 fast Ethernet ports and 1 gigabit Ethernet port
	Novaswitch 12000	Stackable	Blocking	24/12/1	FDDI	802.3x	802.1p and 802.1Q planned	802.1Q planned; port (proprietary), Layer 3	None	$7,945 for ten 10/100 ports and 1 gigabit port
Alteon Networks Inc. San Jose, California 408-574-5500 http://www.alteon.com	Aceswitch 110	Stand-alone	Blocking	8/8/2	None	802.3x	802.1p and 802.1Q planned	802.1Q; port (proprietary)	None	$8,995 for eight 10/100 and two gigabit Ethernet ports
	Aceswitch 160	Stand-alone	Blocking	0/2/6	None	802.3x	802.1p and 802.1Q planned	802.1Q; port (proprietary)	None	Around $15,000 for two fast Ethernet and six gigabit Ethernet ports
Bay Networks Santa Clara, California 408-988-2400 http://www.baynetworks.com	F1200	Chassis	Blocking (switches local traffic within modules)	96/96/12	None	802.3x	802.1p, RSVP, priority queues	802.1Q; Layer 3 (proprietary)	IP	$58,400 for sixty-four 10/100 and four gigabit Ethernet ports
Cabletron Systems Inc. Rochester, New Hampshire 603-332-9400 http://www.cabletron.com	MMAC-Plus	Chassis	Blocking (switches local traffic within modules)	504/168/28	ATM, FDDI, token ring	802.3x on gigabit Ethernet modules; planned for 10/100 switch ports	802.1p and 802.1Q planned	802.1Q planned; port (proprietary); MAC (proprietary); Layer 3 (proprietary)	IP and IPX, proprietary	$169,965 for 156 fast Ethernet and 2 gigabit Ethernet ports
	MMAC-Plus 6	Chassis	Blocking (switches local traffic within modules)	216/72/12	ATM, FDDI, token ring	802.3x on gigabit Ethernet modules; planned for 10/100 switch ports	802.1p and 802.1Q	802.1Q planned; port (proprietary); MAC (proprietary); Layer 3 (proprietary)	IP and IPX, proprietary	$70,930 for 60 fast Ethernet ports and 2 gigabit Ethernet ports

TABLE 26.1 Gigabit Switches (Continued)

Vendor	Product	Configuration	Internal architecture	Maximum ports	Other switched connections	Flow control	Quality of service	Virtual LAN support	Layer 3 switching	Price
Digital Equipment Corp. (DEC) Maynard, Massachusetts 508-692-2562 http://www.digital.com	Multiswitch 900	Chassis	Blocking (switches local traffic within modules)	192/64/8	ATM, FDDI, token ring	802.3x	802.1p and 802.1Q planned	802.1Q planned; port (proprietary); MAC (proprietary)	IP and IPX, proprietary	$86,000 for 96 Ethernet and 16 fast Ethernet ports
	Gigabit Ethernet Backbone switch	Chassis	Nonblocking	To be determined	None	802.3x	802.1p and 802.1Q planned	802.1Q planned; port (proprietary); MAC (proprietary)	IP and IPX, proprietary	To be announced
Extreme Networks Inc. Cupertino, California 408-342-0999 http://www.extremenetworks.com	Summit1	Stand-alone	Nonblocking	0/0/8	None	802.3x	802.1p, 802.1Q, RSVP, weighted fair queuing	802.1Q planned; port (proprietary); MAC (proprietary)	IP	$21,995 for eight gigabit Ethernet ports
	Summit2	Stand-alone	Nonblocking	16/16/2	None	802.3x	802.1p, 802.1Q, RSVP, weighted fair queuing	802.1Q planned; port (proprietary); MAC (proprietary)	IP	$11,995 for sixteen 10/100 and two gigabit Ethernet ports
Foundry Networks Inc. Sunnyvale, California 408-731-3800 http://www.foundrynet.com	Fastiron	Stand-alone	Nonblocking	16/2/2	None	802.3x	802.1p, 802.1Q, RSVP planned	802.1Q planned; port (proprietary); MAC (proprietary)	IP, IPX	$6,490 for eight 10/100 ports and one gigabit Ethernet port
	Netiron	Stand-alone	Nonblocking	8/2/2	None	802.3x	802.1p, 802.1Q, RSVP planned	802.1Q planned; port (proprietary); MAC (proprietary)	IP, IPX	$12,490 for eight 10/100 ports and one gigabit Ethernet port

Company	Form	Blocking	Ports	Protocols	802.3x	VLAN	802.1Q	Routing	Price
Gigalabs Inc. Sunnyvale, California 408-481-3030 http://www.gigalabs.com									
Gigastar 100	Stackable	Blocking (switches local traffic within modules)	32/32/4	ATM, FDDI, proprietary	802.3x	Proprietary; 802.1p and 802.1Q planned	802.1Q planned; port (proprietary); MAC (proprietary)	None	$29,575 for eight fast Ethernet ports and one gigabit Ethernet port
Gigastar 3000	Chassis	Blocking (switches local traffic within modules)	144 Ethernet/144 fast Ethernet/9 gigabit	ATM, FDDI, proprietary	802.3x	Proprietary; 802.1p and 802.1Q planned	802.1Q planned; port (proprietary); MAC (proprietary)	IP planned	$228,600 for 288 fast Ethernet and 4 gigabit Ethernet ports
Gigastar 3000s	Chassis	Blocking (switches local traffic within modules)	144/144/9	ATM, FDDI, proprietary	802.3x	Proprietary; 802.1p and 802.1Q planned	802.1Q planned; port (proprietary); MAC (proprietary)	IP	$235,600 for 288 fast Ethernet and 4 gigabit Ethernet ports
Gigastar 3000t	Stackable	Blocking (switches local traffic within modules)	32/32/4	ATM, FDDI, proprietary	802.3x	Proprietary; 802.1p and 802.1Q planned	802.1Q planned; port (proprietary); MAC (proprietary)	None	$155,000 for 256 Ethernet and 8 gigabit Ethernet ports
Nbase Communications Chatsworth, California 818-773-0900 http://www.nbase.com									
Megaswitch II	Stand-alone	Blocking	8/4/2	ATM	802.3x	Proprietary; 802.1p and 802.1Q planned	802.1Q planned; port (proprietary)	None	$10,460 for eight 10/100 ports and one gigabit Ethernet port
Megaswitch E	Stand-alone	Blocking	16/32/4	ATM	802.3x	802.1p and 802.1Q planned	802.1Q; port (proprietary); MAC (proprietary); Layer 3	IP, proprietary	$15,000 for twenty-eight 10/100 ports and one gigabit Ethernet port
Gigaframe	Chassis	Blocking (switches local traffic within modules)	48/48/12	ATM	802.3x	802.1p and 802.1Q planned	802.1Q; port (proprietary); MAC (proprietary); Layer 3	IP, proprietary	$30,000 for 12 gigabit Ethernet ports

TABLE 26.1 Gigabit Switches (Continued)

Vendor	Product	Configuration	Internal architecture	Maximum ports	Other switched connections	Flow control	Quality of service	Virtual LAN support	Layer 3 switching	Price
Plaintree Systems Inc. Waltham, Massachusetts 617-290-5800 http://www.plaintree.com	Waveswitch 9200	Chassis	Blocking (switches local traffic within modules)	384/128/16	None	802.3x	802.1p and 802.1Q planned	802.1Q planned; MAC (proprietary); Layer 3 (proprietary)	IP planned	$71,960 for 112 fast Ethernet and 2 gigabit Ethernet ports
Prominet Corp. Westborough, Massachusetts 508-870-5570 http://www.prominet.com	P550 Cajun switch	Chassis	Blocking (switches local traffic within modules)	120/120/24	None	802.3x	Proprietary; 802.1Q planned	802.1Q planned; port (proprietary); MAC (proprietary); Layer 3	IP planned	$36,000 for 60 fast Ethernet and 2 gigabit Ethernet ports
	F600	Chassis	Blocking (switches local traffic within modules)	48/48/6	None	802.3x	802.1p, RSVP, priority queues	802.1Q; Layer 3 (proprietary)	IP	$31,200 for thirty-two 10/100 and two gigabit Ethernet ports
	F200	Stand-alone	Blocking (switches local traffic within modules)	32/32/4	None	802.3x	802.1p, RSVP, priority queues	802.1Q planned; Layer 3 (proprietary)	IP	From $21,000 for twenty-four 10/100 and two gigabit Ethernet ports

	Model	Type	Blocking	Ports	Media		VLAN	Trunking	Routing	Price
3Com Corp. Santa Clara, California 408-764-5000 http://www.3com.com	Superstack II 9000	Stackable	Nonblocking	0/0/8	None	802.3x	Proprietary; 802.1p and 802.1Q planned	Port (proprietary)	IP, via route server	$19,998 for 8 gigabit Ethernet ports
	Superstack II 3000	Stackable	Blocking	0/12/1	None	802.3x	Proprietary; 802.1p and 802.1Q planned	Port (proprietary)	IP, via route server	$9,290 for twelve 10/100 ports and one gigabit Ethernet port
	Superstack II 1000	Stackable	Blocking	0/24/1	None	802.3x	Proprietary; 802.1p and 802.1Q planned	Port (proprietary)	IP, via route server	$6,545 for 24 Ethernet ports and 1 gigabit Ethernet port
	Corebuilder 5000	Chassis	Blocking (switches local traffic within modules)	408/68/2	ATM, FDDI	802.3x	Proprietary; 802.1p and 802.1Q planned	802.1Q, Layer 3	IP, via route server	From $35,250 for 24 fast Ethernet ports and 1 gigabit Ethernet port
	Corebuilder 7000	Chassis	Blocking (switches local traffic within modules)	144/64/8	ATM	802.3x	Proprietary; 802.1p and 802.1Q planned	802.1Q	IP, via route server	$109,100 for 64 fast Ethernet and 2 gigabit Ethernet ports

RSVP, Resource Reservation Protocol
SOURCE: *Data Communications* magazine.

26.7   Classy Chassis

Chassis-based switches are headed for the corporate net courtesy of Bay Networks, Cabletron Systems Inc. (Rochester, New Hampshire), Digital Equipment Corp. (DEC, Maynard, Massachusetts), Gigalabs Inc. (Sunnyvale, California), Nbase, Plaintree Systems Inc. (Waltham, Massachusetts), Prominet Corp. (Westborough, Massachusetts), and 3Com. These devices often boast very high port counts. Cabletron's MMAC-Plus maxes out at 504 Ethernet or 168 fast Ethernet or 28 gigabit interfaces. Plaintree's Waveswitch 9200 can field 384 Ethernet or 128 fast Ethernet or 16 gigabit. And Prominet's P550 Cajun also racks up the ports, with 120 fast Ethernet or 24 gigabit. However, the biggest switch as of this writing was still the Corebuilder 9000, with 504 fast Ethernet ports or 126 gigabit Ethernet ports.

Extensive fault tolerance also distinguishes chassis-based contenders from their smaller kin. Almost all can accept dual hot-swappable power supplies. Prominet's P550 can be loaded with an extra crossbar switching element that is brought up automatically if the first element fails. And Plaintree's Waveswitch 9200 can be configured with redundant switching modules.

What really sets this gear apart, though, is the mix of media it accommodates—making it easier to integrate chassis into existing networks. Cabletron's MMAC-Plus and MMAC-Plus 6 and Digital's Multiswitch 900 all field token ring, FDDI, and Asynchronous Transfer Mode (ATM).

26.8   The Inside Story

Vendors are eager to talk up their high port counts and high-speed uplinks. Corporate networkers who want to know what a gigabit switch can really handle without choking should look inside. Basically, it comes down to blocking versus nonblocking architectures—and net managers are catching on to nonblocking nirvana (see Fig. 26.1).

A little applied math can work wonders here. For a nonblocking architecture, the internal capacity of a switch must be greater than the sum of the capacity of its ports. And experts say a nonblocking architecture is a real benefit on heavily loaded networks, ensuring that none of the packets get thrown away at the switch.

For right now, switch architecture may not be the most pressing criterion for choosing a switch; gigabit pipes are unlikely to be filled anytime soon. But net managers who do the math today can be sure they've deployed the right switch tomorrow, when gigabit traffic starts to climb (as it inevitably will).

Alteon's stand-alone Aceswitch 110 features eight 100-Mbit/s ports and two gigabit ports, for a total capacity of 2.8 Gbit/s. According to the vendor, its shared-memory scheme can field 2.5 Gbit/s—so the switch looks as if it can stand up to almost a full load.

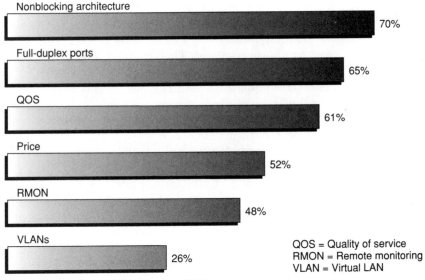

Source: Infonetics Research Inc. (San Jose, Calif.)

Figure 26.1   What's in the Box? What's really important in a high-speed LAN switch? A recent survey of 130 corporate networkers ranked nonblocking architecture as the most critical feature, followed closely by full-duplex ports and QOS.

But there's a catch—a big one. Most switches run full-duplex. Thus, total port capacity on the Aceswitch could actually hit 5.6 Gbit/s, which means the switch could be swamped on heavily loaded connections.

Nbase's stand-alone Megaswitch II also would run into trouble on fully loaded gigabit Ethernets. Configuring it with ten 100-Mbit/s switch ports and one gigabit port gives it a total full-duplex capacity of 4 Gbit/s. Its internal bus tops out at 1 Gbit/s.

Extreme's stand-alone Summit2 comes configured with sixteen 100-Mbit/s interfaces and a pair of gigabit ports, for a total load of 7.2 Gbit/s full-duplex. Its 8.5-Gbit/s shared-memory backplane can handle the load with capacity to spare.

26.9   The Capacity Crunch

The capacity question is even more critical when it comes to chassis-based boxes. But backplane capacity doesn't tell the entire story. On all these boxes it is possible to switch traffic between ports on the same module—without dumping it onto the backplane. Thus, making the most of the switch architecture really comes down to savvy network design: for example, keeping local servers and their associated end users on the same switching module helps reduce the load on the backplane.

Still, there's a limit to how much heads-up design can help. Corporate Intranets are rewriting the tried-and-true 80:20 rule, which posits that 80 percent of traffic typically stays local while the remaining 20 percent is shipped to another subnet. This, in turn, means that it's going to get harder and harder to keep traffic off the backplane.

That's why it still pays to do the math. The goal in this case isn't to determine whether the architecture is blocking or nonblocking—all the chassis-based switches are blocked when fully loaded—but to see how much headroom has been built in.

Prominet's P550 is a capacity leader. Fully loaded, it can accommodate 24 gigabit Ethernet ports, for a full-duplex load of 48 Gbit/s. Its crosspoint matrix maxes out at 41.6 Gbit/s. Gigalabs also does well: its Gigastar 3000 series can handle nine gigabit ports—no problem for its 18-Gbit/s crosspoint matrix. Where the 3000 runs into a capacity crunch, however, is when fully loaded with 144 fast Ethernet interfaces—for a full-duplex load of 28.8 Gbit/s. Bear in mind, though, that each switching module boasts 16 fast Ethernet ports, which means at least some of the traffic can be kept off the backplane.

26.10 All Backed Up

Conversely, Cabletron could be facing a capacity crunch on its MMAC-Plus and MMAC-Plus 6. Right now, Cabletron isn't promoting either MMAC as a full-blown gigabit switch. But it says it will bump up the clock speed of its INB so the bus can churn through far more packets. And it's been talking about a 75-Gbit/s backplane upgrade for some time.

One reason for the squeeze could be that the MMAC-Plus got its start as a lower-speed switching hub. There's nothing wrong with retrofitting per se, but net managers who need products that can stand up to full gigabit loads should probably look at boxes built from the ground up to do the job.

26.11 Go With the Flow?

If there's a chance that a switch can be swamped under heavily loaded conditions, then flow control is a critical concern. All the vendors either support or have pledged to support 802.3x, which, despite its aforementioned flaws, is the only officially sanctioned mechanism.

Implementing 802.3x means deploying new media access control (MAC) chips. The requisite silicon is easy to add to new switches. But vendors retooling older gear are only likely to integrate the chips on their gigabit switching modules. After all, they can hardly expect customers to dump modules they've already bought just to get flow control. 3Com is taking this tack with its current Superstack IIs and its

Corebuilder 5000 and 7000; DEC is doing the same with its Multi-switch 900.

Here's the problem: if there's no flow control on fast Ethernet links, there's no way to shut these connections down when congestion builds up.

Acacia, Cabletron, and Gigalabs are the exceptions to the retro-fitting rule. Cabletron and Gigalabs are using daughter cards to up-grade their switching modules for 802.3x. Thus, net managers who want flow control only have to swap out a card rather than ditch the entire module.

Acacia builds its switches around digital signaling processors (DSPs). The vendor says the chips can be upgraded in firmware; it says it plans to ship 802.3x code. Bay is the only vendor already shipping new switches that has decided to limit 802.3x to its gigabit modules.

26.12 Priority Push

If gigabit networks are going to be congested, then net managers need to worry about more than just flow control. Flow control stops everyone on the link. What network managers really want is prioritization, which lets unimportant packets get buffered while important packets get sent through.

All the switch vendors have lined up behind IEEE 802.1Q and 802.1p. Only those with integrated routing capabilities are concerned with IETF RSVP. Here again, vendors who are retrofitting older gear could run into trouble. For 3Com and DEC it's pretty much the same story as with 802.3x: they'll implement the spec on new switching mod-ules. Cabletron and Gigalabs will handle upgrades with daughter cards; Acacia will do the deal in firmware.

3Com already has a proprietary QOS scheme for its Superstack II Models 1000 and 3000 and its Corebuilder 5000. Priority Access Con-trol Enabled (PACE) works between the switch and the end station and requires software to be loaded on all Ethernet adapters. It offers two priorities: one for delay-sensitive traffic, the other for ordinary data. 3Com has published PACE and Prominet has implemented it in its P550 Cajun switch.

Some vendors argue that quality of service is the weapon that will let Ethernet kill ATM. Thus far, though, this claim is more rhetoric than reality. The prioritization capabilities of gigabit products have yet to be tested, and ATM's are proven.

26.13 Virtual Reality

IEEE 802.1Q also addresses virtual LANs—the key to keeping broad-cast storms from overwhelming large switched networks. Thus far, the spec only addresses port-based VLANs, which enable net managers to group ports on different switches onto the same virtual LAN.

But the standard doesn't cover protocol-based VLANs (built around either specific protocols or particular IP/IPX network addresses) and MAC-based VLANs (grouped by end-station addresses). That gives gigabit switch vendors plenty of leeway, and they've been quick to do their own thing. Proprietary schemes are the order of day, and interoperability is not generally in the picture.

26.14   The Layer 3 Look

VLANs also mean something else to corporate networkers—routers. That's the only way to shunt traffic between virtual LANs. In theory a gigabit pipe can deliver better than 14 million 64-byte packets per second (pps). Cisco's top router, the 7500, can process 2.1 million at best. That's why some vendors are building Layer 3 switching into their gigabit gear, claiming that this lets them route as fast as they switch.

Extreme, for example, claims that its Summit1 and Summit2 can forward 11.5 million pps. Granted, Extreme only supports IP, while Cisco's 7500 is a full-blown multiprotocol device.

Besides speed, the big benefit to Layer 3 switching is cost. In essence, vendors are giving away supersonic routing with their boxes. Bay, for instance, sells its F1200 chassis-based switch for $58,400 (complete with 64 fast Ethernet and 4 gigabit ports). A Cisco 7500 with six fast Ethernet ports (and no gigabit interfaces) costs at least $55,000. Extreme and Bay are not alone. Six other vendors now offer Layer 3 switching. All these schemes are IP only, except Foundry's, which can field IP and IPX.

It's important to realize that not all vendors are basing their Layer 3 schemes on conventional routing protocols like Routing Information Protocol (RIP) and Open Shortest Path First (OSPF). 3Com, for instance, relies on route servers that implement the IETF's Next Hop Routing Protocol (NHRP). True, NHRP is a standard, but there's no guarantee that other vendors will implement it. Cabletron and DEC, meanwhile, are shipping proprietary routing technologies that force net managers to reconfigure end-user IP stacks.

26.15   Management Makeovers

Managing switched networks is a problem. Managing switched gigabit networks is shaping up to be a gigaproblem. Some management experts argue that the conventional SNMP schema—using Remote Monitoring (RMON) probes to capture data and send it back to a central console to be analyzed—is impractical at gigabit speeds. They reckon smart agents are a better option. These agents monitor local segments and only send information back to the console if there is a problem. Some agents can even take corrective action, such as shutting down a switch port that's propagating bad packets onto the network, on their own.

All the gigabit vendors say they'll offer agents on their switches that can monitor the four main RMON groups (statistics, history, alarms, and events) on all ports. 3Com goes further. Its Superstack II 1000 and 3000 and Corebuilder 5000 and 7000 support all nine RMON groups, although the vendor warns that running all groups on all ports simultaneously will take a big bite out of performance. Some vendors also offer port mirroring. This allows data from one port to be sent to another port connected directly to a probe for packet capture and analysis.

26.16 Retail Revelations

And what can corporate networkers expect to shell out for these bit blasters? Not much compared with other technologies: FDDI switches average $3900 per port, 155-Mbit/s ATM switches, $1985—for about one-sixth the capacity. In contrast, Extreme's gigabit-only Summit costs $2750 per gigabit port.

And what about switches that support a range of speeds and technologies? One way to gauge price/performance is to work out per-megabit costs. At face value, the $235,600 that Gigalabs wants for its Gigastar 3000s is enough to give any net manager sticker shock. But the money buys 288 fast Ethernet and 4 gigabit ports—or 32,800 Mbit/s of capacity. That works out to roughly $7 per megabit. 3Com's Corebuilder 7000 lists for $109,100, which buys 64 fast Ethernet and 2 gigabit ports—for a total capacity of 8,400 Mbit/s. Run the numbers and it comes out to roughly $13 per megabit.

26.17 The Hub Club

Things are a lot simpler when it comes to hubs, since there are far fewer features to evaluate (see Table 26.2). Net managers need to consider the number of ports, connections to existing networks, flow control, RMON groups, and price.

Hubs will typically find two homes—on server farms and small, high-performance workgroups—according to Packet Engines Inc. (Spokane, Washington), which is shipping the 12-port FDR gigabit hub. XLNT Designs Inc. (San Diego, California) also is shipping an eight-port gigabit hub module for its Millennium 4000, a 10/100 Ethernet switch with optional FDDI uplinks. At press time, the Network Products Division of

TABLE 26.2 Gigabit Hubs

Vendor	Product	Gigabit ports	Connection to switched network	Flow control	Makes network adapter	Price
Network Products Division, Cabletron Systems Inc. Rochester, New Hampshire 603-332-9400 http://www.cabletron.com	GBE Repeater	8	Undecided	802.3x planned	Yes	To be announced
Packet Engines Inc. Spokane, Washington 509-922-9190 http://www.packetengines.com	FDR	12	Ethernet, fast Ethernet	802.3x	Yes	$17,995
XLNT Designs Inc. San Diego, California 619-487-9320 http://www.xlnt.com	Millennium 4000	8	Ethernet, fast Ethernet, FDDI	802.3x	No	To be announced

SOURCE: *Data Communications* magazine.

Cabletron also had a gigabit hub in the works—the GBE Repeater. All three vendors have decided to develop full-duplex buffered distributors rather than half-duplex repeaters.

Buffered distributors, by definition, have buffers on every port. CSMA/CD is used within the box to arbitrate access to the pool of bandwidth. Since arbitration is done in the hub, CSMA/CD is disabled on the network. This allows the buffered distributor to work in full-duplex mode—sharing 2 Gbit/s among end users and servers.

Flow control is a big deal with hubs—just as it is with switches—only in this instance it's asymmetric flow control that is needed. Asymmetric flow control only works one way on the link: only the hub can tell end stations to stop sending. If the end stations could tell the hub to shut down, the hub would stop sending to all attached nodes—bringing down that segment of the network. All of the vendors support 802.3x, which implements symmetric and asymmetric flow control.

Flow control is also an issue if a gigabit hub can connect to a slower-speed network—something that XLNT and Packet Engines can do. With Packet Engines' FDR, net managers set thresholds on the 10/100 port. If the load on the interface gets too high, the port is automatically disabled and the net manager is notified. XLNT says that its Millennium 4000 is built to prevent port overloads, but won't disclose any technical details.

26.18   In the Cards

Interoperability is a very big issue for any net manager who plans on running gigabit Ethernet anytime soon. One way to take some of the pressure off is to buy switches and network adapters (or hubs and adapters) from the same vendor. Alteon, DEC, Gigalabs, Packet Engines, and 3Com all sell gigabit cards or have them in the works.

But one-stop shopping is really a short-term strategy until interoperability issues are ironed out and the big adapter vendors—like 3Com and Intel—start shipping products in volume. Packet Engines alone among gigabit vendors admits this and indicates that it doesn't plan to be in the adapter end of the business for very long.

It's not hard to see why most smaller suppliers don't want to compete with companies like 3Com. The vendor has already announced that its gigabit cards will be loaded with all of the drivers now found on its Ethernet and fast Ethernet adapters—more than 50 all told. And 3Com's huge manufacturing facilities let it drive down prices to the point where smaller outfits can't compete.

Not all vendors are scared off, though. Alteon is looking to carve out a niche for itself in the server market. It wants to front-end server farms with its Aceswitch 160 and use its Acenic adapters to blow through bottlenecks. One way it's pumping up performance is with its proprietary Jumbo packets—at 9 Kbytes they're far longer than Ethernet's 1518-byte legal limit. Bigger frames mean less processing, freeing server CPUs for other tasks. The illegal frames are restricted to switch-to-server links.

26.19 Conclusion

So is gigabit a go? Corporate networkers are definitely intrigued by the idea of a one-size-fits-all transport. But they need to realize that gigabit can be as dangerous as it is powerful. Those who don't want to get blown away should play it safe by using the following guidelines.

- *Do the math.* For stackables and stand-alones, backplane capacity should be greater than the sum of the ports. Chassis-based boxes can switch local traffic within modules to keep it off the backplane.

- *Go with the flow.* Buy switches that support 802.3x but only activate the flow-control protocol on links to end stations—not other switches. This prevents the protocol from shutting down uncongested segments.

- *Take it slow.* Wait for results of interoperability testing. Rely on independent evaluations, not vendor demos at trade shows.

The First Gigabit Ethernet Products

The Staff of *Data Communications* Magazine

27.1 Introduction

Every new networking technology has its pioneers. Gigabit Ethernet has been no exception. What is surprising is the speed with which vendors (both established players and start-ups) have rushed the first gigabit Ethernet–capable products to market. During the course of 1996 and 1997, *Data Communications* magazine published in-depth analysis of the most significant gigabit Ethernet hardware as it was announced. The material in this chapter is based on those assessments.* Six products are covered:

- The Summit2 switch from Extreme Networks Inc. (Cupertino, California)

- The IP9000 Gigabit Router from Torrent Networking Technologies Corp. (Landover, Maryland)

- The Corebuilder 3500 switch from 3Com Corp. (Santa Clara, California)

- The Corebuilder 9000 switch from 3Com Corp. (Santa Clara, California)

- The Alteon/NIC and switch from Alteon Networks Inc. (San Jose, California)

- The Gigastar 100 switch from Gigalabs Inc. (Sunnyvale, California)

---

*Vendor offerings may have changed since the information was first published.

These products target a wide variety of applications, including router replacement, data center switching, and switch-to-server hookups. But they all have one thing in common: used tactically, they can be powerful tools for net managers seeking to ramp up the performance of their corporate networks.

27.2 Extreme's Summit2 Departmental Switch

Product Summary

Summit2

Extreme Networks Inc., Cupertino, California, 408-342-0999

http://www.extremenetworks.com

Pros: routes IP at switch speeds

Cons: limited network and protocol support; not available until July 1998

Today's high-speed LANs demand high-speed IP routing. Extreme Networks Inc. says there's no reason why one product can't deliver the best of both those worlds. It claims its Summit2 gigabit Ethernet switch routes IP as quickly as it switches other protocols, thanks to a nonblocking architecture and hardware routing on each port. The switch also boasts a variety of prioritization features and can be used to create virtual LANs (VLANs).

But while the Summit2 is built for speed, it's also built for IP only—managers of multiprotocol nets probably won't see any performance benefits. Further, it offers only Ethernet, fast Ethernet, and gigabit Ethernet connections.

27.2.1 Speed Racer

The Summit2 has sixteen 10/100-Mbit/s Ethernet ports and two 1-Gbit/s ports for connection to servers or to a backbone gigabit Ethernet switch (see Fig. 27.1). It runs Routing Information Protocol (RIP) and Open Shortest Path First (OSPF). Extreme says the Summit2 gets around the routing table lookup problem thanks to an 8.5-Gbit/s switching matrix—enough capacity to handle all incoming and outgoing traffic streams simultaneously without delays. Routing ASICs on every port chip in by examining the routing information in packet headers and furnishing lookup. And although that infor-

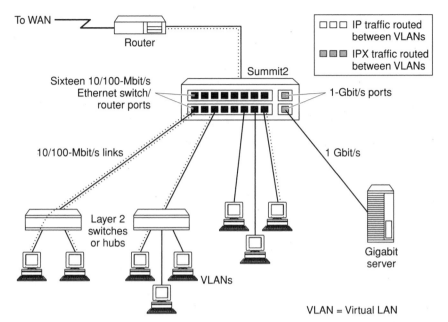

To WAN

Router

☐☐☐ IP traffic routed
       between VLANs

■■■ IPX traffic routed
       between VLANs

Summit2

Sixteen 10/100-Mbit/s
Ethernet switch/
router ports

1-Gbit/s ports

10/100-Mbit/s links

1 Gbit/s

Layer 2
switches
or hubs

Gigabit
server

VLANs

VLAN = Virtual LAN

Figure 27.1 The Summit2: a route through the switch.

mation is stored in a central cache, it can be accessed by all ports simultaneously.

So what's the peak speed of Summit2? Extreme says 5 million packets per second—or more than twice the maximum throughput that Cisco Systems Inc. (San Jose, California) claims for its 7500 router. Extreme also says that switch latency is 6 µs at worst—whether a packet is switched or routed. And with the ASICs taking care of the IP routing, the vendor guarantees the Summit2 won't lose packets even when operating at full-duplex with all ports fully loaded.

When acting as a router, the Summit2 can implement weighted fair queuing, a scheme for managing the buffers so that low-volume traffic isn't starved for bandwidth. It also runs Resource Reservation Protocol (RSVP)—which permits the reservation of bandwidth between switch connections and end stations—and offers IP multicasting. All are selectable on a port-by-port basis.

27.2.2 The switch side

As for switching capabilities, the Summit2 also features packet prioritization. It can assign up to eight levels of priority based on the IEEE's 802.1p draft standard for traffic class and dynamic multicast filtering services in bridged LANs. For each level, net managers can set a maximum amount of bandwidth on a per-port basis to ensure that one type of traffic does not exceed a certain percentage of bandwidth. It's also possible to set minimum bandwidths and maximum delays.

On the management side of things, the Summit2 comes with SNMP Management Information Bases (MIBs); a Web management application that can be used to configure VLANs according to the IEEE's 802.1Q VLAN tagging specs; support for the first four groups of RMON; and a port mirroring capability, so that a probe can be attached to gather traffic data for further analysis. The MIBs and Web application also can be used to configure routing as well as to set quality of service parameters based on such factors as time of day or duration of traffic flow.

27.2.3   Some competition

Of course, Extreme isn't alone in offering such a product. The Netiron from Foundry Networks Inc. (Sunnyvale, California) and the F200 from Rapid City Communications Inc. (Mountain View, California) are both capable of switching and IP routing at wire speed, according to their vendors. But its range of quality of service features is what sets the Summit2 apart. Net managers also could skip IP routing altogether and choose IP switching instead. But not every IP switching scheme works with gigabit Ethernet networks.

The Summit2 costs $14,995—comparable to Foundry's Netiron gigabit switch router ($12,490 for eight 10/100-Mbit/s ports and one gigabit Ethernet uplink) and Rapid City's F200 ($21,000 for sixteen 10/100-Mbit/s ports and two gigabit Ethernet ports). It will be available in July 1998.

27.3   Torrent's IP9000 Gigabit Router

Product Summary

IP9000 Gigabit Router

Torrent Networking Technologies Corp., Landover, Maryland, 301-918-7187

http://www.torrentnet.com

Pros: routes at wire speed

Cons: routes IP only

Like Extreme, Torrent Networking Technologies Corp. also has a strong routing story to tell. It's targeting its IP9000 Gigabit Router at net managers who are fed up with investing in more and more processing power for their conventional backbone routers—only to find that the routers are still running out of puff.

Torrent start-up says its IP9000 Gigabit Router can route IP at wire speed, since it keeps a copy of the routing table on every port. It also boasts service-based (Layer 4) switching for application flows, and it packs a broad range of protocols—from Open Shortest Path First (OSPF) to RIP to border gateway protocol version 4 (BGP#4). BGP#4, Torrent says, is key for companies planning to set up Extranets with trading partners. And though the IP9000 is more expensive than Layer 3 switches, the vendor claims it still has a lower price tag than the carrier-class gigabit IP routers its design is based on.

Sounds good, but prospective buyers should keep in mind that—like Extreme's Summit2—the IP9000 is an IP-only option. Of greater concern is Torrent's decision to use what it calls a "loosely coupled background CPU" for route processing. Translation: the route processor is connected to the switching fabric via a single fast Ethernet connection—which could make for a bottleneck.

27.3.1 Box basics

The IP9000 comes in two versions: an 8-slot chassis with 10-Gbit/s throughput, and a 16-slot chassis with 20-Gbit/s throughput. And there's no shortage of switching options: a partial list of modules that both boxes can handle includes eight-port 10/100 autosensing Ethernet, one-port gigabit Ethernet, and four-port 155-Mbit/s (OC3) ATM. The router has three basic elements: a centralized route processor, which runs the routing protocols and distributes routing updates to the ports; the core switching fabric, which handles per-flow queuing and multicast switching at Layer 3; and ASIC-based forwarding engines at each port, which perform per-packet route lookup.

Torrent says the big advantage to this design is speed. The IP9000 distributes a complete copy of the entire routing table—including flow-based Layer 4 information—to every port (see Fig. 27.2). Route lookup, packet classification, and packet forwarding all take place at wire speed, no matter how large the network.

Torrent also says the IP9000 can handle up to 200,000 routing entries (a typical core Internet router handles about half that) or 100,000 routing entries and their respective Layer 4 User Datagram Protocol (UDP) or TCP port numbers. That helps differentiate the router from many Layer 3 workgroup switches, according to the vendor.

27.3.2 Packet path

One of the biggest bottlenecks on a router is the header processing and how to find a route, according to Torrent. But the vendor says it

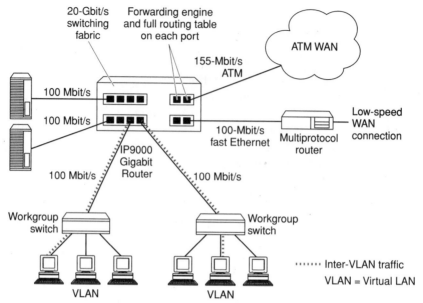

Figure 27.2 The IP9000: a table on every port.

addresses this problem via a proprietary algorithm for matching prefixes, which it runs on the forwarding ASICs.

Here's how the IP9000 puts everything to work. It receives a packet at the media access control (MAC) port and looks up the IP header in the local routing table while the packet is still entering the router. It then identifies and tags the appropriate priority and queue information based on IP destination, source address, or application flow. Flows can be identified and configured statically by using UDP or TCP port numbers, or dynamically via Resource Reservation Protocol (RSVP). The device then performs regular IP routing functions and forwards the packets according to the priority information. And just like a regular IP router, the IP9000 also can be used for inter-VLAN routing. As for management features, the IP9000 complies with SNMP, features Telnet access and a command-line interface, and comes with a Web-based Java interface for configuration. Pricing for the eight-slot gigabit router (which Torrent plans to ship by the end of 1997) starts at $600 per fast Ethernet port; a fully loaded version with 64 fast Ethernet ports is $55,000. The carrier-class GRF 400 router from Ascend Communications Inc. (Alameda, California), by comparison, costs about $100,000 when fitted with 32 fast Ethernet ports.

27.4   3Com's Corebuilder 3500 LAN Switch

Product Summary

Corebuilder 3500

3Com Corp., Santa Clara, California, 408-764-5000

http://www.3com.com

Pros: multiprotocol Layer 3 switch

Cons: some protocols and transports not available in first release

The solutions from Extreme and Torrent are great—provided net managers are running IP-only nets. But many corporate networkers live in a multiprotocol world. It will be years before they are able to pare down their traffic to TCP/IP alone. 3Com believes in building boxes for the here and now. Its Corebuilder 3500 is a Layer 3 switch that can field IP, IPX, and AppleTalk—whether they're traveling over Ethernet, fast Ethernet, gigabit Ethernet, FDDI, or ATM.

What's more, the 3500 offers extensive traffic shaping for delay-sensitive applications, as well as heavy-duty policy-based management. And at roughly $30,000 for a fully loaded switch, the 3500 is about half the cost of a full-blown multiprotocol backbone router. Further, the CB3500 routes traffic at full wire speed, right up to the gigabit range. Still, the 3500 isn't the cheapest Layer 3 box in town. In addition, many of the more advanced features are definitely future tense: in late 1997, 3Com said that ATM modules were still a year out.

27.4.1   The inside story

The CB3500 is a four-slot chassis that can switch Layer 2 and 3 traffic at gigabit speeds, according to vendor claims. As indicated, it can deal with an impressive mix of protocols and transports, routing IP, IPX, and AppleTalk over Ethernet, fast Ethernet, gigabit Ethernet, and FDDI (see Fig. 27.3). In the works is a full suite of Asynchronous Transfer Mode (ATM) software, including LAN Emulation (LANE) 1.0 and 2.0 and Multi-Protocol Over ATM (MPOA), which will let the CB3500 blast legacy packets onto a cell-switching backbone.

The CB3500 is built around 3Com's Flexible Intelligent Routing Engine (Fire), a programmable ASIC the vendor says can be upgraded in software to accommodate new capabilities like IP Version 6. The 3500 also comes loaded with two RISC-based processors. The application processor runs the routing protocols and takes care of net management. The frame coprocessor helps with things like Remote

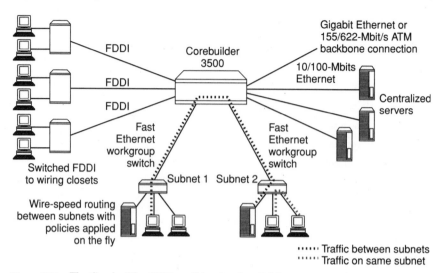

Figure 27.3   The Corebuilder 3500: multiprotocol switching.

Monitoring (RMON) packet capture and filtering. The switch runs Routing Information Protocol (RIP) and Open Shortest Path First (OSPF).

27.4.2 Go with the flow

The Fire ASIC also controls and shapes traffic flows. It has its work cut out for it. For starters, the CB3500 can classify traffic based on protocol type: AppleTalk, Banyan Vines, DECnet, IP, IPX, and SNA. This means net managers can give particular protocols priority through the switch.

The CB3500 also uses 802.1p and 802.1Q—both under development at the IEEE—to identify traffic and prioritize applications. In addition, the switch can identify traffic based on IP source and destination address, multicast address, and TCP and User Datagram Protocol (UDP) port numbers. That enables net managers to pinpoint individual application flows between two end stations—on the fly.

The CB3500 boasts four traffic queues per port. These implement Internet Engineering Task Force (IETF) weighted fair queuing and random early discard (which is like partial packet discard and helps maintain TCP/IP routing at high rates), along with resource reservation protocol (RSVP), to further prioritize delay-sensitive traffic. 3Com's Transcendware software runs on the CB3500 and controls policy management—a way to distinguish which types of traffic get priority treatment.

The CB3500 starts at $9000 for chassis and basic IP software. A six-port 100BaseT module costs $4200; a single-port gigabit Ethernet switching module will cost $3000; a six-port FDDI module will cost about $13,000. Both 155-Mbit/s (OC3) and 622-Mbit/s (OC12) ATM switching modules will be priced at $9000. Additional multiprotocol routing software will cost from $1500 to $3000.

27.5 3Com's Corebuilder 9000 Data Center Switch

Product Summary

Corebuilder 9000

3Com Corp., Santa Clara, California, 408-764-5000

http://www.3com.com

Pros: high density and performance

Cons: some protocols and transports not available in first release

---

The 3500 isn't the only play 3Com is making in the gigabit game. It's also touting another switch, called the Corebuilder 9000. "Bigger is better" seems to be the simple philosophy underlying the 9000's design. How big? How about 504 fast Ethernet ports or 126 gigabit Ethernet ports or 112 OC12 ATM ports?

But the Corebuilder 9000 isn't just the Arnold Schwarzenegger of LAN switches. It also furnishes policy and traffic shaping and offers a choice of Layer 3 switching or cut-through routing using 3Com's Fast IP scheme.

Sounds great, but the 9000 shares the same drawbacks with its smaller cousin. As with the 3500, some of the 9000's most impressive features were MIA in the first release and not due to arrive until the second half of 1998. What's more, 3Com's big switch comes with a big price tag. A 14-slot chassis with a two-port OC12 module costs $51,000.

27.5.1 Going both ways

The Corebuilder 9000 sits in the data center, aggregating traffic from smaller LAN switches (see Fig. 27.4). The switching fabric's initial capacity is 15 Gbit/s; future releases will reach 70 Gbit/s, 3Com claims. And its nonblocking, redundant switching matrix can field cells or frames. Supporting both technologies lets net managers decide whether to go with cells, frames, or a combination of the two, says the vendor.

3Com will ship an eight-port ATM OC3 (155-Mbit/s) module, a two-port ATM OC12 (622-Mbit/s) interface, and a one-port OC48 (2.4-

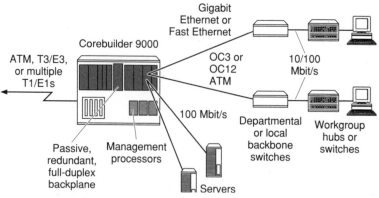

Figure 27.4   The Corebuilder 9000: big-boned.

Gbit/s) ATM module. All 14 slots can be filled; there are 2 separate slots for management modules. This gives the 9000 a total of 112 OC3 or 28 OC12 ports. Later in 1998, 3Com plans to release higher-density modules, boosting the maximum ports to 112 OC12 links.

3Com has equally ambitious plans for all types of Ethernet. Its modules will initially give the 9000 a maximum capacity of 280 RJ-45 fast Ethernet ports, 504 RJ-21 fast Ethernet ports, or 28 gigabit Ethernet ports. Again, 3Com says it will ship higher-density modules later on, taking the maximum capacity to 126 gigabit Ethernet links. All Ethernet modules offer Layer 2 switching and multilink trunking (the ability to inverse-mux traffic over multiple Ethernet channels). They also feature IEEE 802.1p prioritization and 802.1Q virtual LAN standards.

Once 3Com ships all of the components of the 9000, its 126 gigabit Ethernet ports should far exceed the capacity of other high-end LAN switches. For example, the MMAC Plus from Cabletron Systems Inc. (Rochester, New Hampshire) handles 28 gigabit Ethernet ports with its recently upgraded backplane. And Prominet Corp. (Westborough, Massachusetts) piles up just 24 gigabit ports with its switch. Stiffer ATM competition may be on the way from Lannet Inc. (Irvine, California). Lannet plans 70 ATM OC12 and 24 gigabit Ethernet ports on its Meritage switch, due out in 1998.

3Com also promises to offer its Corebuilder 3500 switch as a separate module for the chassis. With it, the 9000 will furnish Layer 3 switched connections to FDDI, ATM, and Ethernet nets. The vendor ultimately plans to add Layer 3 capability to more of the 9000 Ethernet modules.

27.5.2   Fault finder

The Corebuilder 9000 packs some equally prodigious management capabilities. Distributed intelligent agents isolate faults and configure devices. Additionally, the Ethernet switch ports handle RMON 1 and RMON 2 for detailed analysis of traffic flow. The modules also house Java virtual machines, allowing applets for services such as policy management to be written and run on the switch. And the 9000 offers fault tolerance. It can be outfitted with redundant switching fabrics, management module, power supplies, and fans.

A 14-slot chassis with power supply, switching fabric, management controller, and two-port ATM OC12 module is $51,000. A similar gigabit system will run $39,000. Additional dual-port ATM OC12 modules go for $9995 and dual-port gigabit Ethernet modules for $3995.

27.6   Alteon's Alteon/NIC and Switch

Product Summary

Alteon/NIC and switch

Alteon Networks Inc., San Jose, California, 408-574-5500

http://www.alteon.com

Pros: gigabit throughput eliminates bottlenecks

Cons: proprietary; no net analysis tools

Not all gigabit vendors are targeting the sort same sort of mondo backbone applications as 3Com. The products from Alteon Networks Inc. (San Jose, California) are aimed squarely at the other end of the LAN.

Start-up Alteon thinks it can uncork the bottlenecks on switch-to-server and switch-to-switch links with its gigabit Ethernet adapter and switches. Alteon's gigabit gear offers 10 times the performance of 100BaseT (fast Ethernet) for only 2.5 times the cost. The adapter and switch prioritize traffic and offer congestion control. And the switch's dual-processor design keeps things from bogging down.

27.6.1   Solution specifics

Alteon's rack-mountable switch packs eight 100BaseT (fast Ethernet) ports to connect to switches, servers, hubs, or end stations. It also comes with two 1-Gbit/s ports to link to other Alteon switches or servers equipped with the vendor's gigabit Ethernet adapter (see Fig. 27.5).

The switch operates in store-and-forward mode. The faster cut-through mode wasn't an option because the switch has to convert between 100-Mbit/s and 1-Gbit/s speeds. Both 100BaseT and gigabit ports can be set for full-duplex operation. The switch can be managed via SNMP.

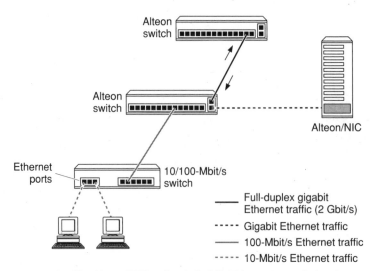

Figure 27.5   The Alteon/NIC and switch: full (Ethernet) speed ahead.

The Alteon/NIC slots into a standard Peripheral Component Interface (PCI) bus. Transmission at 1 Gbit/s might seem far-fetched for standard servers, but PCI buses actually can burst to speeds in excess of 1 Gbit/s.

The switch and the adapter only work with fiber, and drive distances depend on the type of cable. A 50-µm multimode fiber connects devices up to 2 km apart; a 62.5-µm multimode link reaches about 500 m. (As with the rest of Alteon's performance claims, its drive distances have yet to be independently verified.)

27.6.2 Supporting part

Alteon has built support for prioritization, flow control, and virtual LANs into its switch and adapter. The prioritization scheme, which ensures that real-time voice and video applications don't get backed up, is based on the IEEE's 802.1p spec. The flow-control mechanism, which prevents packet loss on busy nets, is based on the 802.3x spec. Alteon's products also adhere to the IEEE's 802.1Q spec, which defines how to implement virtual LANs between 802-compliant LAN switches using port addresses.

27.6.3 Hardware booster

Each port on the switch is outfitted with an ASIC and a 32-bit RISC processor (the same applies to the adapter). The application-specific integrated circuit (ASIC) performs the grunt work of reading addresses, checking for errors, and processing packets. The processor handles tasks like prioritization, flow control, and management. Dividing up the tasks ensures that the overhead incurred by value-added features doesn't lower switch performance, Alteon says.

Despite the tenfold bandwidth boost, Alteon prices its gigabit hardware at about 2.5 times the cost of fast Ethernet gear. The Alteon/NIC costs just under $2000; the switch costs about $1875 per 1-Gbit/s port. Prices for Alteon's eight 100BaseT (fast Ethernet) ports are in line with average 100BaseFX costs—in the $750-per port range—putting the total cost for the switch at approximately $9750.

27.7 Gigalabs' Gigastar 100

Product Summary

Gigastar 100

Gigalabs Inc., Sunnyvale, California, 408-481-3030

http://www.gigalabs.com

Pros: eliminates bottlenecks on switch-to-server links; inexpensive

Cons: proprietary; maximum distance of 25 m over copper

Alteon is not the only company targeting the problem of switch-to-server bottlenecks. Gigalabs Inc. (Sunnyvale, California) is shipping the Gigastar 100 Ethernet switch, which bursts bottlenecks by extending the server's 1.05-Gbit/s PCI bus right to the switch. This technology, known as I/O switching, eliminates the network adapter to deliver fast switch-to-server, switch-to-switch, and switch-to-storage links. The Gigastar also boasts a scalable, stackable design and RMON analysis. And it's easy on the wallet, costing as little as $200 per Ethernet port.

Sounds great, but there are a few drawbacks. In its first release, Gigalabs' I/O switching only runs over copper for 25 m. Also, the Gigapipe technology used in Gigastar is proprietary, which means that cabling and interfaces must be purchased from the vendor.

27.7.1 Star power

The Gigastar 100 is a stackable and chassis-based switch rolled into one. Each chassis sports five slots and a 1-Gbit/s backplane. The slots accommodate any one of several network modules, including eight-port Ethernet, two-port fast Ethernet, single-port FDDI, or single-port 155-Mbit/s ATM. All of the modules work at the MAC layer. But each one serves a different application. The 10/100-Mbit/s Ethernet modules are used for desktop connections. The FDDI and ATM modules link the switch to high-speed backbones. Gigalabs says its switch automatically translates between different packet and cell formats. Up to four switch chassis can be stacked, delivering a total of 160 ports of 10-Mbit/s Ethernet. Switches in the same stack are connected internally via a 1-Gbit/s bus.

For even faster switching, net mangers could use the vendor's three Gigapipe switching modules. Each module takes up a single slot in one chassis. The Gigapipe modules improve I/O performance between switches in different stacks, switch and server, and switch and storage devices (see Fig. 27.6). The server module essentially extends the Peripheral Component Interface (PCI) bus on the server over a cable and into the switch itself. The storage unit delivers a direct SCSI link from the switch to devices like hard drives or CD-ROM drives.

These direct connections into the server or disk drives, Gigalabs claims, deliver significantly better performance than the 100 or 155 Mbit/s of bandwidth offered by fast Ethernet, FDDI, or ATM. Note that while all of the Gigapipe modules have the ability to pump data over the network at 1.6 Gbit/s, PCI and Small Computer System Interface (SCSI) traffic is limited by the speed of the buses to maximum of 1.05 Gbit/s and 320 Mbit/s, respectively.

Not only do the Gigapipe modules run faster than other LAN technologies, they also remove the I/O overhead incurred when data is converted from the PCI bus or SCSI interface into a LAN format. Data is transmitted using native PCI or SCSI signaling, which, Gigalabs claims, lets net managers use the full bandwidth of the PCI bus. The vendor also maintains that Gigastar can handle the new servers that offer PCI buses with 264 Mbytes/s and 64-bit extended PCI architecture.

The box offers a sustained throughput of 400 Mbit/s, according to the vendor. (Actual throughput drops from the theoretical 1.6 Gbit/s due to overhead created when data is copied from the application on the

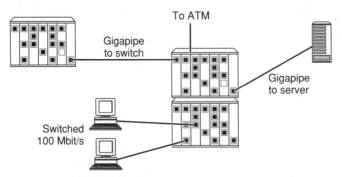

Figure 27.6 The Gigastar 100: stacking up the odds with I/O switching.

server to the operating system's protocol stack.) Gigalabs claims its pipe delivers 4 to 10 times the performance of fast Ethernet, and that net managers are lucky to get 40 to 60 Mbit/s from a fast Ethernet link.

The Gigapipes use a 100-pin 50-twisted-pair copper cable for a distance of 25 m, which can be stretched to 150 m using two Gigalabs repeaters. The vendor also is working on fiber-optic interfaces to extend the switch-to-switch distance to 500 m and 10 km over multimode and single-mode fiber.

27.7.2 Management and price

The Gigastar supports seven Remote Monitoring (RMON) groups per port (capture and filter aren't included). The vendor doesn't recommend performing RMON analysis on all of the ports at once because it could lower switch performance.

The box doesn't support port mirroring, which would allow an RMON probe to view data from any port on the switch. Instead, a probe has to be connected to each Ethernet segment to perform full packet capture and analysis of data coming into a given switch port. A management module that slots into a switch chassis collates RMON stats from a stack of four Gigastar switches.

The Gigastar chassis costs $1595, so a four-switch stack with a maximum of one hundred sixty 10-Mbit/s Ethernet switch port sells for just under $200 per port. The Gigapipe for switch-to-PCI connections costs $2495; the Gigapipe for switch-to-switch link costs $1995; and the Gigapipe for SCSI ports is priced at $4995. The eight-port 10-Mbit/s switch module is $1195; the two-port fast Ethernet module costs $1995. Uplink modules range from $995 for fast Ethernet to $1995 for 155-Mbit/s ATM. The SNMP module costs $1195.

List of Vendors

Gigabit Ethernet owes much of its speedy development and acceptance to the Gigabit Ethernet Alliance (GEA), a vendor consortium that has been working hard behind the scenes to expedite completion of the standard (largely by suppressing the sort of vendor infighting that has held up other networking specs) and also to hasten interoperability testing.

There are now well over 100 member companies in the Gigabit Ethernet Alliance—including every major gigabit Ethernet equipment vendor. Here's a complete list, including the name and e-mail address of each vendor's primary GEA representative.

Acacia Networks Inc.
Alan Raderman
650 Suffolk Street
Lowell, MA 01854
Phone: 978-275-0606
Fax: 978-275-4277
E-mail: araderman@acacianet.com

Acclaim Communications Inc.
Sanjay Sharma
5000 Old Ironside Drive
Santa Clara, CA 95054
Phone: 408-327-0100
Fax: 408-327-0106
E-mail: sanjays@acclaiminc.com

Accton Technology Corporation
Robert Wu
No. 1 Creation Road III
Science Based Industrial Park
Hsinchu, Tawian 300, ROC

Phone: +886-35-770-270, ext. 683
Fax: +886-35-770-771
E-mail: rwu@accton.com.tw

Adaptec Inc.
Eric Brown
691 South Milpitas Boulevard
Milpitas, CA 95035
Phone: 408-957-6645
Fax: 408-262-2533
E-mail:
eric_brown@corp.adaptec.com

Addtron Technology Co. Ltd.
Frank Lee
4425 Cushing Parkway
Fremont, CA 94538
Phone: 888-233-8766
Fax: 510-668-0699
E-mail: at-frank@addtron.com

Alliance Semiconductor Corp.
Vinod Kumar Sutrave
3099 North First Street
San Jose, CA 95134-2006
Phone: 408-383-4900, ext. 210
E-mail: vsutrave@alsc.com

Alteon Networks Inc.
David Callisch
6351 San Ignacio Avenue
San Jose, CA 95119
Phone: 408-360-5540
Fax: 408-574-5501
E-mail: david_callisch@alteon.com

AMD Inc.
Ray Heckman
One AMD Place
P.O. Box 3453
Sunnyvale, CA 94088-3453
Phone: 408-749-5459
E-mail: ray.heckman@amd.com

Amdahl Corporation
Richard Taborek Sr.
1250 Arques Avenue
Sunnyvale, CA 94088
Phone: 408-764-6533
Fax: 408-773-0833
E-mail: rich_taborek@amdahl.com

AMP Inc.
Kirk Bovill
P.O. Box 3608
Harrisburg, PA 17105-3608
Phone: 408-725-4902
E-mail: klbovill@amp.com

Ancor Communications Inc.
Kim Anderson
6130 Blue Circle Drive
Minnetonka, MN 55343
Phone: 612-932-4090
Fax: 612-932-4037
E-mail: kima@ancor.com

Apple Computer Inc.
Wanda Cox
1 Infinite Loop
Cupertino, CA 95014
Phone: 408-974-2178
Fax: 408-974-2691
E-mail: cox.wanda@applecom

Applied Micro Circuits Corp.
Reza Moattar
6195 Lusk Boulevard
San Diego, CA 92121
Phone: 619-450-9333
Fax: 619-450-9885
E-mail: rezam@amcc.com

Asante Technologies Inc.
Chris Tomlinson
821 Fox Lane
San Jose, CA 95131
Phone: 408-894-8401, ext. 297
Fax: 408-272-4582
E-mail: ctomlins@asante.com

Ascend Communications Inc. (Formerly NetStar Inc.)
Dick Kachelmeyer
10250 Valley View Road
Minneapolis, MN 55344
Phone: 612-996-6820
E-mail: dickk@netstar.com

Auspex Systems Inc.
Ashvin Sanghvi
5200 Great America Parkway
Santa Clara, CA 95054
Phone: 408-986-2580
E-mail: asanghvi@auspex.com

Bay Networks Inc.
Jeff Martin
4401 Great America Parkway
Santa Clara, CA 95054
Phone: 408-495-1053
E-mail: jmartin@baynetworks.com

Bell Consulting Inc.
Steve Bell
20111 Stevens Creek Boulevard
Suite 280
Cupertino, CA 95014
Phone: 408-253-2370
Fax: 408-253-4754
E-mail: sbell@bellc.com

Boston Optical Fiber Inc.
Kenneth G. Taylor
155 Glanders Road
Westborough, MA 01581
Phone: 508-347-3309
Fax: 508-836-2722
E-mail: ktaylor@bostonoptical.com

Brooks Technical Group Inc.
Christopher Burke
883 North Shoreline Boulevard
Mountain View, CA 94043
Phone: 415-960-3612, ext. 320
Fax: 415-960-3615
E-mail: cjb@brooks-tech.com

Cabletron Systems Inc.
Steve Augusta
40 Continental Boulevard
Merrimack, NH 03054
Phone: 603-337-5100
E-mail: augusta@ctron.com

Canon Information Systems
Ozay Oktay
3188 Pullman Street
Costa Mesa, CA 92626-3322
Phone: 714-438-7180
Fax: 714-432-1531
E-mail: ozay_oktay@cissc.canon.com

Cisco Systems Inc.
Nathan Walker
170 West Tasman Drive
San Jose, CA 95134
Phone: 408-526-5928
Fax: 408-526-6488
E-mail: nathanw@cisco.com

ComCore Semiconductor Inc.
Bruce Gladstone
27001 West Agoura Road
Suite 210
Calabasas, CA 91301
Phone: 818-880-5192
Fax: 818-880-5193
E-mail: bruce@comcore.com

Compaq Computer Corporation
Scott Johnson
12301 Technology Boulevard
Austin, TX 78727
Phone: 512-433-6756
Fax: 512-433-6399
E-mail: scjohnson@bangate.com

Corning Inc.
Steven Swanson
35 West Market Street
M/S: MP-R3-03
Corning, NY 14831
Phone: 607-974-4252

Fax: 607-974-4941
E-mail: swansonse@corning.com

Cypress Semiconductor Corporation
Greg Somer
3901 North First Street
San Jose, CA 95134
Phone: 408-943-2605
E-mail: gbs@cypress.com

Digi International
G. Ramakrishnan
1299 Orleans Drive
Sunnyvale, CA 94089
Phone: 408-744-2748
Fax: 408-744-2793
E-mail: rama@dgii.com

Digital Equipment Corporation
Patrick Horgan
550 King Street
M/S: LKG1-3/M7
Littleton, MA 01460-1284
Phone: 508-486-6893
Fax: 508-486-7417
E-mail: horganp@mail.dec.com

D-Link System Inc.
Sam Liang
5 Musick
Irvine, CA 92618
Phone: 714-457-6333
E-mail: sliang@irvine.dlink.com

Duke Energy Corporation
Brad Black
401 South College Street
Charlotte, NC 28201
Phone: 704-382-9363
Fax: 704-382-0381
E-mail: bblack@dpcmail.duke-power.com

Edimax Technology Co. Ltd.
Gene Renn
3F, 50
Wu-chuan 7 Road
Wu-Ku, Taipei Taiwan, ROC
Phone: +886-2-299-5648
Fax: +886-2-299-6212
E-mail: gene@edimax.com.tw

Emulex Corporation
Mike Kane
3535 Harbor Boulevard

Costa Mesa, CA 92626
Phone: 714-513-8153
Fax: 714-513-8266
E-mail: m_kane@emulex.com

Essential Communications
Michael McGowen
4374 Alexander Boulevard NE
Suite T
Albuquerque, NM 87107
Phone: 505-344-0080, ext. 308
E-mail: mikemc@esscom.com

E.T.R.I., Systems Technology Section
Hun Kang
161 Kajong-dong
Yusong-gu, Taejon 305-350, South
Korea
Phone: +82-42-860-6113
Fax: +82-42-860-5213
E-mail: hkang@winky.etri.re.kr

Extreme Networks Inc.
Stephen Haddock
1601 South De Anza Boulevard
Suite 220
Cupertino, CA 95014
Phone: 408-342-0982
E-mail: shaddock@ix.netcom.com

Finisar Corporation
Vipul Bhatt
620-B Clyde Avenue
Mountain View, CA 94043
Phone: 415-691-4000
Fax: 415-691-4010
E-mail: vipul@finisar.com

FORE Systems Inc.
Richard Borden
174 Thorn Hill Road
Warrendale, PA 15086
Phone: 412-635-3606
Fax: 412-635-3663
E-mail: rborden@fore.com

Foundry Networks
Drusie Demopoulos
680 West Maude Avenue
Suite 3
Sunnyvale, CA 94086
Phone: 408-731-3820
E-mail: drusie@foundrynet.com

Fujikura America Inc.
Bob Dahlgren
3001 Oakmead Village Drive
Santa Clara, CA 95051-0811
Phone: 408-748-6991
Fax: 408-727-3460
E-mail: bob@fujikura.com

Fujitsu Inc.
Haim Shafir
3545 North First Street
San Jose, CA 95134-1804
Phone: 408-922-9529
Fax: 408-432-9044
E-mail: hshafir@fmi.fujitsu.com

Galileo Technology Inc.
David Shemla
P.O. Box 6374
Hagharoset Street 36
Karmiel, 20100, Israel
Phone: +972-4-988-4259
Fax: +972-4-988-4247
E-mail: david@galileo.co.il

GEC Plessey Semiconductors
Rich Bowers
1500 Green Hills Road
Scotts Valley, CA 95067
Phone: 408-461-6237
Fax: 408-438-5576
E-mail: rich.bowers@gpsemi.com

GigaLabs Inc.
Kon Leong
290 Santa Ana Court
Sunnyvale, CA 94086
Phone: 408-481-3030
Fax: 408-481-3045
E-mail: kleong@gigalabs.com

Hewlett-Packard Company
Rex Pugh
8000 Foothills Boulevard
Roseville, CA 95474
Phone: 916-785-4471
E-mail: rex_pugh@hp.com

Hitachi Cable, Ltd.
Koichiro Seto
3031 Tisch Way
Suite 807
San Jose, CA 95128

Phone: 408-260-2630
E-mail: seto@rop2.hitachi-cable.co.jp

Hitachi Internetworking
J. Felix McNulty
3101 Tasman Drive
Santa Clara, CA 95054
Phone: 408-986-9770, ext. 107
E-mail: f_mcnulty@hitachi.com

HolonTech Corporation
Kelvyn Evans
2039 Samaritan Drive
San Jose, CA 95124
Phone: 408-369-4682
Fax: 408-369-4780
E-mail: kelvyn@holontech.com

Honeywell Inc.
John Bowerman
830 East Arapaho Road
Richardson, TX 75081
Phone: 972-470-4553
Fax: 972-470-4326
E-mail: jbowerma@micro.honey-
well.com

IBM
Michael Taddei
A78/660
600 Park Offices Drive
Research Triangle Park, NC 27709
Phone: 919-486-0550
E-mail:
michael_taddei@vnet.ibm.com

Integrated Circuit Systems Inc.
Charan J. Singh
1271 Parkmoor Avenue
San Jose, CA 95126-3448
Phone: 408-925-9470
Fax: 408-925-9460
E-mail: charan@iscssj.com

Integrated Device Technology Inc.
Robert Napaa
2975 Stender Way
M/S: C3-50
Santa Clara, CA 95054
Phone: 408-492-8632
Fax: 408-492-8469
E-mail: napaa@idt.com

Intel Corp.
Tim Dunn
Network Products Division
5200 Northeast Elam Young Park-
way
Hillsboro, OR 97124
Phone: 503-264-9861
E-mail: tim_dunn@ccm.jf.intel.com

Ipsilon Networks
Tom Lyon
232 Java Drive
Sunnyvale, CA 94089-1318
Phone: 408-990-2000
Fax: 408-743-5675
E-mail: pugs@ipsilon.com

ITRI/CCL
Ben Chao
1590 Centre Point Drive
Milpitas, CA 95035
Phone: 408-946-3015
Fax: 408-946-3019
E-mail: itrica@ix.netcom.com

Jato Technologies
Peter Rauch
3520 Executive Centre Drive
Suite 100
Austin, TX 78731
Phone: 512-342-0770, ext. 203
Fax: 512-342-0776
E-mail: prauch@jatoech.com

Kingston Technology Co.
Steven Chen
17600 Newhope Street
Fountain Valley, CA 92708
Phone: 714-437-3311
E-mail: steve_chen@kingston.com

LANart Corp.
Yongbum Kim
145 Rosemary Street
Needham, MA 02194
Phone: 617-444-1994
Fax: 617-444-3692
E-mail: ykim@lanart.com

Lanoptics Ltd.
Dono Van Mierop
P.O. Box 184
Migdal, Ha-Emek, 10551, Israel

Phone: +972-6-44-9944
Fax: +972-6-54-0124
E-mail: dono@lanoptics.co.il

LANQuest Group
Gail James
47800 Westinghouse Drive
Fremont, CA 94539
Phone: 510-354-0940
Fax: 510-354-0950
E-mail: gjames@lanquest.com

Level One Communications
Kirk Hayden
9750 Goethe Road
Sacramento, CA 95762
Phone: 916-854-2871
E-mail: khayden@level1.com

LSI Logic Corp.
Shekar Rao
1525 McCarthy Boulevard
Milpitas, CA 95035
Phone: 408-954-4625
Fax: 408-433-8980
E-mail: shekar@lsil.com

Lucent Technologies
John Bestel
555 Union Boulevard
Room 23R-060
Allentown, PA 18103
Phone: 610-712-7790
Fax: 610-712-4216
E-mail: bestel@lucent.com

Macronix America Inc.
Chang-chi Liu
1338 Ridder Park Drive
San Jose, CA 95131
Phone: 408-451-3819
Fax: 408-453-8488
E-mail: ccliu@macronix.com

Madge Networks Inc.
Mike Salzman
2310 North First Street
San Jose, CA 95131-1011
Phone: 408-952-9814
Fax: 408-955-0970
E-mail: msalzman@madge.com

MediaWise Networks Inc.
Krishan Viswanadham
3160 De La Cruz Boulevard
Suite 102A
Santa Clara, CA 95054
Phone: 408-235-8031, ext. 200
Fax: 408-235-8030
E-mail: krishna@cellswitch.com

MMC Networks
Henry Hsiaw
1134 East Arques Avenue
Sunnyvale, CA 94086-4602
Phone: 408-731-1620
E-mail: henry@mmcnet.com

Motorola Semiconductor
Gary McGibbon
8601 Six Forks Road
Suite 201
Raleigh, NC 27615
Phone: 919-870-4365
Fax: 919-870-4350
E-mail: rvaj50@email.sps.mot.com

Myricom Inc.
Chuck Seitz
325 North Santa Anita Avenue
Arcadia, CA 91006
Phone: 818-821-5555
Fax: 818-355-7825
E-mail: chuck@myri.com

National Semiconductor Corp.
Michele Holguin
2900 Semi Conductor Drive
Santa Clara, CA 95052
Phone: 408-721-2880
E-mail: michele.holguin@nsc.com

NBase Communications Inc.
Yechiel Kurtz
8943 Fullbright Avenue
Chatsworth, CA 91311
Phone: 818-773-0900
Fax: 818-773-0906
E-mail: kurtz@nbase.com

NEC Electronics Inc.
Pronay Saha
475 Ellis Street

M/S: MV4136
Mountain View, CA 94043-2203
Phone: 415-966-5435
Fax: 415-965-6752
E-mail: pronay_saha@el.nec.com

Neo Networks Inc.
Hemant V. Trivedi
10800 Lyndale Avneue South
Suite 295
Bloomington, MN 55420
Phone: 612-884-6844
E-mail: hemant@neonetworks.com

NeoParadigm Labs
Michael Yang
1735 North First Street
Suite 108
San Jose, CA 95112
Phone: 408-451-1247
Fax: 408-452-8665
E-mail: myang@nplab.com

Netcom Systems Inc.
Mark Fishburn
20500 Nordhoff Street
Chatsworth, CA 91311
Phone: 818-700-5100
Fax: 818-708-7991
E-mail: mark_fishburn@netcomsys-
tems.com

NetVantage, Inc.
Hava Zernik
201 Continental Boulevard
Suite 201
El Segundo, CA 90245-4427
Phone: 310-726-4130
Fax: 310-726-4131
E-mail:
hava.zernik@netvantage.com

NetVision Corporation
Robert J. Zecha
One Comac Loop
M/S: #1A
Ronkonkoma, NY 11779
Phone: 516-737-2363
Fax: 516-737-2372
E-mail: r.zecha@is.netcom.com

Network Appliance Inc.
Cheena Srinivasan
2770 San Tomas Expressway
Santa Clara, CA 95051
Phone: 408-367-3203
Fax: 408-367-3100
E-mail: cheena@netapp.com

Network General Corporation
Cheryl Haines
4200 Bohannon Drive
Menlo Park, CA 94025
Phone: 415-473-2627
Fax: 415-321-0855
E-mail: cheryha@ngc.com

Network Peripherals
Oliver Szu
1371 McCarthy Boulevard
Milpitas, CA 95035
Phone: 408-321-7399
Fax: 408-321-9218
E-mail: oszu@npix.com

Optical Data Systems Inc.
Jerry Pate
1101 E. Arapaho Road
Richardson, TX 75081
Phone: 972-301-3605
Fax: 972-234-4059
E-mail: jerry@ods.com

ORNET Data Communications
Technologies
E. Littwitz
P.O. Box 323
Carmiel 21613, Israel
Phone: +972-4-998-1314
Fax: +972-4-998-1315
E-mail: eilon@ornet.co.il

Packet Engines Incorporated
Bernard Daines
P.O. Box 14497
Spokane, WA 99214
Phone: 509-922-9190
Fax: 509-922-9185
E-mail:
bernardd@packetengines.com

Panasonic
Ren Franse
1 Panasonic Way
M/S: IE4
Secaucus, NJ 07094
Phone: 201-392-4466
Fax: 201-348-7942

PlainTree Systems Inc.
Alan Greenfield
Prospect Place
9 Hillside Avenue
Waltham, MA 02154
Phone: 613-831-8300
Fax: 617-965-2466
E-mail: agreenfield@plaintree.com

PMC-Sierra Inc.
Richard Cam
105-8555 Baxter Place
Burnaby, BC V5A 4V7, Canada
Phone: 604-415-6022
Fax: 604-415-6207
E-mail: cam@pmc-sierra.com

Prominent Corporation
Menachem Abraham
100 Nickerson Road
Marlborough, MA 01752
Phone: 508-303-8885, ext. 202
Fax: 508-303-8161
E-mail: mabraham@prominet.com

Rapid City Communications Inc.
Bert Armijo
1215 Terra Bella Avenue
Mountain View, CA 94043
Phone: 415-237-8668
Fax: 415-237-7388
E-mail: bert@rapid-city.com

Richard Hirschmann GmbH
Harald Baums
Stuttgarter Strasse 45-51
D-72654 Neckartenzlingen, Box 1649
D-72606 Nurtingen, Germany
Phone: +49-7-127-14-1544
Fax: +49-7-127-14-1561
E-mail: hbaums@nt.hirschmann.de

Rockwell Semiconductor Systems
Howard Chan
4311 Jamboree Road

Irvine, CA 92660-3095
Phone: 714-221-5834
Fax: 714-221-6544
E-mail:
howardchan@nb.rockwell.com

Samsung Information Systems
Andrew Shieh
8 New England Executive Plaza
Burlington, MA 01803
Phone: 617-270-5353, ext. 127
Fax: 617-272-5281
E-mail: shieh@cnl-samsung.com

SEEQ Technology Inc.
Robert Frostholm
47200 Bayside Parkway
Fremont, CA 94538
Phone: 510-226-2904
Fax: 510-657-2837
E-mail: bfrostholm@seeq.com

Shiva Corp.
Mike Allan
Shiva Park
Stanwell Street
Edinburgh, FH107, Scotland
Phone: +44-1-31-561-4134
E-mail: mike@europe.shiva.com

Siemens AG
Harvey Sherman
Hofmannstrasse 51
Munich, D 81359, Germany
Phone: 408-492-6844

Siemens OED FO
Shelto Van Doorn
19000 Homestead Road
Cupertino, CA 95014
Phone: 408-725-3436
Fax: 408-725-3435
E-mail: schelto.van-
doorn@sci.siemens.com

Signal Consulting Inc.
Howard Johnson
16541 Redmond Way
Suite 264
Redmond, WA 98052
Phone: 425-556-0800
Fax: 425-881-6149
E-mail: howie@signal.com

Silicon Dynamics Inc.
Stefan Wurster
255 Santa Ana Court
Sunnyvale, CA 94086
Phone: 408-245-6600
Fax: 408-245-6644
E-mail: smw@sidynamics.com

Silicon Graphics Inc.
Joe Gervais
2011 North Shoreline Boulevard
M/S: 08L-855
Mountain View, CA 94043-1389
Phone: 415-933-7479
Fax: 415-964-0811
E-mail: gervais@engr.sgi.com

Silicon Image Inc.
David Lee
1032 Elwell Court
Suite 222
Palo Alto, CA 94303
Phone: 415-964-9482
Fax: 415-964-9435
E-mail: dlee@siimage.com

S-MOS Systems Inc.
Hillol Sarkar
150 River Oaks Parkway
San Jose, CA 95134-1951
Phone: 408-922-0200
Fax: 408-433-0554
E-mail: hsarkar@smos.com

Spike Technologies Inc.
Nikhil Modi
500 Calaveras Boulevard
Suite 206
Milpitas, CA 95035
Phone: 408-945-0354
E-mail: nikhil@spiketech.com

Standard Microsystems Corp.
Tom Grasmehr
6 Hughes
Irvine, CA 92718
Phone: 714-707-4803
E-mail: tom.grasmehr@smc.com

Sumitomo Electric Industries
Satoshi Hagihara
1-3-12, Motoakasaka
Minato-ku

Tokyo, 107, Japan
Phone: +81-6-466-5609
Fax: +81-6-466-5732
E-mail: hagihara@rcom.sei.co.jp

Sun Microsystems Inc.
Alan Dobbs
10 Network Circle
M/S: UMPK10-106
Menlo Park, CA 94025
Phone: 415-688-9569
E-mail: alan.dobbs.eng@sun.com

Super Highway Co.
Mason Ju Jr.
10 Bennett Avenue
Suite 3D
New York, NY 10033
Phone: 212-923-8551
Fax: 212-923-8551

Symbios Logic Inc.
Margit Stearns
4420 Arrows West Drive
Colorado Springs, CO 80907-344
Phone: 719-533-7482
Fax: 719-533-7480
E-mail: margit.stearns@symbios.co

Synergy Semiconductor Corp.
Dona Flamme
3450 Central Expressway
Santa Clara, CA 95051
Phone: 408-980-9191
E-mail: dona.flamme@synergy.com

Texas Instruments Inc.
Greg Waters
8505 Forest Lane
M/S: 8650
Dallas, TX 75243
Phone: 214-480-2605
E-mail: glw2@msg.ti.com

3Com Corporation
Bob Roman
5400 Bayfront Plaza
M/S: 1317
Santa Clara, CA 85052-8145
Phone: 408-764-5237
Fax: 408-764-5001
E-mail: bob_roman@3mail.3com.com

3M
Tad Szostak
6801 River Place Boulevard
Austin, TX 78726
Phone: 512-984-1800
Fax: 512-984-3417
E-mail: tszostak1@mmm.com

Torrent Networking Technologies
Gordon Saussy
8181 Professional Place
Suite 160
Landover, MD 20785
Phone: 301-918-7187
Fax: 301-918-7193
E-mail: gsaussy@torrentnet.com

UB Networks
Ervin Johnson
5 Corporate Drive
Andover, MA 01810
Phone: 508-691-6421
Fax: 508-687-4442
E-mail: ejohnson@ub.com

UNI
Sergey Malyshev
5/1 Novovagankovski Lane
Moscow, Central 123242, Russia
Phone: 7-095-956-9037 or 956-6444
Fax: 7-095-941-4634
E-mail: mss@uniinc.msk.ru

Vitesse Semiconductor Corp.
John Schaefer
741 Calle Plano
Camarillo, CA 93012
Phone: 805-388-7571
Fax: 805-987-5896
E-mail: schaefer@vitesse.com

Vixel Corporation
Stan Swirhun
325 Interlocken Parkway
Bldg. A
Broomfield, CO 80021
Phone: 303-464-2276
Fax: 303-466-0290
E-mail: sswirhun@denver.vixel.com

VLSI Technology
Craig O'Sullivan
1109 McKay Drive

M/S: 21A
San Jose, CA 95131
Phone: 408-434-7673
Fax: 408-434-7866
E-mail:
craig.osullivan@sanjose.vlsi.com

Wideband Corp.
Dr. Roger E. Billings
26900 Pink Hill Road
Independence, MO 64057
Phone: 816-229-0017
Fax: 816-220-0300
E-mail: billings@science.edu

Winbond Electronics Corp.
Wilbur H. Chuang
No. 4 Creation Road III
Science Based Industrial Park
Hsinchu, Taiwan 30077, ROC
Phone: +886-3-579-2558
Fax: +886-3-579-2647
E-mail: whchuang@winbond.com.tw

XaQti
Samba Murthy
1630 Oakland Road
Building #A214
San Jose, CA 95131
Phone: 408-487-0803
E-mail: samba.murthy@xqati

Xircom Inc.
Jim Soriano
2300 Corporate Centre Division
Thousand Oaks, CA 91320-1420
Phone: 805-376-6880
E-mail: jim_soriano@xircom.com

XLNT
Robert Grow
15050 Avenue of Science
Suite 200
San Diego, CA 92128
Phone: 619-487-9320, ext. 232
E-mail: bob@xlnt.com

Xylan Corp.
Anees Narsinh
26679 West Agoura Road
Calabasa, CA 91302
Phone: 818-878-4672
Fax: 818-880-3560
E-mail: anarsinh@xylan.com

Xyratex
Nigel Brownlow
P.O. Box 6
Langstone Road
Havant, Hampshire PO9 1SA, UK
Phone: +44-1-705-443138
Fax: +44-1-705-498158
E-mail: nigel_brownlow@uk.xyra-tex.com

ZNYX Corp.
Alan Deikman
48501 Warm Springs
Boulevard
Suite 107
Fremont, CA 94539
Phone: 510-249-0800
Fax: 510-656-2460
E-mail: alan@znyx.com

Index

ABOUT THE AUTHOR

Stephen Saunders is executive editor of *Data Communications* magazine, the world's premier networking technology publication. He is widely recognized as an authority on high-speed networks. Mr. Saunders is a four-time recipient of the Jesse H. Neal editorial achievement award, which is widely recognized as the Pulitzer Prize of the business and technology press. He lives in New York.